Unconventional Aerial Phenomena

in the Hudson and Wallkill River Valley of New York

Bruce Cornet, Ph.D.

Copyright © 2019 Bruce Cornet

Unconventional Aerial Phenomena

All rights reserved. No part of this publication may be reproduced or transmitted in any form or by any means, electronic or mechanical, including photocopying, recording, or by any information storage and retrieval system, without permission in writing from Kindle Direct Publishing. Reviewers may quote brief passages.

Diagrams are by Bruce Cornet unless otherwise noted. Cover picture shows the Wallkill River Valley with the Shawangunk Mountains in the distance, Orange and Ulster counties, NY, over which many observations of UAP were made between 1992 and 1997. Cover design assisted by Scot L. Stride and Robert Morningstar.

ISBN: 9781075968075

Printed in the United States of America

DEDICATION

This book is dedicated to all the humans and non-humans and spirits who have helped guide me through my life to the completion of this assignment.

Unconventional Aerial Phenomena

Contents

Introduction 1
Chapter 1 Characteristics of UAP Energy Emission Signatures 13
The Experiment 16
The Experiment Continued 32
Discussion 38
Chapter 2 Other Types of Unconventional Craft 43
The Orange-Ulster-Sullivan County Mini-Flap 44
The Hudson Valley Mega Ships 49
The Manta Ray reveals rotating headlights 62
First encounter with the Manta Ray 66
Another Manta Ray Encounter 75
Swamp Gas 92
The Second Event 96
Chapter 3 The Trickster 110
Y Pine Bush 112
The Ships Ellen Observed 116
The Night Siege Ships 118
The Triangles 119
A Shift in UAP Sightings to the East 120
NIDS Involvement 128
Chapter 4 Visitors Studying of Earth and Humans 139
The "Teacher" 140
Ancient Hybrid Civilizations: Lemuria and Atlantis 141
A Biaviian Poem 143

Do we see the Light?	145
Chapter 5 The Movement of Lights	152
Fire in the Sky	156
Light Control and Telepathy	166
Electromagnetic Force Fields	175
Mechanical Sounds and Synthetic Sounds	186
Reversed Doppler	191
Another Reversed Doppler Spectrograph	193
Encounter with Reversed Doppler	198
Introspection	236
Chapter 6 Signs	239
Invisibility	244
Contactee Information: Rules of Contact	257
The Brooking Report Effect on Government Policies	258
Inflation Theory and ET Visitation	265
DNA Symbols	270
Holographic Projections - The Golden Pyramid	277
The Green Temple	283
The Vortex Photo Revisited	295
More than a Coincidence	304
Chapter 7 The Awakening	306
Mimicry in Black	318
Other UAP-Plane Sightings	337
The Landings	338
The Transformation	345
Chapter 8 Why a Geological Investigation?	352

A Magnetic Focusing System from an Underground Robotic Probe 356

Chapter 9 Summary and Conclusions 376

References 383

Unconventional Aerial Phenomena

ACKNOWLEDGMENTS

I wish to acknowledge the following people who have helped make this book possible:

Barbara Hartwell, Ben Field, Billy McNamara, Bob Tarantino, Bob Wisch, Bryan Williams (Sargel18), Budd Hopkins, Chris Burns, Dawn Ley, Dinah M. Bertran, Dr. Ellen Crystall, Dr. John Mack, Ernie Kittner, Evelyn Brock, Fred Brock, Fred Max, George Filer, John Macedo Jr., Kythea Love, Marc Whitford, Mindy Gerber, Niles Leisti, Norma Smith, Patricia Huff, Phil and Lynn Martin, Richard C. Hoagland, Rich Pascorella, Robert Morningstar, Dr. Colm Kelleher, Robert Bigelow, Dr. Eric Davis, Sal Cirami, Sam Sherman, Scot L. Stride, Sharon Cunningham, Stephen Tzikas, Steven Johnson, Sue Mann, Thomas Valone, Tommy Hawksblood Sinisi, Vince Valenti, Vinney Polise,

Unconventional Aerial Phenomena

Introduction

Before we begin, some definitions are needed. Conventional aerial phenomena is that which we understand and agree upon based on observational studies of nature, or of aerial vehicles that we have built.

UFO and UAP are used interchangeably in this book, with UAP preferred over UFO. UFO appears in quotes and in book titles.

Unconventional aerial phenomena is that which does not conform to anything known in nature, and is difficult to identify as being constructed by humans.

Unconventional Aerial Phenomena (UAP) is anything observed in the sky which is initially deemed unnatural and not conforming to any known aerial vehicles.

A flap is the repeated observation in one area of many "Unidentified Flying Objects" (UFOs) over a period of time. Dr. J. Allen Hynek, in his book Night Siege (1986/1998), with co-authors Philip Imbrogno and Bob Pratt begin their research findings with the statement: "Something truly extraordinary happened not long ago in the Hudson Valley [1982-1986] just a few miles north of New York City. Hundreds, and probably thousands, of astonished people looked up in the sky and saw something they had never seen before.

"It was enormous, awesome, and spectacular. It was seen not just once but many times over a period of more than three years. No one knew what it was. No one yet knows." (Hynek et al., 1998: xi). The first edition was published in 1986, the year Hynek died. The second edition was published 12 years later, and was expanded and revised to include the beginning of a second in Hudson valley between 1992 and 1995, which lasted longer to the west of the Hudson River until 1997 (in the region of Pine Bush, NY). My study and field research document this second flap in Orange and Ulster counties, New York. The Middletown Times Herald Record, however, has many accounts and stories on file of sightings in the area dating back to the early 1960s.

Hynek et al. (1998) gives detailed accounts and witness descriptions for over 62 individual sightings and some group sightings of the thousands of reports filed with them and police stations across Dutchess, Orange, Putnam, Rockland, and Westchester counties in New York, and in Fairfield and New Haven counties in Connecticut (see map). Their files contain records of 7,046 sightings in the Hudson Valley region from 1982 to 1995. Preston Dennett provides an extensive overview in "UFOs Over New York, A True History of Extraterrestrial Encounters in the Empire State" (2008).

Dr. Ellen Crystall also did an extensive ten year study of this phenomena in the area of Pine Bush, NY (northern Orange County), which she published in 1991 in her book "Silent Invasion, The Shocking Discovery of a UFO researcher." Her research extends sightings and paranormal activity 30-35 miles west of the activity documented by Hynek et al. (1998) for the area north of New York City. Why there should be a second area of concentrated sightings, paranormal activity, and abductions will be explored later in Chapter 3. It is Dr. Crystall's work and book that made "Pine Bush" a household name for those interested in UFOs. Because of her book and willingness to educate others in the field, I became involved in doing research with her and others. I subsequently conducted a five year, ground-based magnetic study in the Wallkill River Valley running through the Pine Bush region. That magnetic study, documented in Chapter 8, resulted in a geologic map of magnetic anomalies over a 24 square mile area showing the locations of unexplainable magnetic hot spots, some of which turn on and off periodically when certain star constellations are overhead.

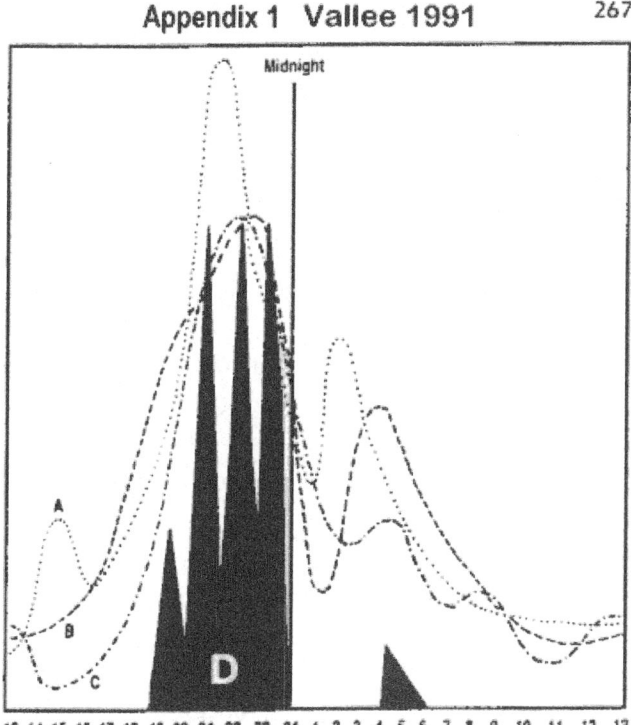

Figure 1. Frequency of close encounter reports as a function of time of day.

A: 362 cases prior to 1963, all countries
B: 375 cases between 1963 and 1970, all countries
C: 100 cases from Spain and Portugal only
D: 70 cases for 1992-1993, Pine Bush

Dr. Jacques Vallee published a graph in Appendix 1 of his book, Revelations: Alien Contact and Human Deception, 1991, p. 267, which shows UAP activity worldwide. When the Pine Bush activity recorded by Cornet for just the years 1992 and 1993 are plotted against his curves, a remarkable similarity occurs. It is clear this phenomena is not unique to Pine Bush, but represents global activity by Earth's Visitors.

Whereas the Hudson Valley sightings began on December 31, 1982 in Kent, Carmel, and Lake Carmel, Putnam County, NY (east of the Hudson river), Ellen Crystall was photographing large triangular craft, rectangular-shaped craft with windows, small boomerang-shaped craft, and pod (dome)-shaped craft with captured images of humanoid beings inside and outside those craft more than two years earlier at or near the town of Pine bush, NY (July 1980). Whereas the Hudson Valley sightings tapered off in 1986 after spreading north to Kingston, NY, and south to southern Long Island, NY, and northern New Jersey, Crystall and other witnesses with her spotted or observed anomalous craft near Pine Bush into late June of 1988. See also "In The Night Sky, Hudson Valley UFO Sightings From The 1930s To The Present" by Linda Zimmermann (2013) for previously unrecorded testimony of sightings to the north and northwest of New York City.

A year after Dr. Crystall's book appeared in book stores in 1991, sightings of Triangular and Boomerang-shaped craft were being made on 5 February 1991 in Williamsport, PA, and surrounding areas. At least 9 sightings based on 13 eye-witness reports were published in the June 1992 issue of MUFON UFO Journal (Number 290), in an article written by late investigator Dr. Samuel Greco (aerospace engineer) with an introduction by Stan Gordon (PASU). Even an article by Vallee entitled, The Hybrid Question appears in that same issue. Sightings were reported in Williamsport, Cogan Station, Lewisburg, Linden, and Selinsgrove, PA. These sightings of the same types of craft seen around Pine Bush 185 miles to the east, extend the corridor at least that far west. These sightings occurred just after Crystall's book was published, and just before a two-part, four-hour miniseries, "Intruders," appeared on CBS-TV in May of 1992. This mini-series is based on Budd Hopkins' book by the same name. This mini-series is what got me interested in the UFO-Abduction phenomena (Chapter 7 The Awakening).

In the early 1990s Dr. Steven Greer, an ER doctor working in North Carolina, took an active interest in studying UFOs, and formed

CSETI, the Center for the Study of Extraterrestrial Intelligence (Greer, 1999: Extraterrestrial Contact). In 1993 he led a Rapid Mobilization Investigative Team (RMIT) to central Mexico to study the ongoing wave of UAP sightings there. He felt due to the frequent sightings his team had an opportunity to capture ETVs (Alien Vehicles) on film, and have scientifically trained personnel present to witness and describe what they saw.

While he and his team were documenting the phenomena, around an 18,000 foot high volcano named Popocatepetl, Cornet and Crystall were documenting the phenomena in Orange and Ulster counties around Pine Bush, NY. This area also had frequent sightings, which Crystall documented in her book, Silent Invasion (1991). Thus, in the same year research teams were in the field on opposite sides of North America and Central America.

The following is the article written by Dr. Joseph Burkes:

"The Center for the Study of Extraterrestrial Intelligence (CSETI), is a scientific research organization dedicated to the better understanding of Extraterrestrial Intelligences (ETI) and their civilizations. CSETI believes the extraterrestrial hypothesis is not just a theory, but is a fact of life. CSETI is attempting to perform real time research of UAPs and ETs by employing specific techniques and prearranged methods of attraction, or as CSETI's director Dr. Steven Greer calls it "vectoring in an ET's spacecraft to a carefully chosen research site." At these sites, investigators attempt to peacefully interact with an extraterrestrial spacecraft and its occupants by using a variety of techniques. This process includes using powerful lights, meditation and playing tape recorded sounds obtained during sightings of ET craft.

"CSETI members use these techniques in locations where there have been waves of UFO sightings. So far, the results have been quite impressive. Dr. Greer started this international organization after having a transformational encounter of his own. Dr. Greer calls these human initiated contacts with the ETI, Close Encounters of the Fifth Kind (CE-5), and has successfully facilitated such interactions on numerous occasions not only in the U.S. and Europe, but also in Mexico.

"I [Dr. Joseph Burke] was selected to join the Mexican investigation, because I speak Spanish and more importantly, because I have been leading a group of Los Angeles based CSETI researchers since August of 1993. As coordinator for the L.A. group, I have been practicing my hiking skills by regularly leading our team into remote locations in the Southern California area to make contact with ETI.

"These skills came in very handy during our investigation in Mexico. Upon arriving in Mexico City, we immediately drove to the base of an active volcano and set up camp. The team consisted of five members. For security reasons, the exact location of our research sites cannot be given at this time. In 1992, during an investigation being conducted in England, hordes of tabloid journalists and uninvited UFO enthusiasts invaded the private farm where Dr. Greer was leading a team effort. Some of these unruly intruders had to be forcibly removed and several arrests were made. Because the Mexico UFO sightings are very active, and an important part of CSETI's ongoing research, the term "volcanic zone" will be used for our research location.

"The first night after driving up the mountain road on the slope of a volcano, we set up our camera equipment, radar detector (radar detectors pick up microwaves and can serve as inexpensive UFO alarms), and sleeping bags in a clearing on a hillside surrounded by pine trees. From this site we had a good view of the town below. We watched as one UFO appeared, then possibly two, over the town below. They were amber-colored glowing globe-shaped objects which behaved as though they were intelligently controlled. When

Dr. Greer aimed his half a million candlepower light to signal the object, it appeared to signal back.

"The sighting of a globe-shaped UFO had been described to local investigators several weeks prior to our visit. We later obtained a copy of a home video which showed just such an object taken at twilight from a local restaurant during a wedding party. During our first night on the volcano, the temperature dropped into the low thirties and Dr. Greer was suddenly hit by a powerful amber-colored beam of light which appeared to come from the sky although we could see no identifiable source. We were very surprised by this flash of light and no ill effects were felt.

"The next day, we traveled to a new location where according to local authorities, sighting of UFOs during both day and night are a common occurrence. They told us 80% of the population has seen a UFO and many of these sightings are close enough to see details of the UFO's structure. We found a research site in a field that was in full view of an 18,000 foot high volcano named Popocatepetl." (Burke and Greer, 1993: Mexico City Sightings – Large Triangle).

In the years following the 1982-1986 "Siege" (i.e. from 1986 to 1995) over the central Hudson Valley, Imbrogno reports sporadic sightings in the lower Hudson Valley (Westchester and Putnam counties, east to Danbury and New Fairfield, Connecticut), with additional sightings to the north from Newburgh to Kingston, NY (Hynek et al., 1998). In their book, "Celtic Mysteries," Imbrogno and Horrigan (2000) identify numerous Celtic stone chambers dated to ~2,000 B.C.E. (by an obsidian dagger blade using water loss method) in Putnam, Westchester, and Putnam counties, NY, that seem to mark the locations of unusual Earth magnetic anomalies, paranormal activity, strange hooded beings, and angular craft of various sizes witnessed by dozens of people. Their maps tracking UFO activity and paranormal activity imply interdimensional portals or vortexes may be connected to the magnetic anomalies, and show that many of the sightings during the "Siege" originated and later ended near those same magnetic anomalies and associated stone chambers.

In this book I will show a similar correlation to anomalous Earth magnetic anomalies in the Wallkill River Valley between the towns of Wallkill, Walden, Montgomery, and Pine Bush to the northeast, and Pine Island, Warwick, and Edenville to the southwest near the New York and New Jersey state boundary (Sullivan, Orange, and Ulster counties). The Wallkill River Valley runs northeast-southwest across lower New York State, and gets its name from the Wallkill River, which flows from its headwaters in northern New Jersey to the northeast, where it joins the Hudson River at Kingston, NY. Whereas the magnetic anomalies due north of New York City appear to be associated with very ancient metamorphic rocks dating back to the Precambrian Era more than 550 million years ago, those in the Wallkill River Valley appear to be associated with granitic masses which were intruded into younger Precambrian rocks. The host rocks were then eroded before the Cambrian, 550 million years ago, which exposed those granitic intrusions as mountains on a very ancient landscape of the North American craton.

During the Cambrian and Ordovician periods at the beginning of the Phanerozoic, 420 to 550 million years ago, New York State was covered by an epicontinental sea. Those hard granitic mountains were high enough to be exposed as islands in that sea until the end of the Ordovician, when subsidence and sea-level rise submerged them. Alien mining probes landed on those islands, and remained there. Due to recent erosion during the Ice Ages, the Ordovician and Cambrian black shales and carbonates once completely surrounded and covered those mountains were partially removed. Some of the mountains now poke up through their once entombing, softer oceanic sediments, or can be detected just below the surface using magnetic mapping. Mount Adam and Mount Eve near the New Jersey border just south of Pine Island, for example, represent two of those ancient granitic mountains, which you can see and climb today.

Anomalous UAP activity is closely associated with these magnetic mountains to the west of the Hudson River. These granitic intrusions contain exotic minerals and elements (cf. Crystall, 1991), which have been extensively mined at Franklin, NJ, just south of the border with New York State (see The Franklin Mineral Museum and Nature Center, 32 Evens Street, Franklin, NJ 07416). These intrusions were formed when the Earth's mass was still reorganizing, and sialic Crust (i.e. Continental Plates) was just beginning to form billions of years ago, and the Earth's Mantle was excluding and ejecting those elements into the newly forming Crust, elements which did not fit into the crystalline mineral structure of the Mantle. Those elements and minerals have been sought not only by miners during the last several hundred years (especially concentrated iron ores), but also by visitors coming to Earth hundreds of millions of years ago when those granitic islands were easily accessible during the Cambrian and Ordovician. This book provides evidence for those ancient mining probe from other worlds, some of which are still there trapped on those islands, entombed in Ordovician black shales and mudstones. We have found them where they landed. Some of them are still functioning, because they are nuclear powered and nearly indestructible.

The Hudson Valley UAP Corridor, which stretches from Pine Island to Long Island, is still attracting alien probes, because the extraction probes left behind 420 million years ago are still sending signals out into space. That is why this area is a UAP hotspot, a kind of stargate. Crystall hypothesized there is an alien base underground in the Wallkill River Valley, and she is in part correct, but not a recent one. These ancient probes may form a functioning interdimensional-intergalactic stargate, with Earth as a portal to the Universe. Just as the secret about this stargate started to be investigated by the government, UAP activity around it seemed to wane. Cheryl Costa, writer for the Syracuse Newtimes, and manager of the New York Skies, A UAP Blog online, provides sighting information for a 20 year window from 1995 to 2015, reporting 496 sightings in Orange and Suffolk counties of New York. The importance of her data is not that

Unidentified Flying Objects are more common over Long Island and its surrounding waters, but that the activity in and around Pine Bush rapidly dropped off in 1997, and moved elsewhere.

So why did UAP activity almost stop around Pine Bush in 1997 (read The Pine Bush Phenomenon by Vincent Polise, 2005)? Orange County is located in the western part of the UAP Corridor, while Suffolk (eastern Long Island) is located in the eastern part of the Corridor. Although Costa under reports the number of sightings in Orange County for the years 1995 to 1999, she documents a surge in UAP sightings in Suffolk County or Long Island, NY, for the years 2000 to 2015, with the highest number of sightings between 2008 and 2015. Is this shift in UAP activity to the East significant?

There is much we do not know, and very few people who want to uncover the truth. It is through the dedicated work of Hynek, Imbrogno, Crystall, Pratt, Horrigan, Polise, Zimmermann, Costa and others that a picture and pattern is beginning to emerge which may allow scientists to test hypotheses, dig up artifacts, and make predictions about our visitors in our quest for answers and understanding.

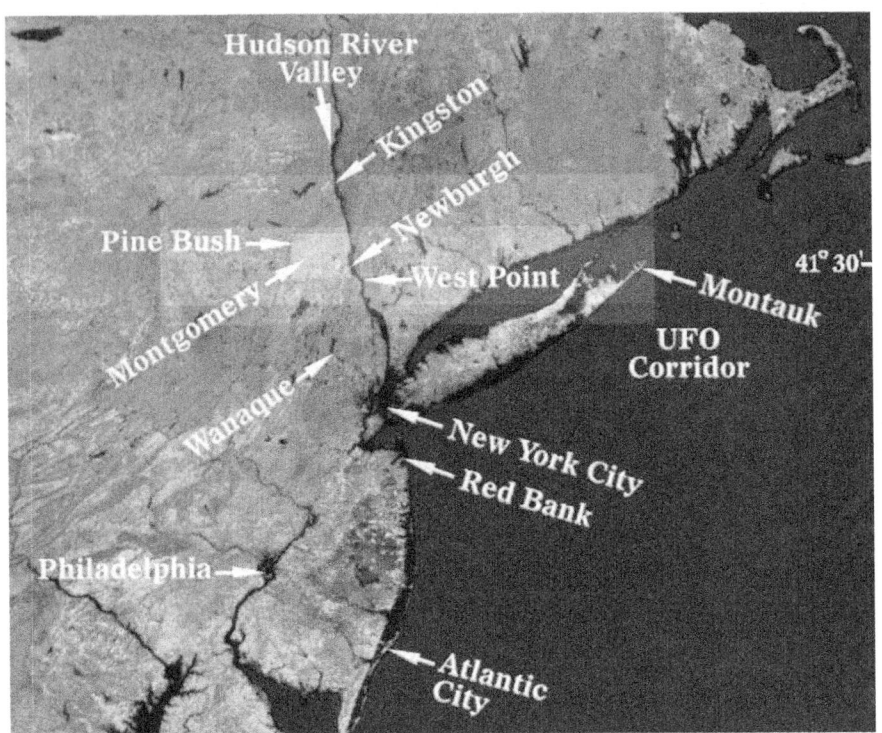

http://www.sunstar-solutions.com/AOP/Appendix_I/APPENDIX_I.htm

Chapter 1 Characteristics of UAP Energy Emission Signatures

By now the reader should accept that something very unusual happened in the Hudson Valley region during the 1980s and 1990s, extending from Long Island north and west to Connecticut, to as far north as Kingston, New York, and as far west as Pine Island, New York, through the Wallkill River Valley. This region became known as the UAP Corridor, and has a history of strange activity going back to the 18th Century, to Salem, New York, and strange tales such as the one about the headless horseman. Hundreds of unusual stone chambers were built across Dutchess, Putnam, and Westchester counties as long ago as 4,000 years, dating to pre-Stonehenge (Imbrogno and Horrigan, 2000). But what did people see, photograph, film, videotape, and record between 1980 and 1997 in this corridor? The most recent books by Dennett (UFOs Over New York, 2008) and Zimmermann (In The Night Sky, 2013) will bring you up to date on eyewitness accounts.

One of the earliest sightings in New York State of a large triangular object occurred in Plymouth Township. The night was 23 November 1977. This UAP wasn't silent, as if it wanted to be noticed and call attention to the arrival of a new phenomenon. A farmer and his wife reported it thundered so loudly, it shook their house when it hovered almost directly overhead. The whole house vibrated. Other sightings of triangular craft were reported across the United States that year from Memphis, Tennessee, to Carroll County, Arkansas. 1980 turned out to be a particularly significant year, with many reports of triangular craft, and hundreds of sightings in November in north-central Missouri. The Age of the Black Triangles (also known as Flying Triangles) had been born (Hynek et al., 1998).

Dr. Ellen Crystall was the first person in the Pine Bush region to photograph these anomalous and strange craft, the best pictures of which she took in July of 1980. Even though she and her companion, Harry Lebelson (UAP editor for Omni magazine in 1980), saw these craft either flying low over their heads silently, or landed near them, the photographs they took with her Zenite 35 mm SLR or his twelve hundred dollar Leica did not turn out as expected. Lebelson's pictures were all blank, while Crystall's pictures showed concentrated masses of light and emission sparks and trails. Even though their eyes saw clear shapes of structured craft, which did not resemble airplanes or helicopters, only Crystall's film picked up any images. She would later learn through her photography professor at Rutgers University that her film had been exposed to short wave radiation, and the clusters of emissions and bluish clouds in her pictures came from x-ray and ultraviolet emissions (Crystall, 1991). But this explanation didn't explain why Lebelson's (Harry's) Leica produced blank, apparently unexposed frames of film.

Only a few pictures taken by many witnesses during the 1980s show the lights of UAPs clearly; none of those show the types of radiation and emissions captured by Crystall's camera. Pictures taken by me and others during the 1990s continued to clearly show lights; only a few showed evidence of emissions, glows around the object, and indications of structure (i.e. reflections) next to the lights. Something changed in the way this phenomena presented itself to humans. Similar to the trend in UAP history from 1) alleged contact with alien beings (some very human-like) prior to the 1960s and back at least to the 1920s, 2) followed by mostly sightings of disc-shaped craft through the 1970s, and 3) sightings mostly of angular craft (i.e. triangular and boomerang shapes) in the 1980s to the present, the Pine Bush phenomenon followed an identical but abbreviated pattern (The Pine Bush Phenomenon, by Vincent Polise, 2005). Crystall (1991) was even able to reveal through enlargements what appear to be alien beings standing next to their circular and dome-shaped craft, or visible through windows in those craft.

The frustration of observers who could not capture images of what they visually saw in the field was a problem that concerned me. How could people standing near or next to one another get very different results on their photographs? This is a serious problem which threatens the reliability of all photographic evidence if the images can be manipulated by the source. How can people in a group who supposedly witness the same event give substantially different accounts of what they saw? Let us put this into perspective: If differences can be scientifically documented, and can be repeated, then conventional explanations of aircraft, rocket launches, etc., can be completely ruled out. In fact, any known human technology cannot account for such differences, especially when two adjacent cameras and observers witness and record completely different phenomena. The problem in recording (not just observation) needs to be addressed before eye witness descriptions can be given. It should become clear to the reader that the intelligence behind this phenomena is giving humans only the information it wants to give, and there is a definite pattern and objective in the interaction between humans and the unknown intelligence. Even Crystall states that she felt like a "lab rat" in an experiment, but one that was allowed to look outside the maze. Even though we have limited ability to know much about these experimenters (because of our technological limitations and their lack of permission to know who they are or why they are experimenting with us), we still can understand what is happening, and come to plausible, rational explanations for why it is happening.

The information given below is based on 11 years of field study (1992-2003) and the capture on film and video, and/or through multiple witnesses, over 130 personal close encounters: Appendix I.

The Experiment

On 3 June 1993 Ellen Crystall and I were positioned at our favorite observation site at the dogleg bend along west Searsville Rd. We had our SLRs on tripods next to one another (no more than four feet apart). The road was behind us about 50 feet, and a large open field stretched out in front of us to the east, bordered by a distant tree row oriented north-south, which defined the western boundary to another farm field. On the eastern side of the distant field is a circular mound (200-300 feet) rising about eight feet above the field. I have dubbed it the "Indian Mound," because it may have been constructed by Native Americans (perhaps the ancestors of the Mohonk or Ramapo tribes) to mark an area of anomalous nighttime light activity not unlike what we are witnessing today. The lights come up out of the ground and dive down into the ground around this mound. The Indians may have regarded the lights as spirits and the area as sacred.

When we had the opportunity to fly over the Hotspot in a helicopter on 10 May 1994, we discovered an Indian Mount and permanent crop circle defined by trees. This unknown archaeological site resembles Native American Sweat Lodges with a sacred path oriented East to a mound or Unchi where a fire was built.

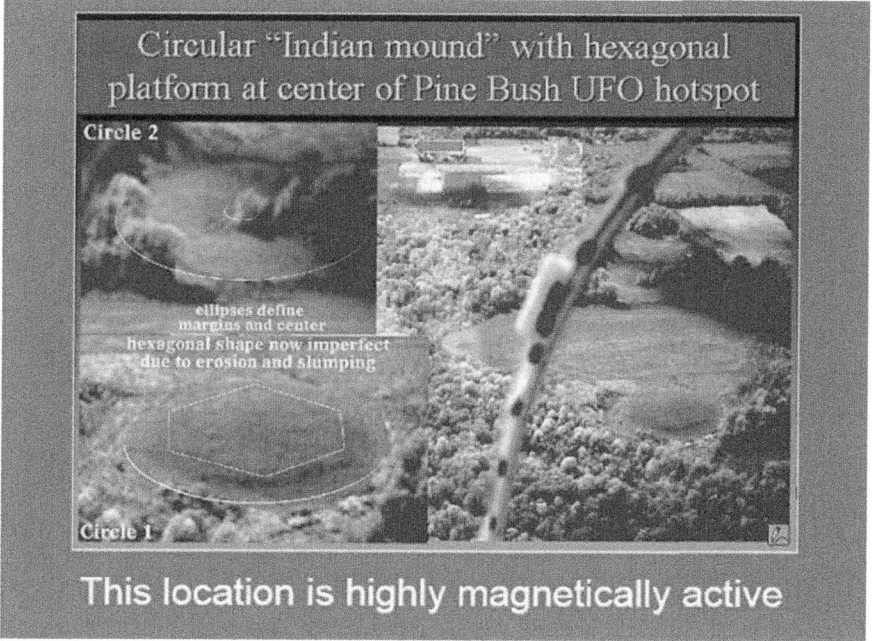

The arrow at the top points to the station along West Searsville Rd. where most sky watching took place in this area.

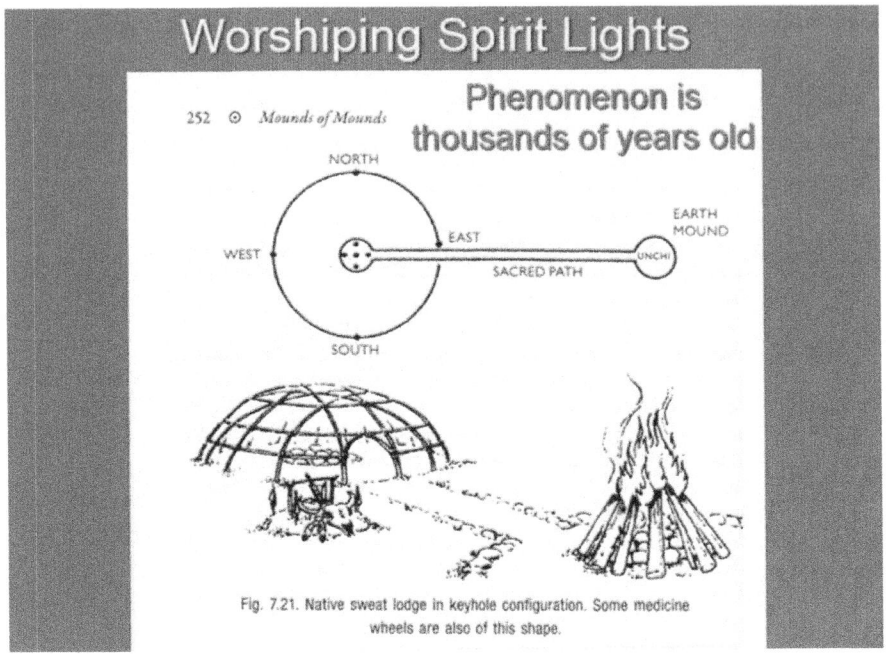

Fig. 7.21. Native sweat lodge in keyhole configuration. Some medicine wheels are also of this shape.

Above illustrations from The Lost History of the Little People, by Dr. Susan B. Martinez (2013).

On 31 May 1995 while photographing a plasma light to the East, there was a flash of light coming up from the area of the Indian Mound. When that time exposure was printed, an image of a golden pyramid was present rising up just above the tree line across the field. A golden splay of light was seen shooting up through this pyramid, symbolically representing what a pyramid means: Fire (pyro) in the mid (middle). See Appendix I for color picture. That hologram represents the fire built on the mound (Unchi), located due East of the circle, which may represent the boundary of a huge sweat lodge that once stood there.

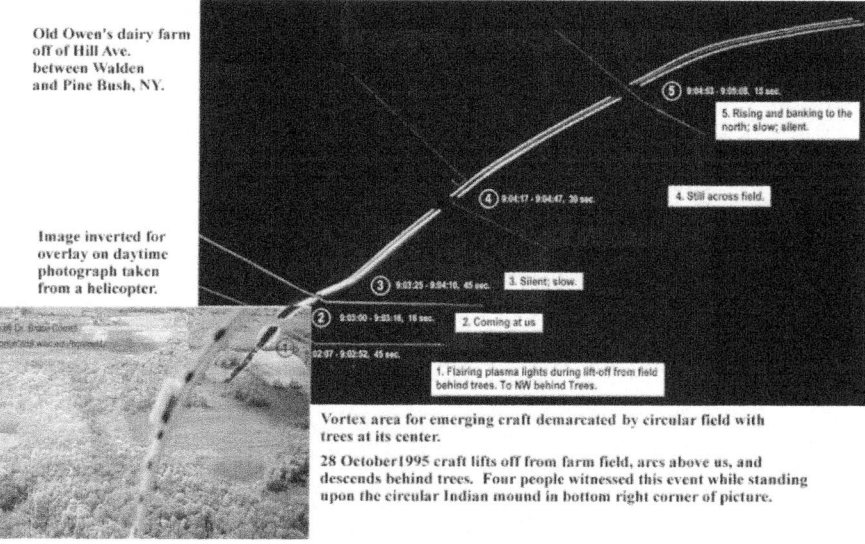

Black Triangle takes off from circle where sweat lodge was located.

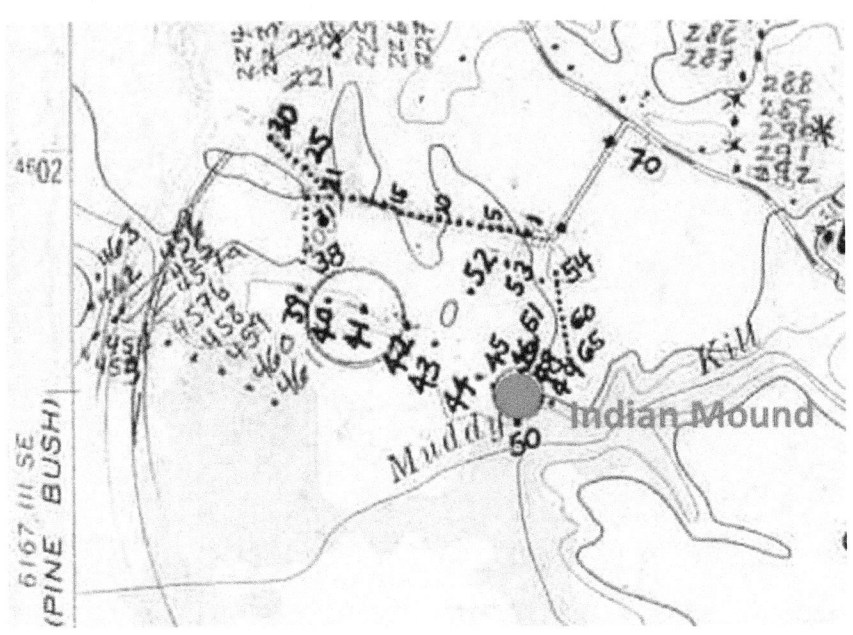

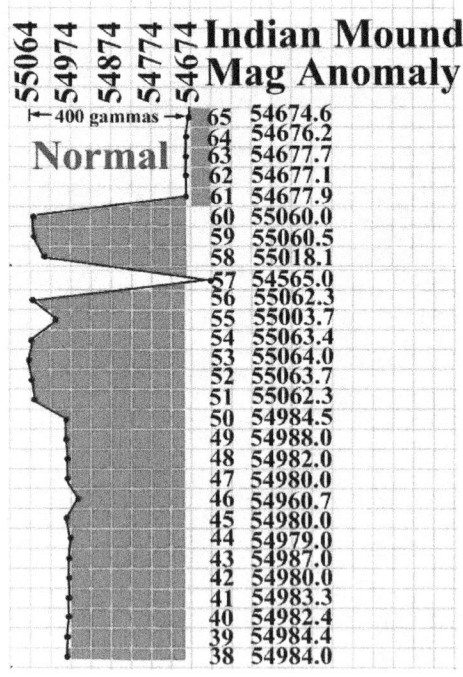

When the UAP Hotspot area was surveyed with a Proton Precession Magnetometer in 1995, a significant magnetic anomaly as much as 400 gammas above background was recorded around that hotspot. As measurements were being recorded (see graph below), the magnetic anomaly suddenly stopped, accounting for the abrupt drop in magnetic readings to background levels.

Back to The Experiment. At 11:11 pm on 2 June 1993 Ellen and I witnessed a light blink on near the tree top level over the "Indian Mound." We both opened the shutters of our cameras at about the same time. That exposure was the first of four time exposures I took over a period of three minutes 50 seconds, as this light slowly rose up into overhead clouds, disappeared momentarily inside the clouds, then appeared again as it dropped just below the clouds as it traveled south. Crystall was situated to Cornet's right and photographed the same light rising up into the cloud.

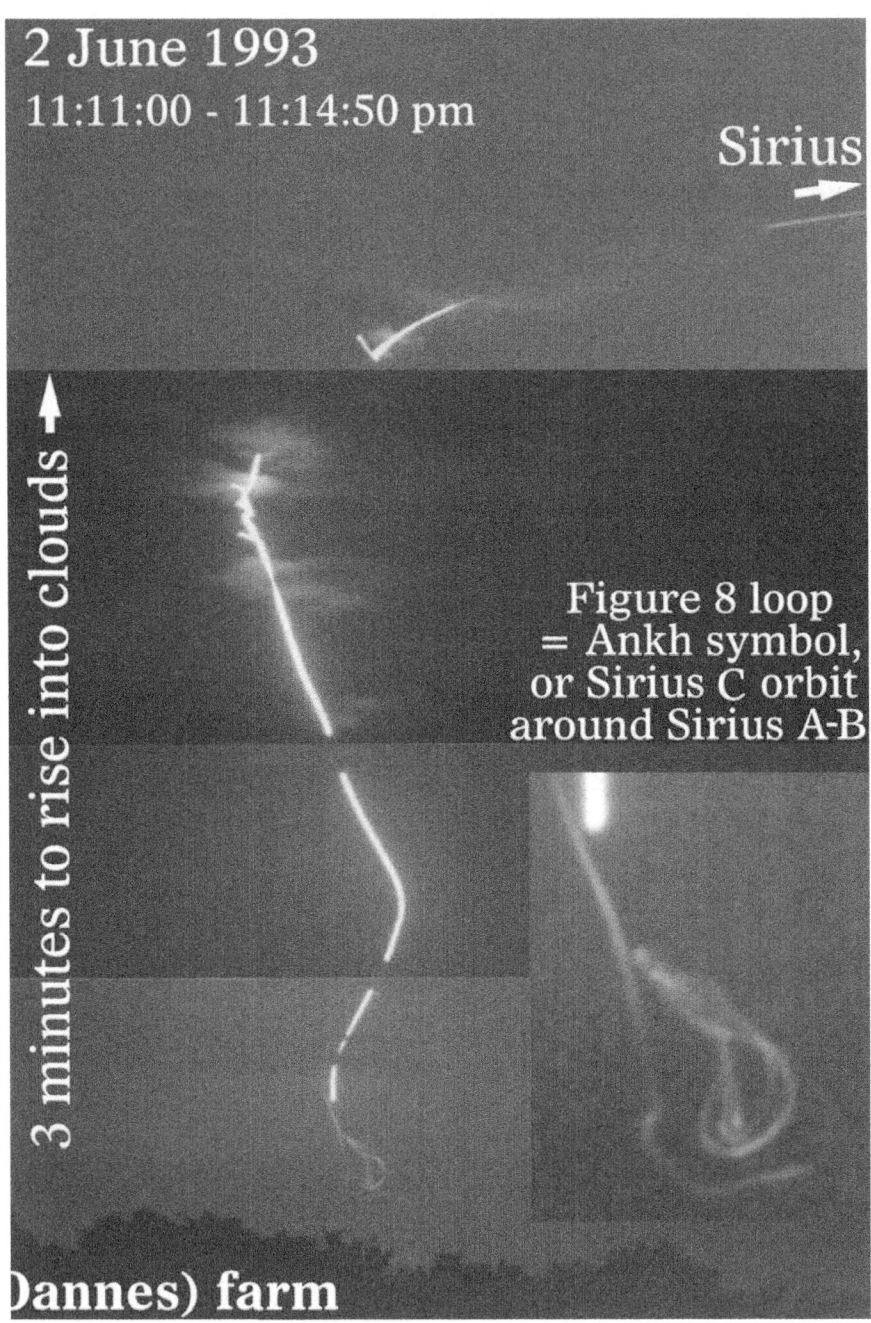

What is evident in the time exposure 3 below are lateral movements of the light just before the UAP enters the cloud. The original of Crystall's image (not shown above) does not contain those lateral motions. Cornet wanted to know why these images are not identical.

Contactee Vol. 6 No. 2 Fall 1993

Ellen's picture (hard to see due to glare) shows no sideways movement, although the amount of brightness (ambient light reflection) around the light trace was greater in her photograph. Question is, which image was deliberately altered at the source, and how was it done without influencing the other camera no more than four feet away?

The overall shape of the light's flight path was a lazy 'S' shape. When I got my pictures back, I was surprised to see a red light (orb) emerging from the main white light just before the end of the first time exposure (top of photograph one). We saw this red orb emerge and drop towards the ground, but it happened so fast we couldn't imagine (see) the complex figure eight pattern it produced just above the tree line. See Appendix I for additional pictures.

Keep in mind the red orb emerged within a millisecond of the time exposure ending as the reflex shutter closed (from top to bottom across the film). The entire movement of the red light occurred within a millisecond as the shutter closed! How did the intelligence behind this performance (and it was obviously a performance for our cameras) know when the shutter on my camera was closing? Crystall experienced similar synchronicities during the 1980s, and attributed them to advanced alien technology. Experiments I conducted in the field (e.g. 1 July 1993 and 9 May 1994), and photos I took in a helicopter over the valley on 10 May 1994 of a tetrahedral-shaped object that buzzed the helicopter in broad daylight, demonstrate both advanced technology and mental telepathy.

Study the images below:

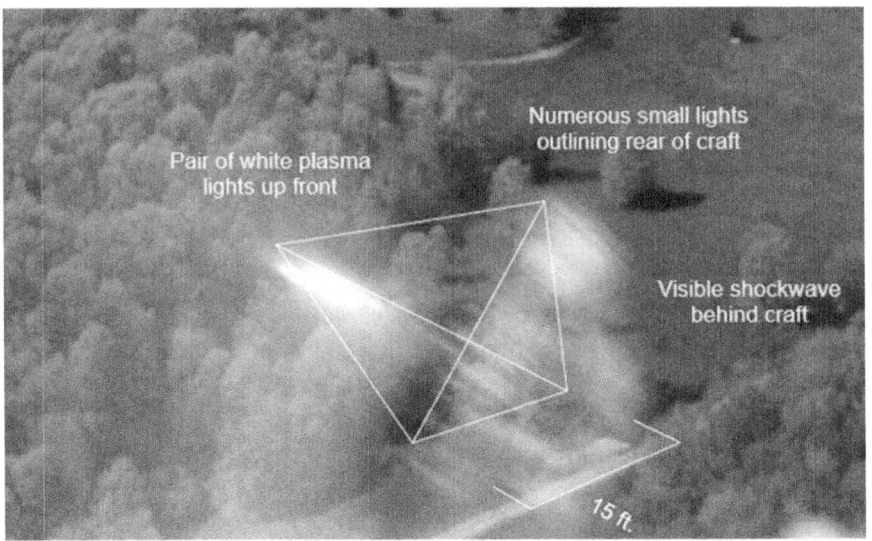

http://www.sunstar-solutions.com/AOP/Experiment/experiment.htm

That image of a craft was made in broad daylight (about 2:00 pm on 10 May 1994) as the helicopter flew northwest from Stewart International Airport. We had just passed over the Wallkill River, headed towards Pine Bush, NY. I was seated between the pilot on my left and a Japanese cameraman who was holding a beta camera out the open door of the canopy so he didn't have to film through the canopy. I was holding a Minolta XG7 SLR and photographing through the canopy. Its shutter speed was set for 1/1000 second. I was waiting for us to get close enough to a large geographic feature on the ground so it filled my viewfinder frame. As we reached that point, I went to press the shutter button. To my surprise, a clear unmistakable voice in my head said, "Wait!" I looked to see if the cameraman or pilot had spoken to me, but both were preoccupied. I reframed the area on the ground, and went to push the shutter button again. Then that same voice repeated, "Wait!" I instinctively listened and waited for him to say "Now." And sure enough, he said, "Now," and I pushed the shutter button.

At that very moment, there was a flash about 300 feet in front of and below the helicopter. Almost immediately the helicopter was hit by a shockwave. It bounced the helicopter enough to make the pilot moan. We didn't know what had happened. All we saw was a flash of brilliant light traveling from lower right to middle left. This same scenario happened two more times flying over other areas on that same flight. All the pictures show a craft defined and outlined by lights in the photographs. Its body is not visible, because the film was not fast enough (400 ISO). Based on estimating the size of the object and measuring the length its lights traveled in 1/000 second, I calculated that a tetrahedral-shaped craft had buzzed our helicopter (three times) traveling a Mach 7! Impossible you say? If the craft was only 10 feet wide and not 15 feet, it still would have been traveling at over Mach 4. We have no aircraft that can travel that fast in the lower atmosphere. The brilliant plasma at its nose contoured the air around it so it would not produce a sonic boom. There is a detectable shockwave directly behind the craft, however, which blurs the lights behind the shockwave, and it is that energy wave which bounced our helicopter. And yes, the beta camera also recorded the flyby, but it was so fast all there is on one video frame is a streak or blur.

Let me describe how we discovered deliberate interference or manipulation of our camera images through the use of high technology.

The second time exposures taken by each of us on 2 June 1993 show the light rising into an overhead cloud (see images above). Crystall's photograph shows a straight vertical ascent (published in Contactee, her news flyer published on 10 November 1993). Cornet's photograph, however, shows sideways digressions of the light several times as the light quickly moved to the left or right, then back to its original vertical ascent. My camera was mounted on a stable tripod, and the shutter was operated by an electronic cable, precluding sideways movement. Once we recognized these differences, it became apparent that something else was controlling the characteristics of the light reaching each of our cameras, independent of the other.

Almost a year later, on 9 May 1994, I had the opportunity to conduct an experiment with the help of a friend, John Macedo, Jr. I wanted to set up two cameras on tripods only two feet from each other. If my interpretation was correct, I wanted to repeat the results of 2 June 1993 where the distance between cameras was about four feet. But in order for that to happen, I needed the cooperation of the intelligence behind the phenomena. I came to the conclusion after two years and about 80 sightings and encounters (see Appendix I) that I was being tested and mentally probed. Frequently in the presence of other observers, a light would turn on low to the trees or below tree top in an adjacent field behind a row of trees, brighten substantially as it rose above the trees headed in our direction. I recall vividly how Crystall (recording an event with her Canon camcorder) would speculate about the light with other observers who were present. As the light came closer it rose higher in altitude silently, sometimes changing the brightness of one or the other paired headlights. "It's just a plane," someone would state. Crystall might say it is traveling too slow to be a plane. The lights would dim and navigation lights (red and green) would appear at the outside edges of the object. "It's a plane," someone would insist, even though the object was now closer than 1,000 feet from us, and too low for FAA regulated minimum altitude. Then at about 500 feet distance, a jet-like sound would be heard that grew steadily louder. Even I would be deceived, sometimes saying, "It's a jet." Then as it flew over us we would see a Black Triangle, not an aircraft fuselage with wings and tail assembly.

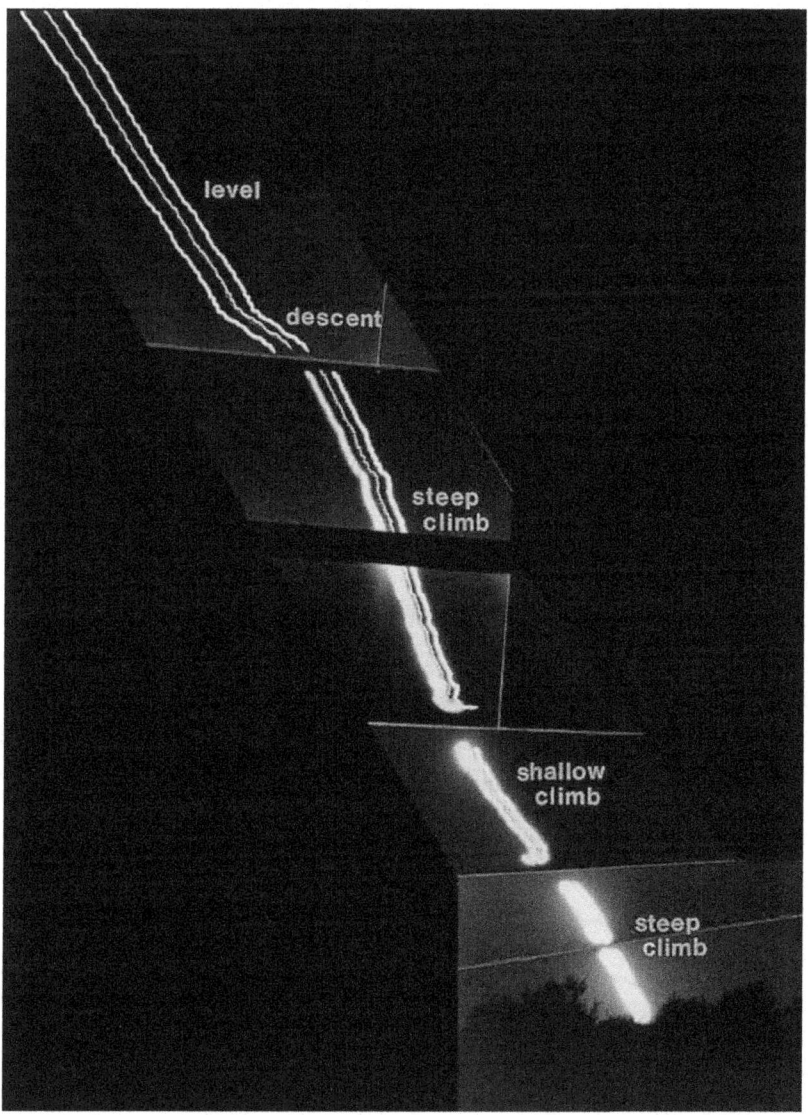

Above: On July 13, 1992, at 11:00 pm at dogleg bend in South Searsville Rd., next to Wilde farm, six time exposures were taken using infrared black & white film. Black Triangle, craft lifted off silently from a farm field behind tree row, brightened its lights with an ionization of the atmosphere, then dimmed them quickly; as Black Triangle approached, it began to level off its climb, then brightened its right light produced ionization of atmosphere; it then reduced that light's brightness slowly to match the brightness of the left light; craft then dropped slightly in altitude and leveled off; first jet-like sound heard was about 600 feet away; initially lights rapidly spiraled; spiraling slowing down as craft approached observers; average speed 17 mph; sound analysis reveals synthetic jet sound exhibiting Reversed Doppler effect [Bruce Cornet, Ellen Crystall, Ralph, and Steve were present].

Skeptics would argue that it was only a conventional aircraft based on the navigation light configuration and color, and on the jet-like sound. But after careful examination of my sequential time exposures, Crystall's video, and two separate audio recordings, which were analyzed with spectral analyzers, major anomalies appeared that conflicted with a conventional interpretation. These anomalies will be discussed in more detail in Chapter 5. We are being given mixed signals in order to confuse us, as if we were either enlisted for quality control of camouflage techniques and mimicry, or being tested for our intelligence, or both. Then again, our visitors could have been just jerking our chains for their own amusement.

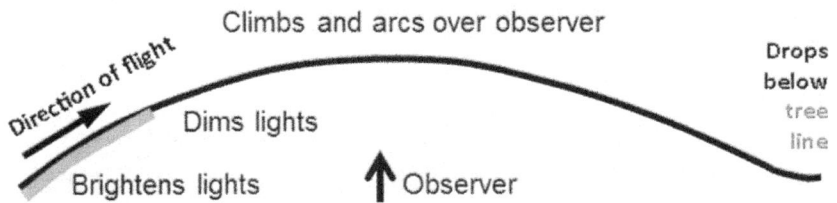

Changes in Light Intensity during Transformations

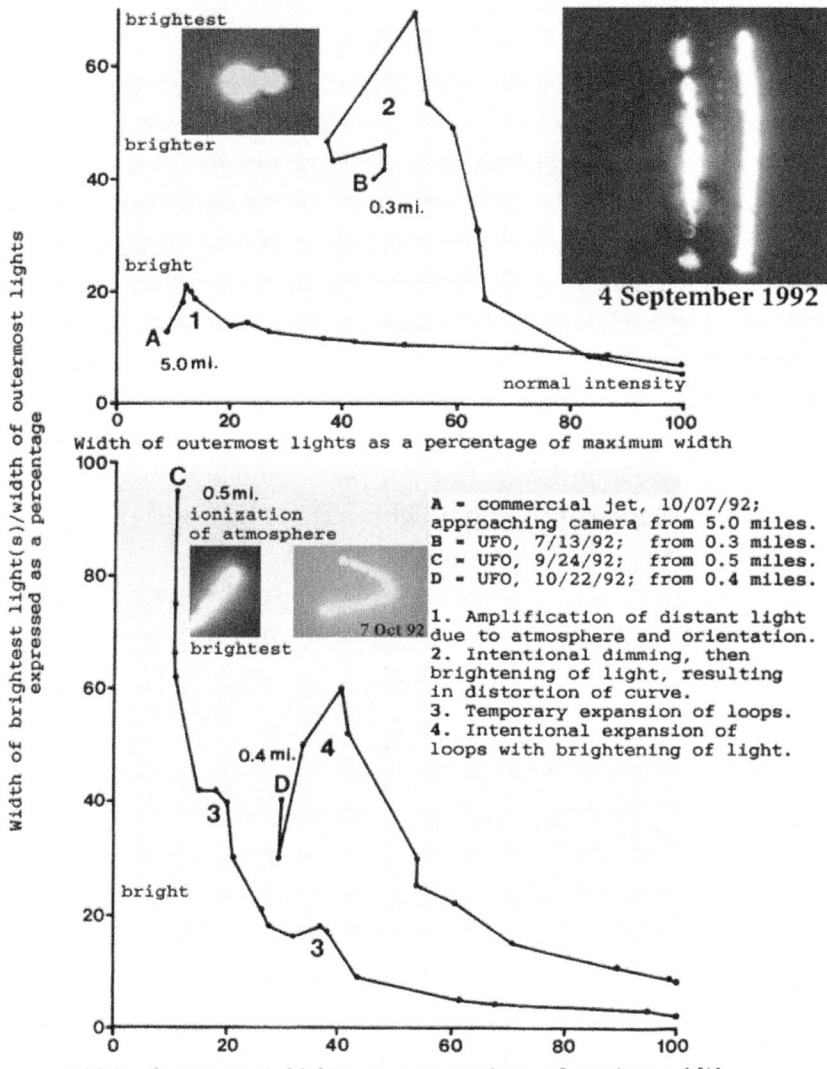

A = commercial jet, 10/07/92; approaching camera from 5.0 miles.
B = UFO, 7/13/92; from 0.3 miles.
C = UFO, 9/24/92; from 0.5 miles.
D = UFO, 10/22/92; from 0.4 miles.

1. Amplification of distant light due to atmosphere and orientation.
2. Intentional dimming, then brightening of light, resulting in distortion of curve.
3. Temporary expansion of loops.
4. Intentional expansion of loops with brightening of light.

The Experiment Continued

The evening was cool and dry. The sky had a high cloud cover. Cornet and Macedo set up their cameras next to one another. At 9:22 pm Cornet spotted a pair of lights approaching from the west at an estimated altitude of about 1,500 feet. Macedo turned on the camcorder and began tracking the lights. Cornet took a series of time exposures as the AOP flew overhead and seemed to descend rapidly to the east, where it disappeared below the horizon. What Cornet saw approaching him was a pair of "very brilliant" yellowish white lights. As they approached they flared in typical "performance" fashion, then dimmed. Cornet thought Macedo was seeing the same thing, but got no response to his statement about the brilliance of the right light.

Even though Cornet centered the lights so they would traverse the picture frame beginning at the bottom, the images indicate that cropping had occurred. At least half of the light traces in picture # 1 is missing, while most of the light traces in picture # 2 is missing. Something is wrong with this picture. Somehow what was visible through the lens before the shots were taken was shifted downwards when the shutter opened.

Timing of Photographs

Time Exposure	Camcorder No. in Seconds	Seconds	Counter
1	9:22:18-9:22:23 pm	5 seconds	0
2	9:22:28-9:22:32 pm	4 seconds	13 ¥

3	9:22:38-9:22:44 pm	6 seconds	29
4	9:22:49-9:22:55 pm	6 seconds	43-77 £
5	9:23:47-9:24:00 pm	13 seconds	

¥ Glow of lights in time exposures visible below the lights in video.
£ First sound heard; resembled that of a jet. The sound can be faked.

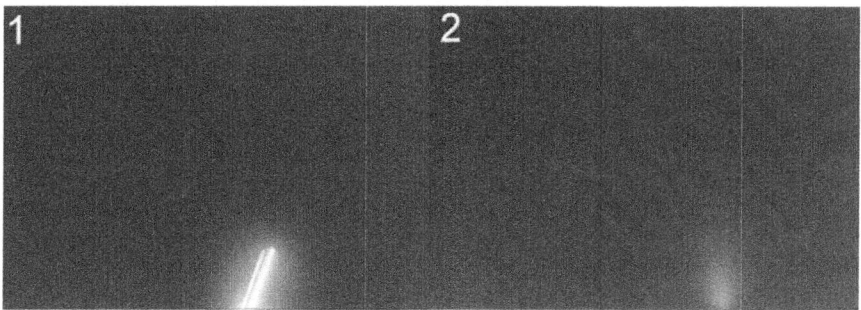

As the AOP flew overhead, Cornet could see additional lights, which were not previously apparent. When he asked Macedo how many lights he saw, he was surprised by his response. Macedo said he could see at least four lights, while Cornet could detect only three with his unaided eyes. After the event was over Macedo said he did not see the right light flaring. He said it appeared smaller than the left light. Cornet was puzzled until he replayed the video, and sure enough, there was no evidence of a flaring light. Even more peculiar, there was only a single light flanked by smaller lights, not two main headlights. But what he soon realized was that Macedo saw and described what was captured on the video, while

Cornet saw and described what would be confirmed by his time exposures. And yet the two of them were standing no more than three feet apart, and their cameras were only two feet apart. How could they see and photograph two completely different sets of lights, when it was clear there was only one AOP that flew over them?

It was also clear from Cornet's analysis of the data that they had seen two very different sets of lights, seemingly coming from only one source or craft. What is more astonishing is Macedo and Cornet came to this location to document such differences between observers on the ground, and the visitors seemed to have known what they were doing. In response, they gave Cornet and Macedo the very evidence they needed in order to document the phenomena.

When Cornet got his prints back from the photographic shop, they showed exactly what he had seen. Because he was calling out times for the beginning and ending of each of his time exposures, the audio portion of the video had a precise record of what time exposures corresponded with what portions of the video. That is when the optical trick was revealed! Only when the right light in his time exposures brightened, did a faint record of that light appear on the video. But its position was many feet (estimate) below the row of multicolored lights that Macedo described seeing and what his camcorder recorded. It was because the brightening of the light caused unpolarized light to form that the camcorder could "see" that light. It became apparent that the lights were so directed and polarized in different directions and for specific targets that two very different sets of lights could be directed from different parts of the same craft to different cameras and observers only two-three feet apart! Mystery solved!

How does one explain two cameras capturing two different sets of lights, and two people seeing only the lights their cameras captured? When one examines the video carefully, something remarkable becomes evident. During the first time exposure, nothing can be detected on the video, but during the second time exposure, the light (image) somehow was bent downwards so that most of it did not get captured on film. When that happened, a faint brownish light or glow can be discerned below the main light on the video (occurs at ¥ in <u>table above</u>. An enlarged video frame in the lower right corner of <u>Graphic No. 2</u> below shows this glow clearly. It soon disappears in the video.

When Cornet saw the right light flare, it produced an orange glow around it, characteristic of Nitrous Oxide gas being produced by an intense plasma gas (Google Search <u>Plasma Orb</u> on the internet). That glow was captured on his time exposure (photo no. 1 above). It's this glow which seems to have been captured on video. Why only the glow? To answer this question we have to consider additional evidence. The results of photographic and video frame analysis are given in the two composite graphics below:

Graphic No. 1

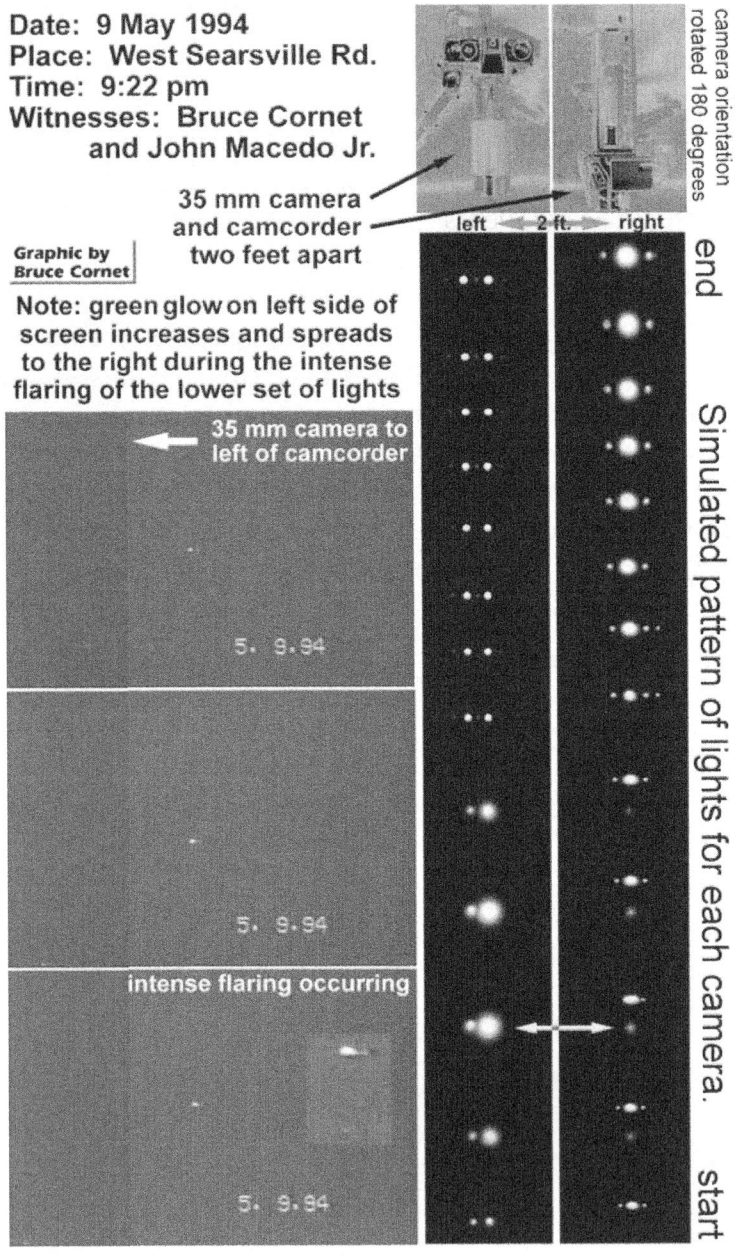

Graphic No. 2

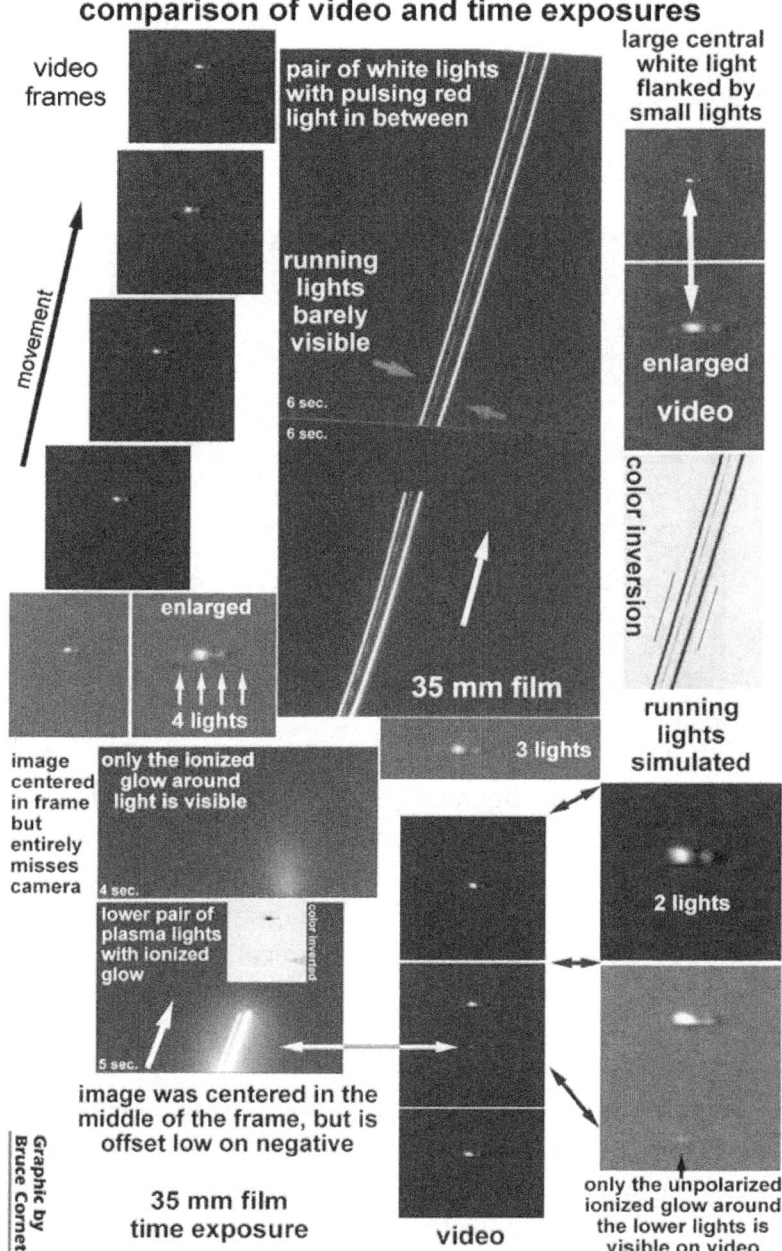

Graphic by Bruce Cornet

Discussion

Look at the three enlarged video frames on the left side of Graphic No. 1. Note the intense (green) glow on the left side of each video frame. Now look at photo no. 1 enhanced (below) in order to bring out a mirror-image artifact (green color distortion) on the right side of the time exposure.

There appears to have been some sort of energy that affected the right side of the film in the SLR camera and on the left side of the videotape in the camcorder, which was positioned just to its right. That energy was capable of causing a magnetic (color) darkening on the videotape and an emulsion (chemical) color darkening on the 35mm film. The darkening is called wave interference.

One possible explanation for this effect is that the images projected to each of the cameras was done using highly coherent beams of energy - not unlike the beams in the Young's experiment (Figure x1). You might say the light was polarized, but not in a conventional sense, because there were no filters on the cameras which could make polarized light visible (Figure x2).

Young's Experiment

A one-color light source passing through a pair of very closely spaced slits becomes two coherent beams of light. Instead of two lines, a pattern of bright and dark bands is observed on a screen far from the slits. Young demonstrated the wave nature of light.

The bright bands happened when the path difference is $n\lambda$, (n = 0, 1, 2, or 3...)
The dark bands happened when the path difference is $(n+\frac{1}{2})\lambda$, (n = 0, 1, 2, or 3...)
The band is bright due to constructive interference, and dark due to destructive interference.

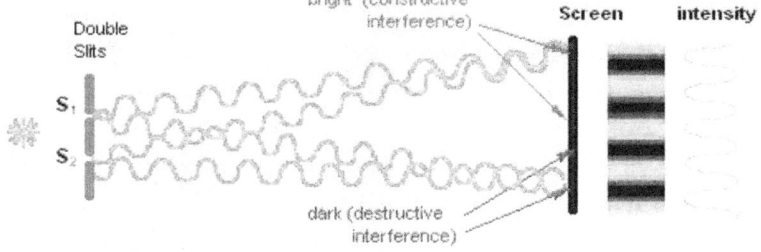

from Science Master: www.supertutor.com

Figure 1. Young's Experiment

Polarization

In an ordinary light, the vibration is in all directions perpendicular to its direction of propagation. It can be resolved into two perpendicular vibrating components. When a polarizing filter is put in the middle of a beam of ordinary light, only those waves vibrating in one plane can pass through and others will be blocked. A polarized light with half its original intensity is then produced.

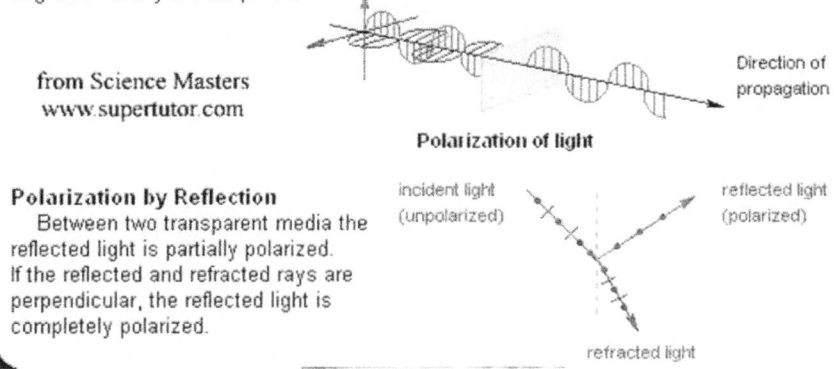

from Science Masters
www.supertutor.com

Polarization by Reflection
Between two transparent media the reflected light is partially polarized. If the reflected and refracted rays are perpendicular, the reflected light is completely polarized.

Figure 2. Polarization

The energy beams were so focused that Cornet could not see what Macedo saw, and Macedo could not see what Cornet saw. Additionally, there was magnetic and/or photonic interference between the beams, since they were so close (magnetic for the video if the influence did not occur through the lens). That interference pattern extended far enough laterally to cause adjacent color distortions on the recording media.

But why then did the videotape pick up only the glow around the flaring lights that Cornet saw? Because the glow or ionization of the air could not be controlled as well as the projected images. It therefore was not focused, and both cameras picked it up. The position of the glow significantly below the main light in the video indicates the position from which the image Cornet saw and recorded was projected. For this feat to have been designed by humans would require the integration of complex optical, infrared, night-vision, steerable electromagnetic beam technology - certainly something that 99.9% of all human-designed aircraft don't have, and certainly don't need.

At that point in Cornet's investigation, it was apparent we were dealing with highly sophisticated projectors of light and sound (and even short wave radiation: Crystall, 1991), designed for the purpose of testing human perceptions in this physical universe. What appeared to be deliberate performances and appearances to selected observers was now supported by objective scientific evidence.

In support of this hypothesis (i.e. that we humans are the subjects of non-human perceptual experiments), I quote observations about the surface structure of these craft by Dr. Ellen Crystall. She said, "I never accepted that theory [that UAPs move in and out of our dimensions from other dimensions]. I still don't. If a person were looking at a fully lit UFO when it suddenly turned off all its lights [at night], it would appear to vanish into thin air. But it would not have dematerialized; it would not be a psychic event. Its disappearance would be due to the operation of advanced technology – a kind of extraterrestrial "stealth" aircraft.

"How does this technology work? The best explanation I can offer at this time involves an analogy to sunscreen glass used for eyeglass lenses... The metallic skin of a UFO probably has a molecular structure giving it properties similar to sunscreen glass. Under some conditions, controlled by electrical current, perhaps, it becomes opaque; under others it becomes clear to the point of transparency." (Crystall, 1991).

Crystall and Lebelson (Harry) had the opportunity in 1980 to view differently-shaped craft up close – within a few feet as they passed or hovered overhead. After that period very few if any other witnesses have described such close encounters with Pine Bush or Hudson Valley craft. On one occasion Crystall was only a few feet from a craft, and it appeared solid. On other occasions she saw windows with glasslike panes. She concludes that "ships" must fly as much during the day as at night, but their invisibility is easier to achieve at night. I have witnessed (on July 1, 1992 after sunset) a craft hovering over Harriman exit #16 and toll booths on the New York Freeway. When I first saw it above the tall highway lights that surrounded the toll plaza as I was approaching my exit from the south, I thought it was a helicopter. I saw a bright headlight with a small red light behind it, as a tail light on a helicopter. I say behind it, because as I took my exit, the lights moved southwest with the bright light in front. However, when it descended and passed directly in front of a high circular ring of street lights on a pole, no physical structure was visible between the lights. No silhouette of an aircraft structure was visible.

After passing through the toll booths, this craft caught up with me, passed directly over my truck, and paced me home on Rte. 17, staying just ahead of me over the trees to the right side of the highway. If I sped up, it sped up; if I slowed down, it slowed down. It even changed its lighting to match the navigation lights of small planes flying to and from Orange County Airport as they crossed over the highway, rising above them. When I got home the white light parked in the night sky in a northwest location where no bright star exists on star charts. I took several pictures of the light, with building roofs in the picture, so I could get an accurate compass direction for star chart comparison.

Crystall relates more information on the stealthy attributes of the craft witnessed in the Hudson Valley east of the Hudson River (cf. Night Siege, 1998; In The Night Sky, 2013). It needs to be pointed out that there is a significant size difference and diversity of shapes reported between areas east and west of the Hudson River. At this time I have no explanation for the smaller sizes and greater diversity of shapes west of that river. How much of this difference is due to experimental parameters and illusions created by the unknown intelligence, and how much is due to the source or sources of these craft is something that investigators in the future need to address. When one considers how much this phenomena has changed over the last 90+ years (since the early 1920s: Alien Bases, by Timothy Good, 1998), and more accepting the public has become of a possible non-human presence, one wonders what the future will hold.

Chapter 2 Other Types of Unconventional Craft

So far I have discussed various shapes of UAP/UFOs and sizes ranging from stealth black triangles about 60-70 feet wide, small boomerang-shaped craft reported by Crystall (1991), large triangles 100 feet or more in width, very large boomerang-shaped ships 300 feet and larger in size, and circular-disk-shaped craft 300 feet in diameter. Even hybrids between angular or V-shapes and disk-shapes have been reported and photographed (Hynek et al., 1998).

Although very large craft have been reported more frequently east of the Hudson River, they have traveled west at least as far as Pine Bush and Monticello (Orange and Sullivan counties, NY). Russel Pardi reported his sighting to me, which appears not to be part of the Hynek-Imbrogno data base, because it occurred well outside their study area.

Pardi wrote to me, saying: "I observed a large delta-shaped device traveling east over Rte. 17 in 1984. I was traveling west from Community General Hospital of Sullivan County [NY} where I had just completed my evening shift as pharmacist. As I drove [under] the Rte. 17 overpass by the hospital I could see two large 'stadium-size' lights approaching over rte. 17. I drove onto the highway and stopped my vehicle on the shoulder, got out and watched as the device flew over me at approximately 15-20 mph; it could not have been more than 150 feet above me. It was certainly a close encounter. I could see into the cockpit of the device: rows of orange-backlit dials behind a helicopter-like clear glass dome on the belly of the front of the device. I heard a very high-pitched turbine-like noise as it passed overhead. No dust or debris were kicked up from the fields surrounding the area. The fuselage was a flat gray-black with small red and green navigation-like lights on the belly. It appeared to be about as wide as the 4-lane highway. Two large and very bright lights were positioned on either end of the delta 'wings.' I could see the countryside illuminated as the device passed me.

"The incident was reported in the Times Herald Record of Middletown. The device was reported to have preceded east bound into Monticello where local police gave chase. I spoke to the Monticello police and they described exactly the same device I had seen. The police also indicated that they had reported the incident to the then Stewart AFB (and naturally no aircraft were supposed to be in the area!). Recently I have seen A&E and Unsolved Mysteries shows depicting the same device in Westchester and Sullivan Counties."

The Orange-Ulster-Sullivan County Mini-Flap

Occasionally the UAP phenomenon occurs in specific localities for an extended period of time. These are referred to as localized "flaps." In other cases two or more UAP/UFOs may appear in the same area for a short period of time. These are called mini-flaps.

On 6 August 1992 something spectacular occurred west of the Hudson River over the towns of Middletown, Maybrook, Montgomery, Walden, Wallkill, Newburgh, and Harriman, New York. A mini-flap happened between 9:30 pm and 10:00 pm on Thursday, and was witnessed in more than five different locations.

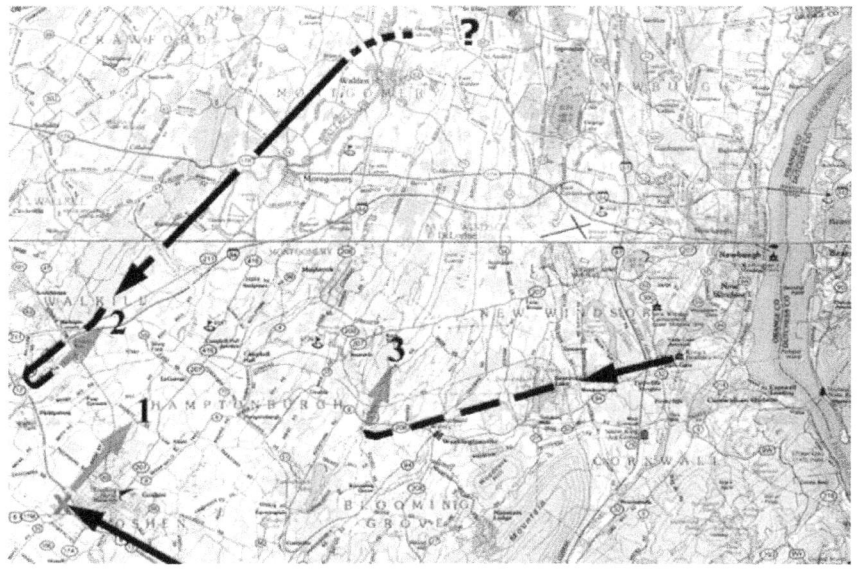

This flap caught the attention of the Sightings TV show and aired on May 21, 1993. Perhaps as many as four different types of UAP/UFOs (at least three types were documented on film and video) were seen traveling west from Beacon, NY, on the east side of the Hudson River, across from Newburgh, NY, to Middletown, Maybrook, and Pine Bush before turning around and returning to the east side of the river again.

Several people near Montgomery and Wallkill saw two craft pass near them. One was large and resembled the "Manta Ray." Another was smaller (about 100 feet wide) and looked like a black triangle (AOP), and its path is given on the map above as UAP/UFO #2.

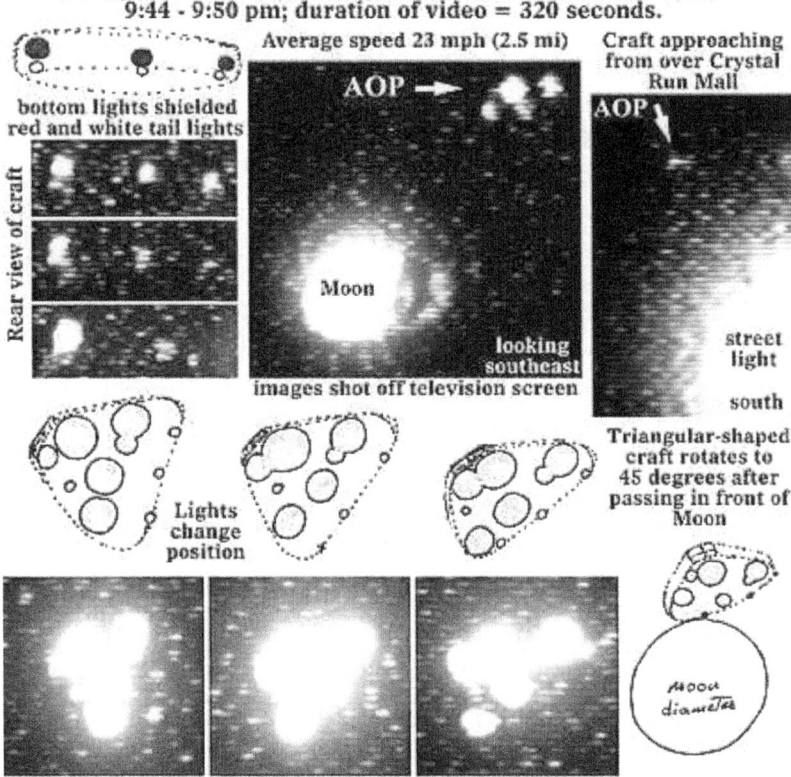

6 August 1992, Goshen Tnpk at Rte 211E, Middletown, NY
9:44 - 9:50 pm; duration of video = 320 seconds.

Witnesses: Bruce Cornet, Patricia Huff, Robyne Leisti, and neighbor.

Slightly to the south, Ellen Crystall was traveling west towards Middletown on Rte. 17 from her home in New Milford, NY, when she spotted a set of lights off to the north traveling in her same direction. When she reached Chester, New York, the lights stopped and hovered about 1000 feet above the highway. She pulled her car off to the shoulder and began videotaping the UAP/UFO from about one quarter mile away. She recorded 12 minutes of a giant craft slowly turning clockwise. In the video you can see internal structure periodically blocking the lights on the far side of the object. When the pattern of lights are compared with those videotaped of the Westchester Boomerang by Pozzuoli on 24 July 1984, they match. Ironically, this UAP/UFO stopped and hovered just West of Chester, NY. Was it the Westchester Boomerang? The collage of images from Crystall's video show the UAP/UFO rotating clockwise. Images #1and #2 below resembles that of the Westchester Boomerang videotaped by Pozzuoli. The path of this UAP/UFO is given in the map above as UAP/UFO #1.

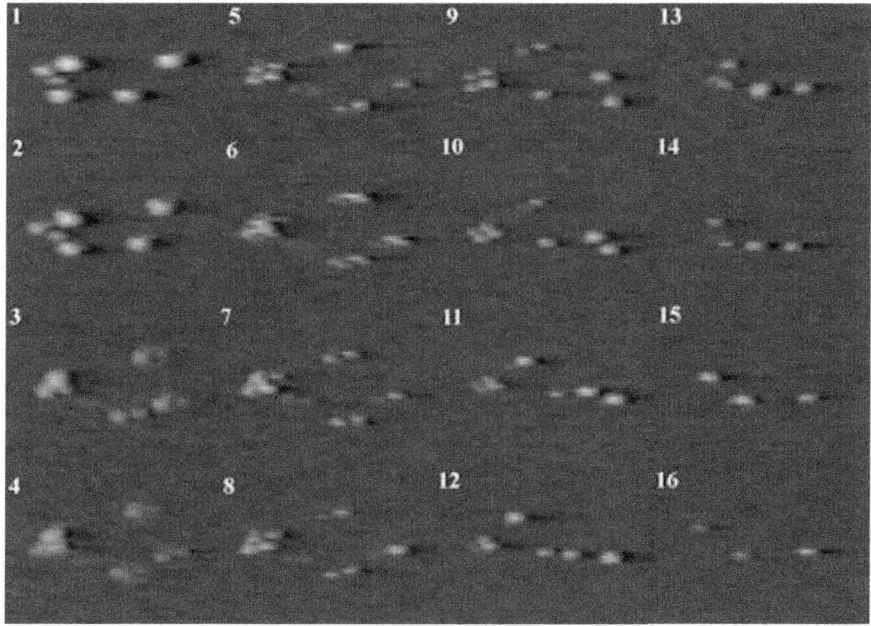

Further south witnesses saw a large disk-shaped craft travel from Vinton near Newburgh, west, where it stopped and hovered above the Crampton house in Maybrook. All the clocks in the Crampton house stopped working at about 9:30 pm. The path of this UAP/UFO is given in the map above as UAP/UFO #3.

At the same time my fiancée and her daughter (Pat and Robyne, respectively) were visiting me at my Middletown condominium, which was located in the village of Scottstown, located at the southeast corner of Middletown, NY. It was 9:44 pm, and I was in the kitchen fixing something to eat when I got a very strong urge to go outside and look east. When I went outside I was greeted by a pair of bright lights approaching my condo from the east, just beyond the complex of parking lots. The lights stopped and turned south (to my right) and moved towards Rte. 211 East that borders the Hillside Village Condominiums. "Knowing" what it was, I ran inside and yelled to Pat and Robyne, who were watching television, to come outside. I grabbed my Sears VHS camcorder and ran outside. I located the UAP/UFO over Rte. 211 East as it traveled west towards the Crystal Run Mall several miles further west. Its shape was distinctly triangular. It had several bright lights up front and more beneath its flat belly. It had several red lights across its rectangular back side. I videotaped it as it moved west, as it became more indistinct among the street lights down the road. Then it stopped, turned around, and started heading back towards us.

By that time Pat and Robyne, and another neighbor girl, were watching this spectacle. The triangle came back and passed us to our south, following Rte. 211 East. It climbed higher in altitude and deliberately flew in front of a nearly full moon! On the video you can see a dark triangular shape with five bright white lights on its belly. The video shows the lights changing position on the belly — something that will be discussed in the following section on lights. The craft then descended and disappeared behind trees in the foreground as it continued to fly east following Rte. 211 East, back the way it had come. With its red tail lights clearly visible blinking on and off in the last images of it captured on the video, it descended to less than 200 feet above the ground. Its speed was slower than the cars traveling on Rte. 211 East, where the speed limit was 40 mph.

This entire event, or mini-flap, was featured on the Sightings TV show, with Crystall, the Cramptons, and Cornet being interviewed in late 1992.

The Hudson Valley Mega Ships

The database assembled by Hynek et al. (1986/1998) includes 7,046 sightings for the Hudson Valley from 1982 to 1995. Most of these sightings occurred in Dutchess, Putnam, and Westchester counties, NYU, on the east side of the Hudson River, while some reports directly across the river in Rockland, Orange, and Sullivan counties are included, as well as some sightings in Fairfield and New Haven counties of Connecticut. Zimmermann (2013) extends the number of sightings with her later interviews of witnesses from this same period of time. The Mega Ships came in various sizes, the largest and most common of which was described as the Westchester Boomerang, because it was first reported in Westchester County. In the study by Hynek et al. (1998), most common shapes reported were Boomerang (33.6%), Triangle (30.0%), and V-shape (24.6%), while disk shape accounts for only 9.9% (Hynek et al., 1998). Most reports indicate the size of the craft was about 100 feet across (59.9%), while large craft (about 300 feet across) accounted for 24.9%, and very large craft (greater than 300 feet) accounted for 15.2%.

In the study by Greco and Gordon (MUFON UFO Journal, No. 290, June 1992) for the Williamsport, PA, wave on 5 February 1992 (western end of the UFO Corridor, 185 miles west of Pine Bush, NY), the most common shape was the Boomerang (76.9%), while Triangle, Pie- or Bell-shape, or other shape were seen only once (7.7% each). The shape of the object most likely varied as a consequence of the angle from which it was viewed. One witness directly under the object reported a true triangular shape. All of the Williamsport witnesses related the size of the craft to distances between trees on their property, that of their house, or that of buildings on the adjoining streets. Most of the reports indicate the size of the craft was between 100-150 feet wide (46.1%) or slightly larger at 150-200 feet (30.8%), while the smallest one was only 50 feet across (7.7%). Those reported wider than 200 feet at 450 feet (7.7%) or 600 feet (7.7%) may be a reference to the Boomerang.

In contrast, most triangular craft observed by me at and near Pine Bush were less than 100 feet in maximum width. Only the large diamond-shaped "Manta Ray" was the size of a jumbo jet (Boeing 747) or larger – over 195 feet in width, while a large disk-shaped craft reported by the Cramptons for the Sightings TV show was estimated to be about 300 feet in diameter. Perhaps because most triangular craft flying over the valley west of Stewart International Airport (at Newburgh, NY) are about the size (width –not length) of conventional aircraft that fly in and out of that airport, they can more easily disguise themselves and mimic the DC-9 (American Airlines) that typically flies out of that airport. Some of the alien craft actually copy the size and shapes of conventional aircraft.

It is because of that similarity in size to conventional aircraft that many people were skeptical of UAP reports in the Pine Bush area. That situation (i.e. misidentification) was not as much of a problem east of the Hudson River, where craft 300 feet in width or larger were more frequently observed, and many of the 100-foot-wide triangular craft were seen hovering silently, sometimes over man-made reservoirs, ponds, and lakes. The triangular craft (FTs) flying west of Stewart International Airport frequently had booms sticking out in front and back of the craft that supported lights, which the DC-9 had in those positions on its cigar-shaped fuselage, and projected synthetic jet-like sounds that when analyzed lack white noise typical of turbofan engines, with their frequencies divided into about 12 discrete frequency bands indicative of "synthetic music."

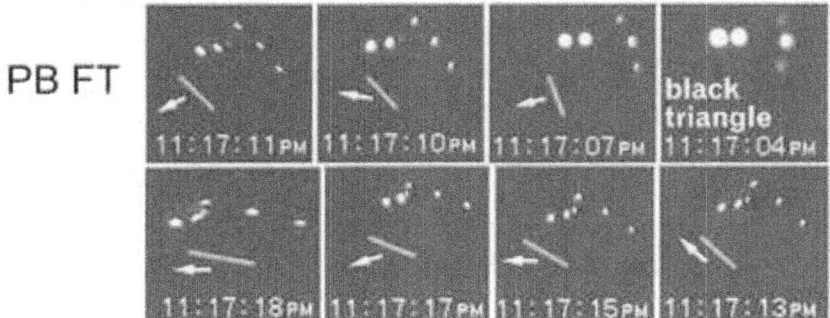

FT lights rotate relative to orientation of lights in back, which supports boom hypothesis with attached lights.
FT videotaped at Pine Bush, NY, on 4 June 1997,
DC-10 videotaped at Sea Bright, NJ, on 6 June 1997.

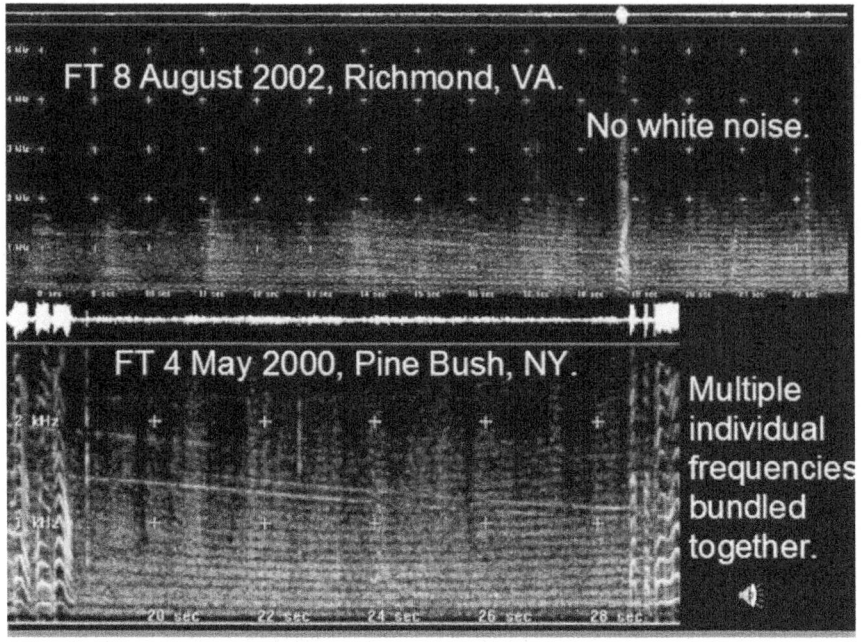

DC-9 showing how tail assembly is illuminated by strobing wingtip lights and reflections off of engine pods behind wings.

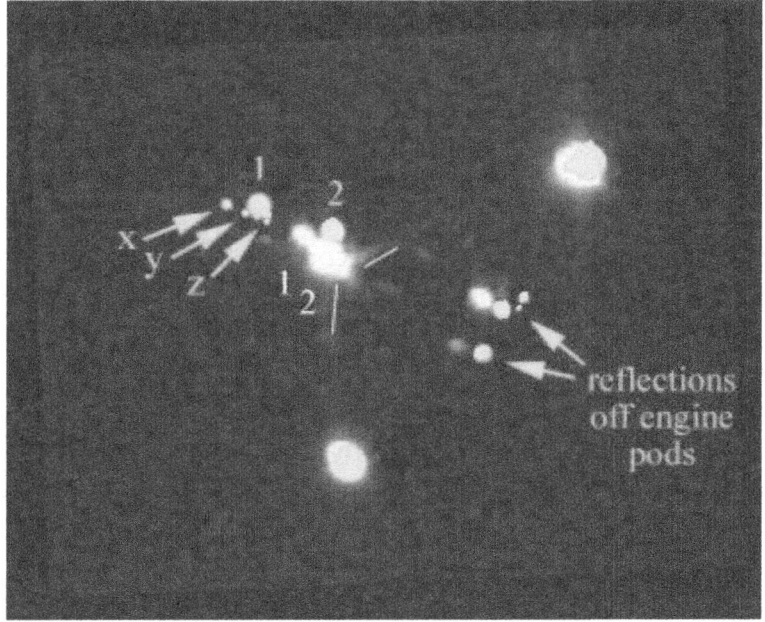

DC-9 showing position of navigation lights and reflections, wingtip strobes, and landing lights, including window lights (x, y, and z) on the side of fuselage.

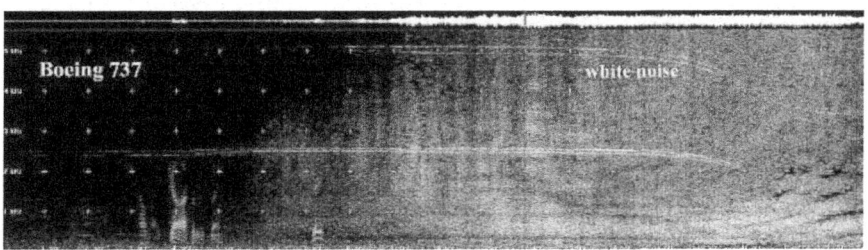

Above: Note the amount of white noise produced by conventional turbofan engines. About five thick frequency bands visible below 3 Khz, unlike UAP/UFO sound with 18 thin frequency bands below 2 Khz.

Another reason why it was easier for people to distinguish UAP/UFOs from conventional aircraft to the east of the Hudson River was a lack of sound (42.8% or a faint humming sound (55.1%):

"Night Siege," 1998. Only 2.2% of those craft produced an engine sound, although not always expected for a large craft if the sound resembled that of a single engine airplane.

The Greco and Gordon study for the Williamsport, PA, wave (MUFON UFO Journal, No. 290, June 1992) reports that a loud rumbling noise (61%) was twice as common as a heavy rumbling sound (30%), while the distinctive rumbling noise of a jet engine (7.7%) was the least common sound. All witnesses agreed that the noise shook the house and rattled the windows (like the experienced I had on 25 September 1995, which caused all the local dogs to bark incessantly). They varied in their interpretations of intensity and frequency, which were probably based on past experiences.

Witnesses east of the Hudson River sometimes were close enough to a craft to observe its structure. In 54.4% of Hynek et al.'s (1998) cases the structure was described as dark and/or metallic. In 34.9% of the cases no structure was observed, but it was not determined if the craft was close enough for structure to be observed. Crystall described in her book, "silent Invasion" (1991), difficulty in seeing structure close up, except around the craft's lights. In Hynek et al.'s book, many people remarked on how little light reflection could be observed off of the craft's structure, and what reflection did exist was very close to its lights. Some witnesses described how the lights appeared recessed into wells or cavities in the structure, which may in part account for so little reflection around the lights.

In the eye witness reports detailed in "Night Siege" (Hynek et al., 1998), several people report a massive network of tubes and grills underneath the object, and tubes associated with some of the lights. Crystall in her book illustrates a drawing of a triangular craft with tubes (presumably for plasma lights) sticking out along the sides of the craft. I received a report from a female police officer who worked in Pine Bush of her seeing an enormous craft hovering and moving slowly over a park (open area) just west of the town. It had a network of tubes and grills on its underside. She said she stood directly beneath the object, and that it looked like the back of a refrigerator. None of these descriptions even vaguely resemble anything manmade. Some of the ships were so large (bigger than a football stadium) that hiding them, especially from satellites (e.g. Google Earth and weather satellites) would be virtually impossible today. And that fact raises an interesting question of timing relative to our rapid technological development in just the past decade. All of my photographs were analog-based, giving high resolution even under a microscope. Such images would have been prohibitively expensive once the relatively inexpensive digital images replaced analog film. In fact, the digital revolution came into being in 1998 just after the Pine Bush flap ended in 1997. Coincidence?

Some witnesses in Hynek et al. (1998) describe an open structure to the craft, where the witness could see into the craft, seeing cross beams and internal supports. Such a description brings to mind Ellen Crystall's description of hull transparency and invisibility. Other Hudson Valley witnesses describe a massive superstructure six stories high above the craft, only made visible when the object passed in front of a full moon. Ellen Crystall (1991) also saw a craft near Pine Bush that had a tall superstructure visible. On May 4, 2000 Barbara Hartwell and I were standing at a gateway to an estate on the hillside just to the west and south of the dogleg bend in west Searsville Rd. where I had watched UAPs hundreds of times during the previous eight years. It was 11:14 pm. We spotted a pair of bright lights coming towards us from due east. The lights were very low to the ground, so low that tree branches (no leaves yet in spring) in the foreground can be seen in front of the lights. I began to videotape the event.

Cornet: "Whoa, whoa, whoa, whoa, what the hell is this thing."
Hartwell: "Bruce, I've seen things like this before. This looks like a real thing. What the hell."
Cornet: "This is huge.
Hartwell: "Look at it. It's so bright!"
Cornet: "It's coming; it's slowly coming towards us."
Hartwell: "Why is it so bright?"

Those branches gave us a reference object for judging distance and altitude. The lights were just above distant trees. They then flared or brightened beyond the capability of landing lights on conventional aircraft. After about 20 seconds they dimmed. As they dimmed, we could see smaller lights between the large lights and just outside them to their left and right. Other lights came on.

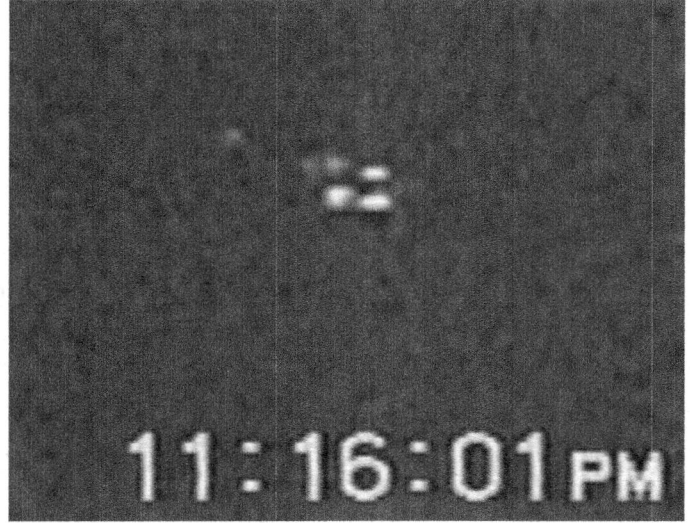

Then the craft turned south without banking. We could see outboard strobes and two rows of large window lights along its side – much too large at a distance of about a mile to be the windows on a Boeing 747, a type of aircraft which does not land at Stewart Airport. Only large C-5 Galaxy Air Force cargo aircraft land at that airport, and they do not have two tiers of windows or lights. As the craft turned slowly east again without banking, we could see a tail light. At no time did we hear a sound coming from the object, but it could have been too far away to hear an engine sound. Crystall (1991) reports seeing a box-shaped UAP pass close to her with a row of three large rectangular windows along its side, and two bright headlights up front. Was this the same craft?

Shortly after that event we spotted what we initially thought was a conventional aircraft on its normal approach to Stewart International Airport across the valley. The second craft flew directly towards and over Cornet and Hartwell at 11:30 pm. At first Cornet thought it was a conventional jetliner, but noted differences in its lights. It had a pair of lights at the nose of a chevron-shaped front, and red and green outboard lights. The nose lights quickly increased to four small lights forming a square. Instead of following the low ridge (Thompson's ridge) to our west and turning east at the red beacon on top of that ridge (which would orient an aircraft towards the main runway at Stewart), it veered towards us and passed directly over us.

As the craft flew over them, it turned on beams of light coming from two bright lights at the outboard margins of its fuselage. Those lights and beams quickly rotated forward and aligned with the other forward facing lights. The craft was no more than 500 feet above them. We could see no forward fuselage, only four small lights demarking the corners of what could have been a large window at the front of a V-shaped or angular craft – similar to what Crystall (1991) shows in her first drawing (A) in her book. As it passed over us I saw its entire shape. It was the "Manta Ray" diamond-shaped craft with a tail light. Although it produced a jet-like sound, later sound analysis showed that the sound was synthetic and violated the Doppler Law (discussed later in Chapter Five). The craft did not travel to Stewart Airport, but disappeared below the tree line to the south. Such an occurrence of two events over a period of 16 minutes, with both craft moving directly towards us and not on conventional flight paths to or from an airport, indicates that we, the observers, were deliberately targeted.

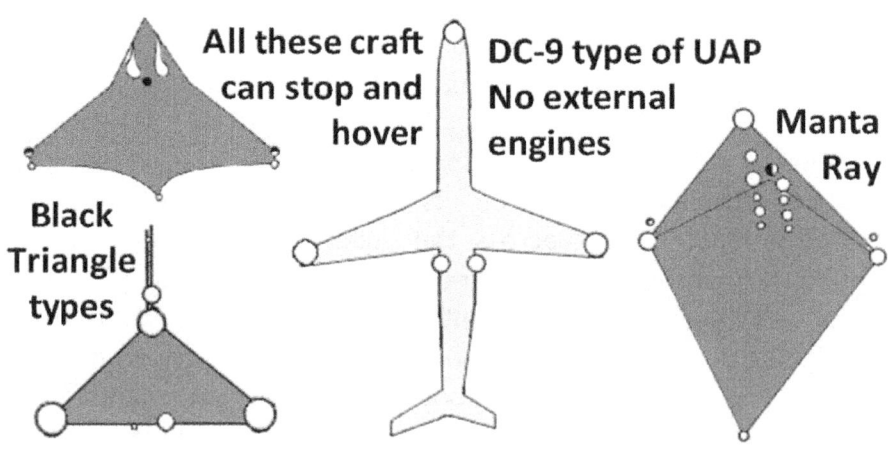

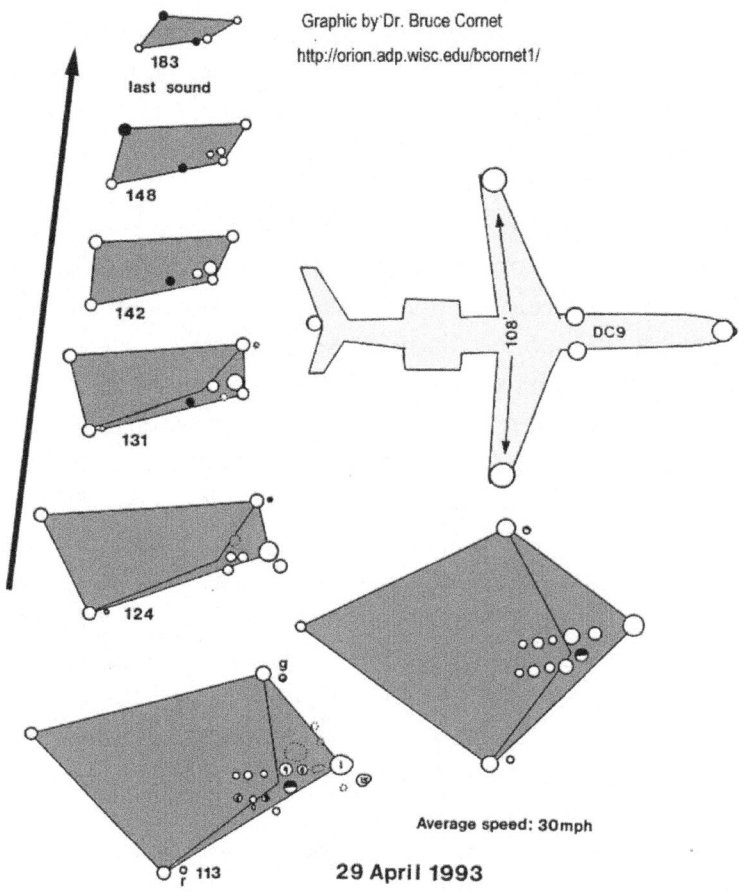

Above: Comparison of the Manta Ray with a DC-9 jetliner, showing lights that appeared on belly of Manta Ray as it turned in front of Sightings TV camera crew, Crystall, Cornet, and two other witnesses.

Mimicry: *Zool*. The superficial resemblance which some animals exhibit to other animals or to the natural objects among which they live, thereby securing concealment, protection, or the like (Webster, 1961).

In order to convince the reader that these craft were not conventional aircraft or that we misidentified them or confused differences in their navigation and landing lights with typical light patterns on commercial or military aircraft, a detailed comparison is given below.

The Manta Ray has been commonly seen in the Wallkill River Valley near Pine Bush and Montgomery, NY (Orange County), and also reported in the Hudson River Valley region east of the Hudson River is a kite-shaped or diamond-shaped craft called here the "Manta Ray" due to its shape and extended tail, which resembles the marine fish by that name. It is a large craft, wider than a Boeing 747 wingspan of 195 feet. It is stealth black, and supports a variety of colored lights of different size on its belly, along its sides, and on its back. It has been videotaped both by Crystall and Cornet flying overhead and changing shape. It was witnessed by a woman standing at night along West Searville Rd. not far south of Hill Avenue as it passed slowly over her head, nearly clipping the tops of the trees on the west side of the road. She saw the tail section telescope into the forward section of the body, and disappear as the craft transformed from kite-shaped to triangular shaped.

The video Crystall took on September 4, 1992 as the Manta Ray flew over the Beth Hillel Jewish Cemetery along Rte. 52 between Walden and Pine Bush (Cornet, Brock, and Wisch were also present) shows forward lights at the nose of its chevron front leading edge changing position, with a top light moving onto its belly. The video I took on 20 August 2003 at Salt Point, NY (northwest of Poughkeepsie, NY, on the east side of the Hudson River) with Billy McNamara as witness, also shows the same craft and forward lights changing position as it slowed to a stop while turning. It did not bank. Again, one forward light moved down and onto its belly and shifted behind a second forward light. That was the last time I witnessed the Manta Ray before moving to the southwestern United States.

The Manta Ray typically displayed bright lights out where wingtips would normally be on fixed-wing aircraft, and when they first turned on they were pointed down towards the ground. These lights turned on just as the craft flew over us, again indicating deliberate targeting. Hartwell spoke out as this happened, "They're looking for you Bruce."

The outer lights were positioned just inside the outermost running lights, and are not landing lights. Landing lights on conventional aircraft are positioned much closer to the midline so that the pilot and copilot can have the runway in front of them illuminated. Landing lights positioned out at the wingtips are illogical and impractical, because they leave the runway directly in front of the aircraft dark. Bright lights at the lateral extremities are a distinct characteristic of anomalous craft witnessed in the Pine Bush area. Therefore, the lights on the Manta Ray are called HEADLIGHTS, not landing lights. Even still, their utility and purpose is in doubt, except as unusual navigation or hazard lights. What where they aligned to illuminate?

It is possible that the outboard headlights of the Manta Ray were functional, and designed to illuminate the walls, floor, and ceiling of a tunnel through which the craft flew to reach its underground base. That explanation supports the observed behavior of all these craft, some of which were actually seen and sometimes photographed passing into the ground after passing through the trees of a forest.

White arrow points to the last visible light of craft before it disappeared into the ground, probably opening up a tunnel in its path. What you see in the photograph is a dense forest extending back hundreds of yards, and not a clearing behind a group of trees.

The Manta Ray reveals rotating headlights

Conventional jet on left; Manta Ray on right.

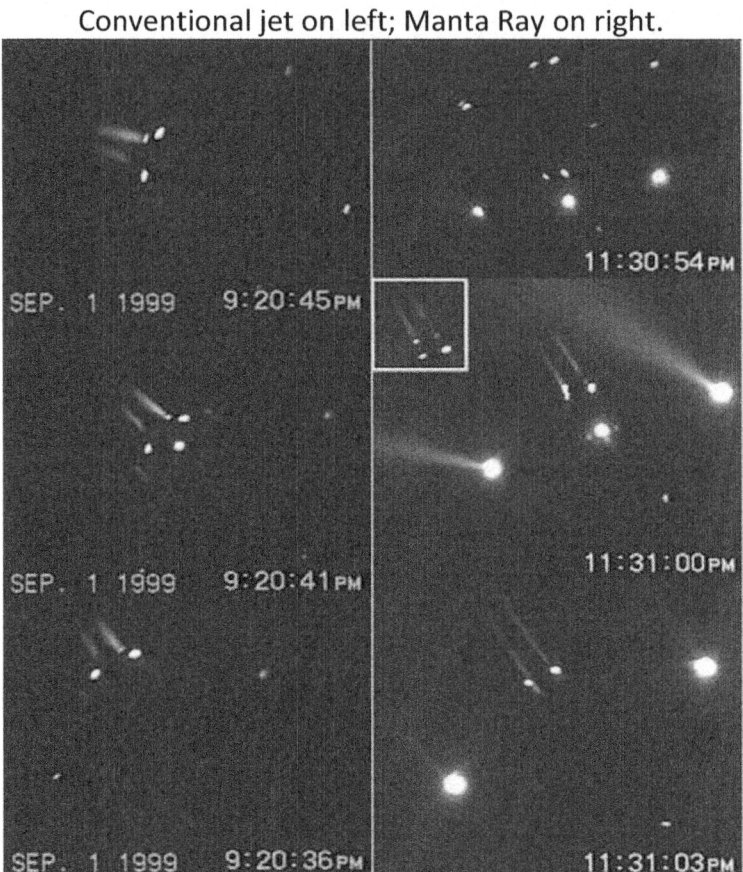

Strobes on conventional jetliners (on pp. **33-34**) and non-stealth aircraft are bright enough to reflect off of portions of the fuselage. The belly strobe on the jetliner (above left) reflects off of the engine pods, while the strobe on the Manta Ray (above right) does not, indicating either shielding and/or the absence of protruding engine pods. Strobes typically reflect off of engine pods on conventional jetliners.

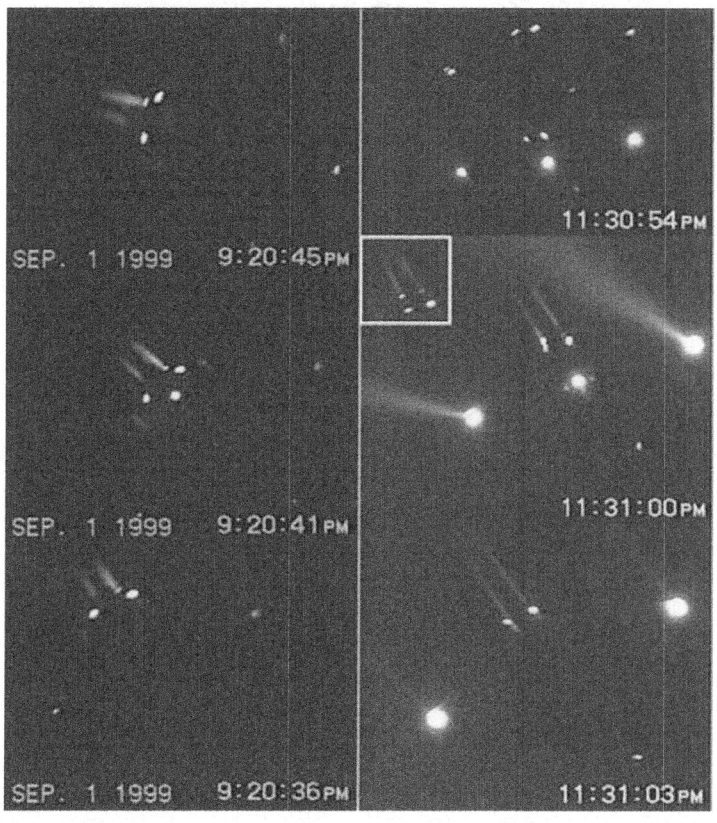

Above: Comparison of Manta Ray lights for September 1992 and May 2000 sightings indicate similar shape and pattern of lights. Note how the paired forward lights are not always aligned. That is because they have been recorded moving around independently, rising, moving sideways, and backwards along the hull of the ship. The pictures below are of the Manta Ray videotaped on 20 August 2003 near Salt Point (Millbrook), NY. They show the movement of the central two lights and panel reflections. This sighting with Billy McNamara was later described in "UFOs Over New York" by Preston Dennett (2008: pp. 223-224).

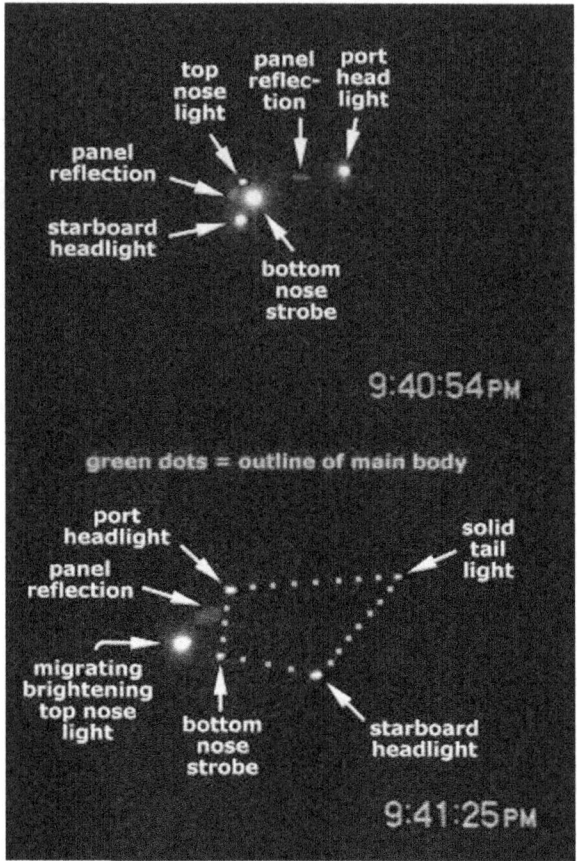

Typically, lights on conventional aircraft (below) are fixed and their wings and structure are made partly visible by navigation lights and strobes. The tail on the jet to the right is clearly illuminated.

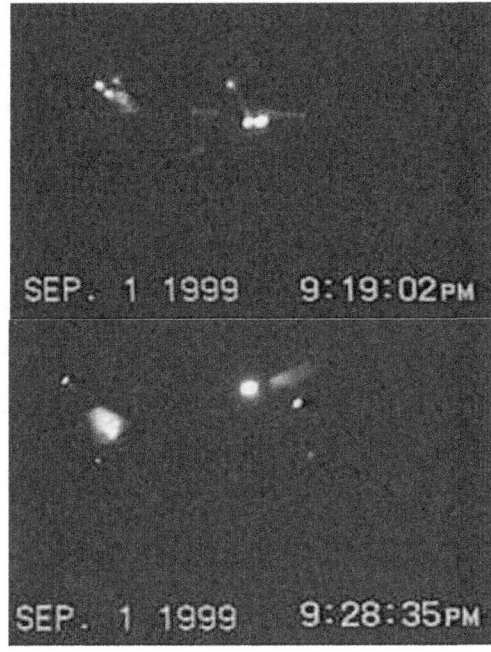

Above: Note how strobes reflect off of tail vertical stabilizer and make it visible on conventional aircraft. No tail assembly is made visible by strobes and lights on the Manta Ray or on the Black Triangles.

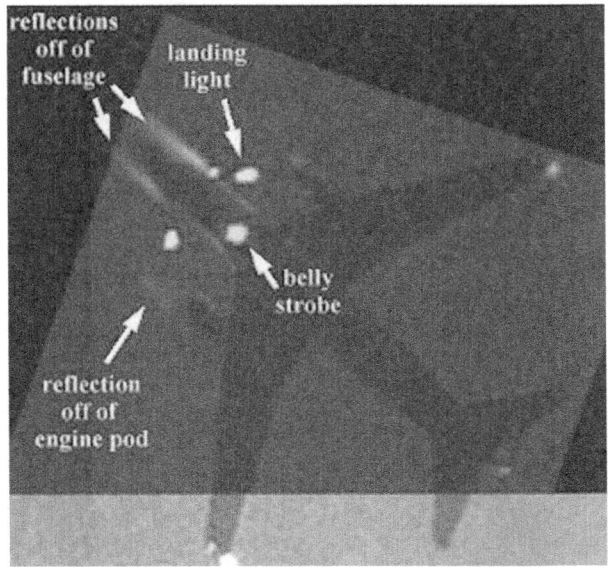

Above: Overlay of Boeing 767 with September 1 jetliner (from above images) to show layout of navigation and landing lights.

Comparison of Manta Ray lights to those of a conventional jetliner indicates significant differences even though there is general similarity (possibly indicating deliberate mimicry). Landing lights on conventional winged aircraft are positioned near the fuselage, not out at the wingtips.

First encounter with the Manta Ray

June 9, 1992 (Tuesday) was the beginning of my field experience with Ellen Crystall, who was with Mike and Chris. As a scientist who avoided the subject of UAPs in the academic world, this was a whole new world of experience. I had recently finished reading "Silent Invasion" by Ellen Crystall, had contacted her, and was asked to join her in the field at night. Because she had hypothesized that an alien base existed underground in the valley, and I had access to professional geological magnetometers, I decided to begin a magnetic survey in order to test her hypothesis. Little did I know what I would discover, and how much it would change my life. My magnetic survey took more than three years to complete and ultimately involved 24 square miles of 1,800 station measurements.

Pat Huff (my fiancée) and I went to the Beth Hillel Jewish Cemetery on Rte. 52 west of Walden, NY, at 12:18 pm in order to take magnetic measurements along that road with an E.G.&.G. Proton Precession Magnetometer, which was on loan from Lamont-Doherty Earth Observatory where I worked. Before sunset Pat headed home to New Jersey where she lived.

That very first evening in the field I would be exposed to stories that seemed to be coming out of Science Fiction novels. After dark I joined Ellen, Chris and Mike, who picked me up at my condominium in Middletown, NY, in Ellen's car. We first went to the Jewish Cemetery on Rte. 52. A party was going on at a house down the road from the cemetery. The music could be heard wafting occasionally across the field to the cemetery. We saw a silent high flying ship, moving very fast from east to west; it made a right angle turn without banking or slowing (indicating to me that it was not a conventional aircraft), and preceded north out of sight. We also saw two UAPs take off from the field behind the cemetery. The second one circled field and then flew east, making an engine-like sound (reciprocating sputter). There was no conventional airport near that location, so I could rule out a conventional explanation.

Then Chris and Mike told us they saw a small bipedal figure running across the field behind the Jewish cemetery fence after hearing an ear-piercing screech. The small figure was only about 3-4 feet tall, and disappeared into the woods. Chris and Mike said they had seen a slow moving "animal," described as a stooped biped on all fours, large, seen moving through field at West Searsville Rd. That encounter was close enough for them to describe "slanted eyes."

These early experiences and sightings told me that something was going on in the valley that was highly anomalous. I would soon be hooked on the mystery, and as a scientist I wanted to collect enough data to be able to answer at least some of the many questions I had. The most important question to me was: Why were my colleagues in academia and the government deliberately ignoring a phenomenon that was so in the face of those residents around Pine Bush who had experienced sightings and paranormal events? I would discover that most residents who had lived in the area for a while did have UAP sightings (Read "In The Night Sky," by Linda Zimmermann, 2013).

The first time I saw the Manta Ray was on July 10, 1992, just one month after I had begun going out with Ellen Crystall and associates for skywatching at night near the Pine Bush-Montgomery-Walden hotspot area. On that night I was alone. I had purchased a roll of Infrared Black & White film, because I wanted to collect data on the relative temperatures of the UAP lights to see if the film captured significant differences from the lamps of conventional aircraft. I drove to South Searsville Rd. just north of Montgomery, NY, from my condominium about 10 miles away. The skywatching location was not far north of Orange County Municipal Airport 2.8 miles away, from which small light aircraft took off and landed. I set up my Minolta XG7 SLR on a stable tripod on the south side of the road at a recess and gate to the fence that bordered a large farm field. Across from me was another farm field that stretched at least half a mile north to the edge of a forest and swampy area. Beyond that forest were more farm fields, barns, and farm houses. South Searsville Rd. Stretched east-west between low hills on the west side of the Wallkill River Valley, and almost to the Wallkill River in the middle of the valley. The Wallkill River flowed north from its watershed area to the southwest of Pine Island, and emptied into the Hudson River near Kingston, NY.

It was 11:53 pm when I spotted a set of lights (two larger headlights with a small red strobe in between) heading south just beyond the forest across the field to the north. My camera was loaded with Infrared Black & White film. I began taking a series of time exposures as the lights slowly approached, headed straight for me. They were too low to the ground for conventional aircraft, which was my first clue that this might be a UAP. By the time the lights reached the field and were half a mile away, I had taken several time exposures, each about 12-15 seconds long.

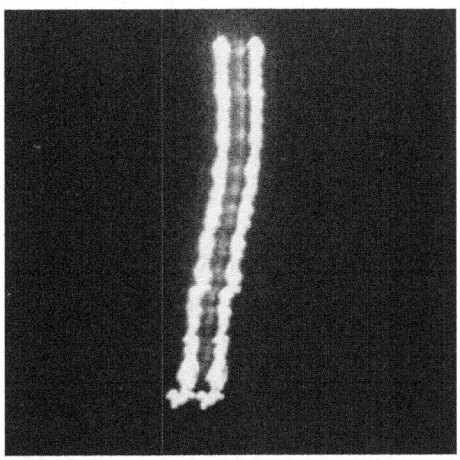

By the time the craft reached the middle of the field, I ran out of film! I thought I could run to my pickup truck parked on the dirt and grass shoulder of the road to my left, get a new roll of film, and reload my camera before the craft passed me out of camera range, because it was traveling so slowly. If this had been a commercial jetliner, it would have been out of sight in 15-20 seconds.

I quickly rolled the IR film back into its canister; popped open the back of my camera; removed the canister; ran to my truck; leaned inside through the driver's door; fumbled for another roll of film as I put the exposed roll into my leather camera bag; ran back to my camera; installed the new roll of film without checking what type of film it was; sighed in relief as the camera sprockets grabbed onto the holes in the film; closed the back of the camera; and looked up. I totally expected to see the craft over me or moving south away from me, but I wondered why I was not hearing any engine sounds.

When I did not see the UAP either above me or behind me, I looked back north to where I had last photographed the craft. To my astonishment, there it was poised in the night sky, stopped, and waiting for me to change the film! What better proof is this that I was not misidentifying a conventional aircraft? As soon as I began refocusing my camera and 300 mm zoom lens on the lights, they began moving towards me again. I took one or two time exposures, but as the craft got closer, I had to reposition and point my camera faster and faster. By the time the craft was over me, I was snapping pictures quickly. As it moved away slowly, my time exposures increased in length again. Then the craft slowly turned east without banking, and disappeared beyond sight past the trees to the east.

When the craft was over me, it was enormous. I could see its silhouette clearly against a lighter black sky. There were no wings. It was kite-shaped with a short tail. Now various colored lights could be seen around its perimeter and on its belly (white, red, green, and even blue lights). I heard no sound as it passed directly over me. It couldn't have been more than 500 feet above me. It was as wide as vertically out-stretched arms.

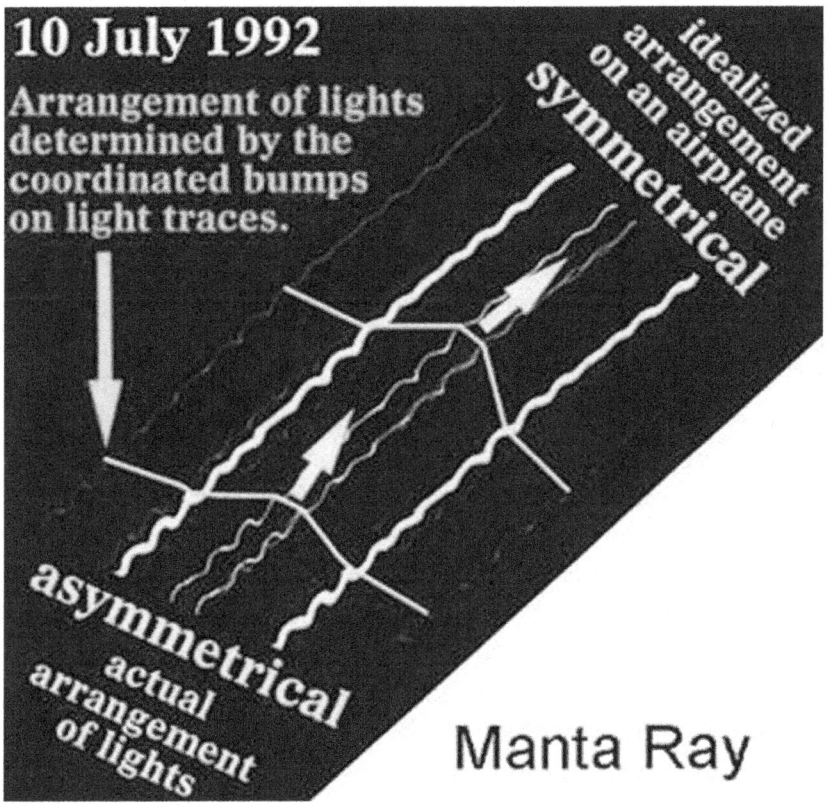

When I removed the second canister of film from my camera, I discovered I had grabbed a roll of Kodak 400 ASA color film. The pictures showed only its colored lights (see Appendix I), but from their positions and pattern, and the position of a single tail light, one can reconstruct its diamond shape. The outboard green and red lights (starboard and port, respectively) were not positioned in a bilaterally symmetrical fashion, indicating that if this had been a conventional aircraft, it would have to have been flying at a skewed angle to its actual direction of flight. That can occur if it did not have a vertical tail stabilizer. The Manta Ray does not have a vertical stabilizer, and resembles the B-2 Bomber in shape and size. That skewed orientation can occur if the airplane was turning or banking, but the light traces on the time exposure indicate that the craft was not turning.

The second time I saw the Manta Ray was on September 4, 1992, less than two months later. At about 11:15 pm Ellen Crystall, Fred Brock, Bob Wisch and I were standing within the fenced area of the Beth Hillel Jewish Cemetery, when someone noticed a set of lights descending rapidly from the clouds to the northwest, and disappearing behind a tree row across a large field filled with seven-foot high corn. Then in a break in the tree line to the right where a farmer's road was located, we saw a pair of lights enter the corn field, flying just above the corn. About halfway to us the lights brightened considerably. Crystall caught most of this initial action on video, while I took a sequence of time exposures. After about 60 seconds [11:16-11:17 pm] the lights dimmed. The total flyover took about four minutes based on Crystall's video time stamp. The maximum distance the craft traveled based on map locations and distance was about 1.4 miles, which would give a speed of about 21 mph. That is way too slow for a fixed-wing aircraft, and a jumpjet would have produced significant noise from down-thrusting engine exhaust (there was no odor of burnt aviation fuel). The craft was much too large for a helicopter, and there was no obvious or forensic sound of rotor blades recorded on the audio.

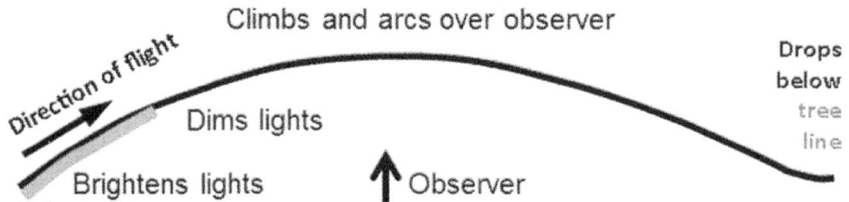

The time exposures I took would reveal something spectacular, something that no other investigator or witness had identified: The lights were not lamps. They were gaseous plasmas controlled by electromagnetic fields and emissions. The evidence for this conclusion will be presented in Chapter Five on plasma lights.

As the pair of bright lights got closer, we could see a blinking red light between them. By the time the craft reached the cemetery, it was at least 500 feet above us, having steadily climbed in altitude. As it passed over us we heard a distinct engine sound that resembled a high-pitch turbine. Crystall blurted out, "Nice Sound," implying that it was synthetic or fake, because she had seen this craft before (and so had I) flying silently. Soon after she said those words, the pitch of the turbine whine increased and then returned to its previous level, as if to acknowledge her comment. Crystall responded with her typical cynicism, "Oh sure!" The nose lights changed position as it flew overhead, with the top light moving down next to the bottom light, then move towards the rear of the craft. Lights on conventional aircraft are fixed and cannot do that.

The Manta Ray leveled off, then began descending rapidly until it disappeared behind a low hill or ridge to the west (Thompson's Ridge), which rose only about 100 feet above the valley floor where we were standing. No normal or conventional aircraft would have made such a dangerous and illegal maneuver, descending behind trees and then climbing to its highest point above us, and then descend back to less than 100 feet above the ground.

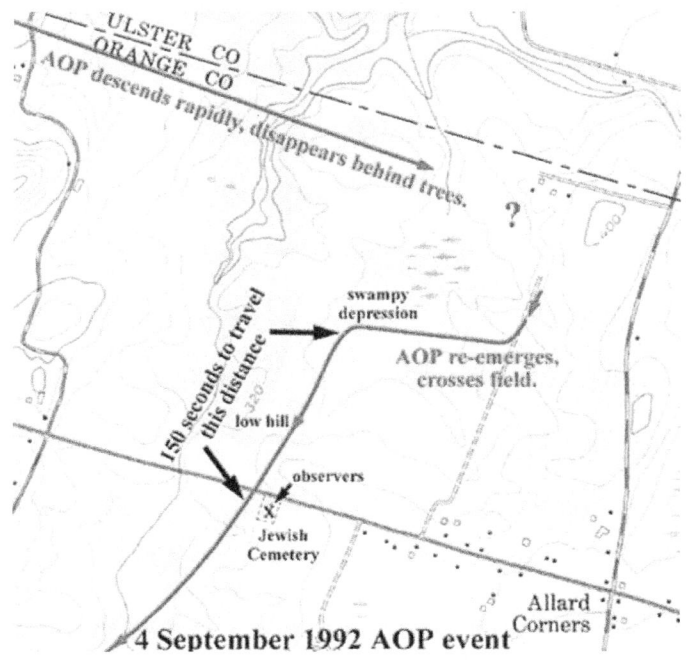

4 September 1992 AOP event

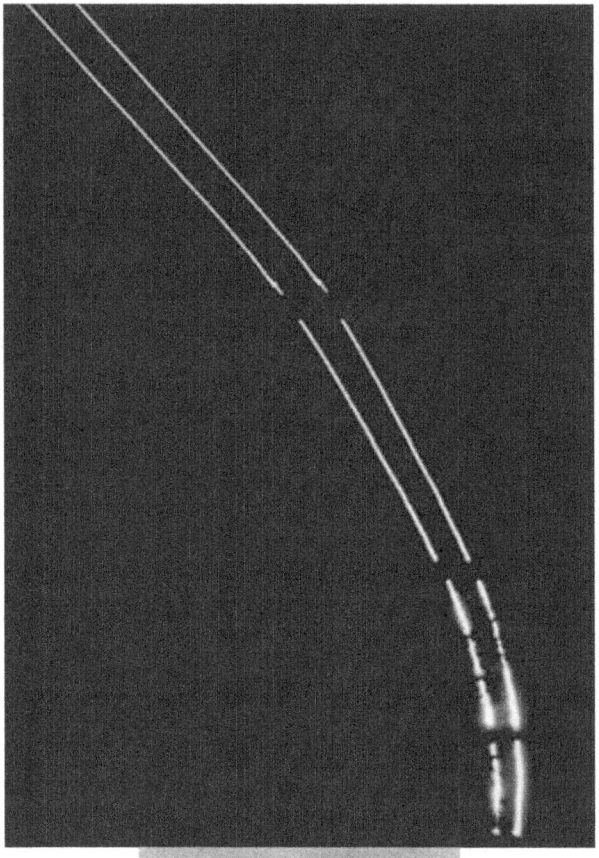

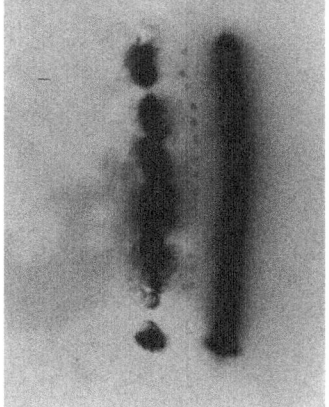

Image of negative showing NO2 fumes coming off plasmas.

When it flew overhead we could see its kite-shape that resembled a marine manta ray. Fred Brock said he thought he saw one engine pod attached to its belly. If that observation is correct, the pod device might account for the turbine sounds. No other description of an engine pod attached to the underside of the Manta Ray has been told to me by other witnesses/ Audio recordings of the Manta Ray sound at other times never duplicated that sound again, except during the performance by this craft for the Sightings TV film crew on April 29, 1993 (related and discussed in Chapter Five: Mechanical Sounds and Synthetic Sounds). No engine pods were visible on the bottom of that craft as it rotated and presented rows of belly lights to the cameras. We heard and recorded turbine sounds again, which changed in pitch as the craft passed low in front of the Sightings TV camera crew and other observers on the ground.

Another Manta Ray Encounter

The second sighting of the Manta Ray was the last one in 1992. There would be three more sightings of this craft in 1993, five in 1994, three in 1995, six in 1996, and three in 1997, for a total of 22 sightings by Cornet and accompanying witnesses during a six year period. Including the additional sightings in May 2000 and in August 2003, there are 15 time exposure sequences, eight video recordings, and nine different audio recordings of the Manta Ray for study and comparison. But before I move on to UAP/UFO lighting, I want to describe five more encounters with the Manta Ray that stand out as examples of extraordinary interplay (interaction or communication) with humans, and that go well beyond any possible overzealous misidentification of conventional aircraft, which violate FAA flight regulations with impunity (see Appendix I).

On April 8, 1993 Pat Huff and her 11 year-old son Niles Leisti joined me for skywatching along West Searsville Rd., Orange County, NY. We went to my main skywatching site at the dogleg bend half way along the length of West Searsville Rd. It was 11:00 pm when we got there and I set up my camera and tripod. At 11:07

pm I noticed a distant light to the east of Walden, NY. It was too low for a conventional airliner on its way to Stewart International airport on the opposite side of the valley. Initially it was traveling west towards us. Then it stopped moving and turned to the south. I began taking time exposures,. As soon as I opened the shutter on my camera (using an electronic cable switch to eliminate mechanical movement and vibration of the camera), the light sped up and then slowed down, then faded and disappeared 23 seconds later. I took a compass reading for the direction of the light.

About 52 seconds later, the light blinked on again, this time to the right of where it had been before. But this time it was slightly brighter than before (larger) and appeared closer to us. I opened the shutter on my camera again (11:08:55 pm), and as before the light sped up moving south as if knowing it was being photographed. Shortly after speeding up, it slowed down and gradually dimmed and went out after 21 seconds. I took another compass reading of its last position.

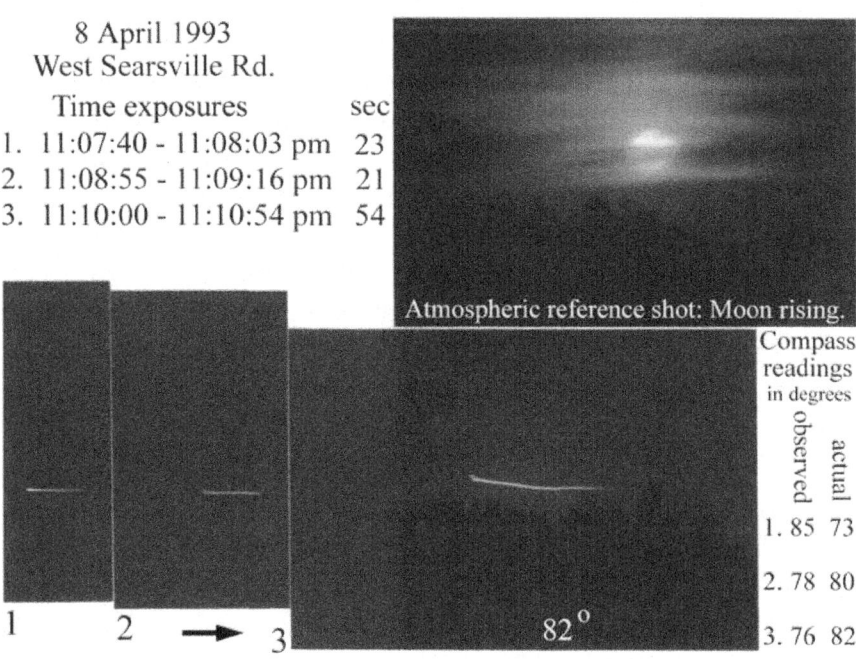

8 April 1993
West Searsville Rd.
Time exposures   sec
1. 11:07:40 - 11:08:03 pm   23
2. 11:08:55 - 11:09:16 pm   21
3. 11:10:00 - 11:10:54 pm   54

Atmospheric reference shot: Moon rising.

Compass readings in degrees

	observed	actual
1.	85	73
2.	78	80
3.	76	82

The same pattern of movement happened again at 11:10 pm, and another time exposure (54 seconds) and compass reading was taken. The light was much closer and nearly due east by this time. I didn't realize it at the time, but my compass readings were being altered from their true direction. I was too busy taking measurements and operating the camera to notice. Compass readings for time exposures: #1 was 85 degrees; time exposure #2 was 78 degrees; time exposure #3 was 76 degrees. Based on treeline directions (silhouettes of trees were compared to daytime pictures for accurate compass readings), the direction of time exposure #1 should have been 73 degrees; time exposure #2 should have been 80 degrees; and time exposure #3 should have been 82 degrees. Do you see the problem? The photographs advance from north to south. Actual compass degrees should have increased from 73 degrees to 82 degrees, but the measurements at the time of the event are reversed: 85 degrees to 78 degrees to 76 degrees. What happened next supports this reversed trend in compass directions, indicating that somehow this craft and invisible support system was altering the local magnetic field around me and my camera.

Then after a period of three minutes and 56 seconds, the light blinked on no less than 3,000 feet away, which is the location of an Indian Mound in a farm field just beyond the field in front of us. This time there were two lights. They were an estimated 800 feet above the ground (appearing to be no more than 300 feet above the tree tops, or five times the width of its headlights), and the lights were so low and bright that they backlit the silhouettes of the leafless tree tops in the foreground. Its lights would not have done that if the craft had been much further away.

Instead of moving quickly from right to left as they had done before, the lights began to form a skyglyph that was captured on six successive time exposures. The paired headlights had a blinking red light between them and outboard green and red navigation lights. But the maneuver this craft made was something no manmade aircraft could do without engines at full throttle. The performance was silent. From the pattern and spread of the lights I recognized this craft as the now familiar Manta Ray.

The craft moved south a short distance, then turned upwards, climbing vertically slowly and silently at least 400 feet. As it did so, it turned off its green and red navigation lights and turned on its axis so that its breadth or width was facing us. The time exposures show full separation of its headlights, giving a measure of its distance from us. Its headlights did several vertical loops as it climbed vertically, as if slipping downwards, then lunging upwards again, only to slip downwards slightly again. In other words, its movement was not smooth and continuously upwards. After climbing about 200 feet, its lights brightened considerably, so much that the plasmas set the atmosphere on fire, creating plumbs of red nitrous oxide gas (recombined nitrogen and oxygen in the atmosphere). With more energy output, the slipping back towards the Earth stopped and the pilot rotated the craft again, this time towards the north as he terminated his ascent and levelled off the Manta Ray to a horizontal course.

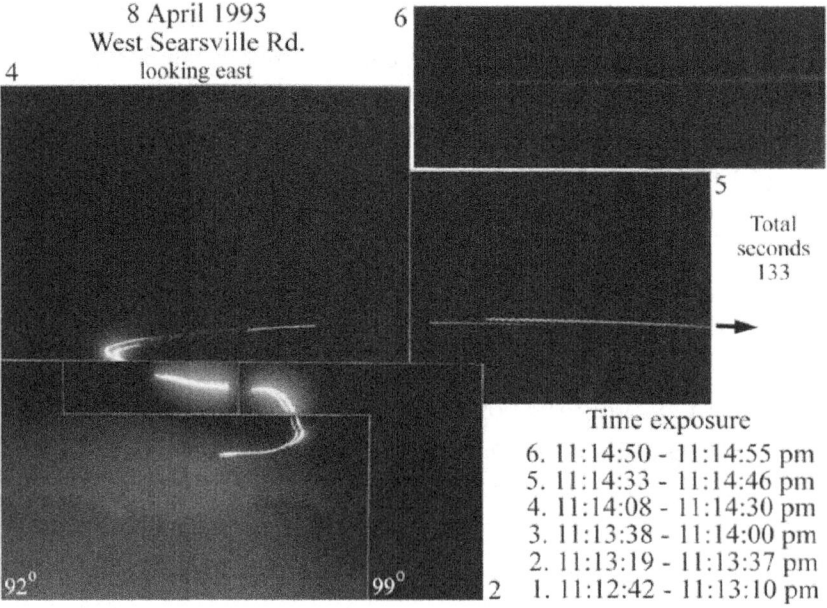

But the headlights remained flared, and throughout most of the second time exposure and all of the third the atmosphere around the craft was glowing bright red: Fire in the Sky. In the fourth time exposure the craft did an inclined U-turn back to the south, as its headlights were dimmed and then extinguished. Its green and red navigation lights again become visible as it turned south. Then only the central flashing red strobe can be seen for a short distance before it too was turned off. See Appendix I for more color images.

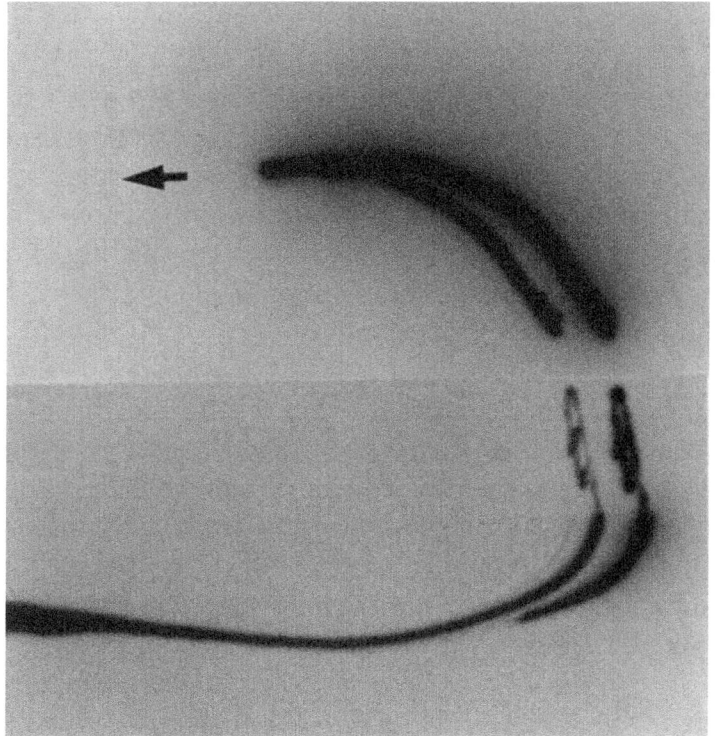

Notice how the outboard red and green running lights turn off before the ascent, then turn back on as the paired headlights began to flare again (arrows). Conventional aircraft cannot do such a climb silently.

With its headlights off, the craft began to accelerate towards the south. About halfway through the fourth time exposure, the pilot turned on the starboard (right) headlight without turning on the port (left) one. That light was gradually dimmed. In the fifth time exposure the pilot turned on his port (left) headlight, as the starboard headlight dimmed out. The Manta Ray departed to the south with only one headlight on, which was then also dimmed. As it departed in time exposure six, only a faint glow can be detected from its headlights. At this time the three witnesses heard a faint whooshing sound, possibly from the air as it accelerated out of sight.

It was apparent at the time that the pilot had given us a signal as he departed. But it wasn't until I assembled the six time exposures that I realized the significance of his aerial maneuver. He was generating a gigantic 'S' sign in the night sky, which looks remarkably like the Superman logo. Was this his intention and meaning? Are we being visited by highly-intelligent, super-powerful beings from another world?

When I took my final compass reading for the direction of the last performance to the east, I was shocked to discover that the compass needle was pointing almost due south! When I checked the other compass readings as the light (craft) got closer to our position, I realized that this was no fluke. The Manta Ray was somehow causing the local Earth's magnetic field (at least around the compass) to reverse. But why?

Through the course of sightings and apparent deliberate performances producing skyglyphs in the night sky, seven 'S'-shaped skyglyphs were created by the Manta Ray and by Black Triangles. One 'S'-shaped ditto was recorded on September 20, 1992, two in 1993, and five in 1994.

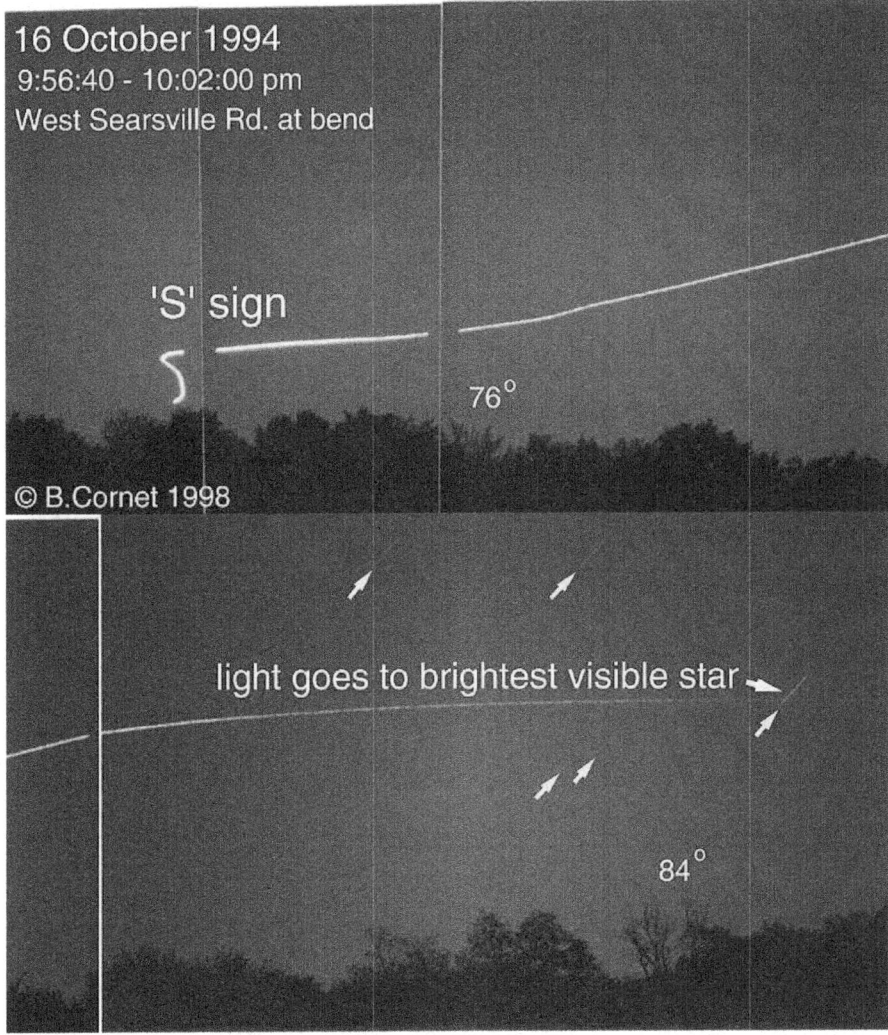

On October 16, 1994 the UAP light created a small but distinctive 'S' maneuver as it rose above the trees, then moved south as it climbed in altitude. The light went straight for one "star" just to the north of due east (75 -84 degrees compass) in the Constellation Torus. But when checked, there is no bright star in that position of the sky. It would appear that the UAP and star were trying to call attention to a particular location in the Milky Way – vey near where the Galactic Center is located.

If this was an attempt to call attention to the Galactic alignment that supposedly occurred on December 21, 2012, when all the planets and Sun will be aligned with the Galactic core (black hole), then the 'S' ditto becomes very significant. In the book, The Mystery of 2012, Predictions, Prophecies & Possibilities (2009), Branden provides a five point outline (p. 14) that includes a possible polar reversal at or soon after that alignment. The Earth's total magnetic field is declining from a peak intensity 2,000 years ago. Values have dropped steadily to the point they are now 38% lower than the peak magnetic intensity, and 7% lower just in the last 100 years. The magnetic poles of Earth flip on an average once every few hundred thousands of years. Based on this average, we are now overdue for a reversal. There has been weakening of the Earth's magnetic field over the past six months, according to data collected by a European Space Agency (ESA) satellite array, called Swarm. Such a flip is not instantaneous, but would take many hundreds if not a few thousand years, as shown in Triassic and Jurassic age cores from the Newark Basin in New Jersey (Kent, Olsen, and Witte, 1995; Kent and Olsen, 1999). "Previously, researchers estimated the field was weakening about 5 percent per century, but the new data reveal the field is actually weakening at 5 percent per decade, or 10 times faster than thought. As such, rather than full flip occurring in about 2,000 years, as was [previously] predicted, the new data suggest it could happen sooner." (Dickerson, 2014, Live Science; Scientific America).

Two facts based on the data recorded for UAP performances impinge on a possible imminent reversal: During the April 8, 1993 Superman 'S'-shaped skyglyph, my compass readings were clearly anomalous. Upon careful examination and the determination of two accurate compass directions based on tree line patterns in time exposures #3 and #4, I discovered that not only was the magnetic field nearly reversed (23 degrees shy of absolute magnetic reversal), but it was also inverted. The results indicated that the local magnetic field was off from an absolute reversal of 180 degrees by just 23 degrees, but the compass directions were inverted from normal. In other words, the measurements were not only off by 157 degrees, but East became West and North became South.

The 'S'-shaped dittos created several times over a three year period could stand for any number of words or meanings. Since most UAP performances ended with the craft flying south, the most likely meaning of 'S' is a reference to South Pole (i.e. future magnetic north). That is complemented by all the craft traveling south once they completed their skyglyphs. The October 16, 1994 performance where a white light made an 'S' ditto and then tagged a light (false star) in the Constellation Torus before going out, clearly relates the 'S' word with a point in the sky coincident with the Galactic Center! The bright "star" that marked this center does not exist on star maps. Unless it was a Supernova event not recorded by astronomers, it could have been deliberately set up as a "sign" of upcoming events.

In summary, the following evidence points to major future event:

- The 'S' sign may be a reference to the South Pole.
- A magnetic pole reversal during an 'S' performance by a UAP.
- Most UAPs and all performing an 'S' skyglyph departed traveling south.
- One 'S' performance ended by tagging a false star marking the location of the Galactic Center, clearly relating these two factors.
- Geologic evidence that the Earth is approaching a magnetic pole reversal, based on magnetic intensity decline over the past 2,000 years.
- Based on new data from the Swarm array of satellites indicates that magnetic intensity is declining 10 times faster in the last seven years than at any time previously.
- The alignment of our planets and solar system with the Galactic Center on December 21, 2012 could be a catalytic event

resulting in magnetic pole shifts for the sun and most of the planets with fluid magnetic cores.

The third encounter with the Manta Ray will now be described in detail. This encounter occurred on September 29, 1994, and was recorded with sequential time exposures and an audio tape recorder. David Ring, a UAP skeptic, was supposed to meet me, but he missed his plane flight to Stewart International Airport. So I went alone to my favorite observation station at the dogleg bend along West Searsville Rd. The Sun had just set, and twilight was turning the deep blue sky to black. At 7:33 pm I spotted a set of white lights about a mile away, flying much too low to the tree line to be a conventional aircraft. I pointed my camera, which was secured to a tripod, and opened the shutter for the first of 10 time exposures. After I closed the shutter and moved my camera to the right just ahead of the slow-moving lights, I didn't expect anything unusual to happen. I opened the shutter again. Almost immediately the craft turned towards me. Then it began to brighten its paired headlights as it slowly rose towards me silently. Then it began a series of turns left and right in a snaking movement – that is, an 'S' shape. The 'S' shape is clearly visible in the time exposures, because the flight path is tilted upwards, becoming higher off the ground the closer the craft came towards me. The lights brightened so much that they produced an orange-red aura of nitrous oxide gas radiating out many diameters from the lights. By the third time exposure the craft was more than half way to me, and its lights dimmed back down. The orange-red aura disappeared. By about 1,000 feet away I realized this was the Manta Ray. I could see red and green navigation lights at its outboard edges. Then a sound could be heard, which grew in volume until it nearly shook the ground as it passed high over my head.

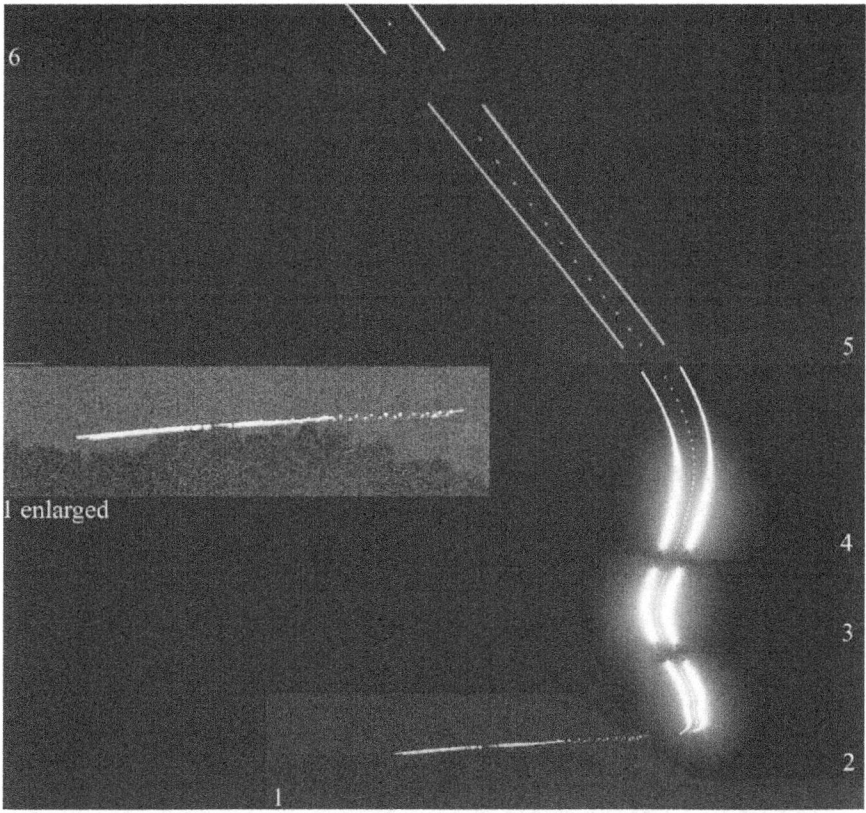

My voice can be heard on the audio recording calling out time for the opening and closing of the shutter. My comments can also be heard as I described the turbine-like sound and its loudness. The sound was that of a jetliner, but something was wrong about it. I heard no high pitch white noise, and it was much too loud before the Manta Ray came close to my location. The loudest part of a jetliner's sound is heard just after the plane passes you, because the sound of its engines is loudest coming out of the back of its engines. A jet aircraft can sneak up on you, but it can't fly away without making a lot of noise.

By the time the Manta Ray reached me, and it headed directly over me, it was at least 500 feet above me. It had been less than 200 feet above the ground when it began its run. I raised my camera as much as my tripod would allow it to tilt, snapping pictures only a second or two long as it quickly passed. But when the Manta Ray was directly over me, it cut its sound. When it did that, my electric shutter control failed. Later I discovered my audio tape recorder stopped recording sound for a few seconds. I banged on my camera several times, hoping it was just a loose electronic cable connection, but to no avail. The Manta Ray then turned on a completely different sound – from a jet-like sound (left) to a harmonic wobbling melody (right) sounding like that a clarinet would make. The differences on frequency spectrogram are major and obvious. Conventional military aircraft cannot do this unless engineered for that purpose. What would be that purpose?

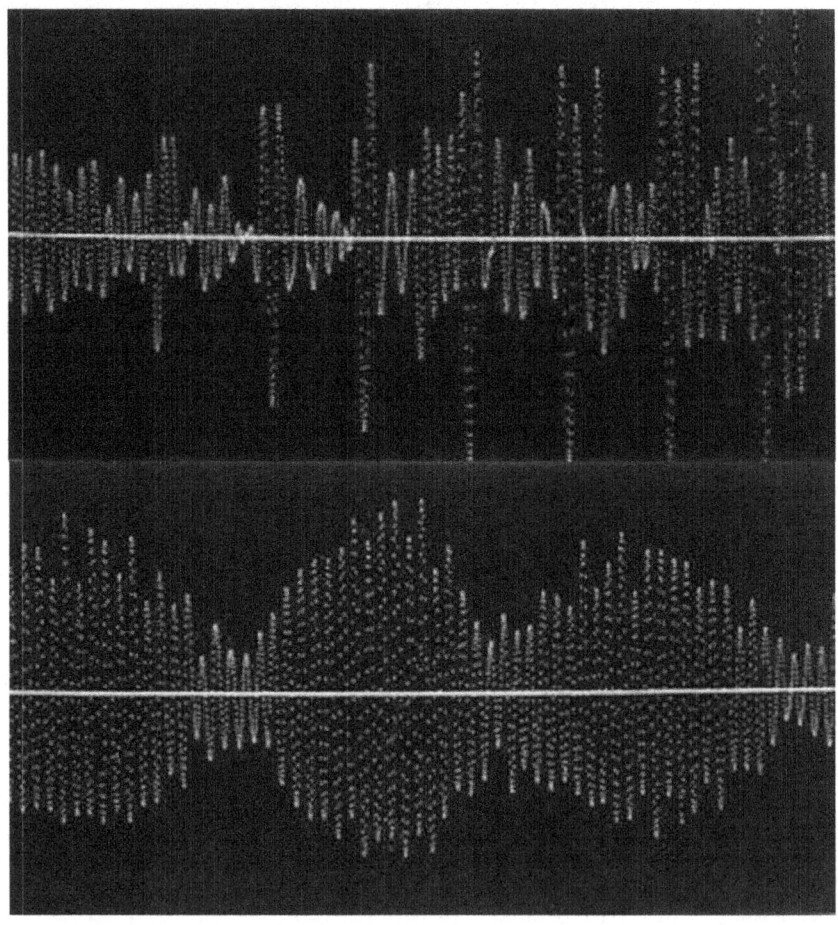

My camera started working again and the motor drive whined away as I captured six pictures of it still climbing rapidly and then diving as it moved away from me towards the west. My audio tape recorder also began working again, recording the change in sound (above), as well as the sound of my voice calling out times for the shutter action.

Path of AOP as it traveled behind hill, coming through gap, and descending to tree top level.

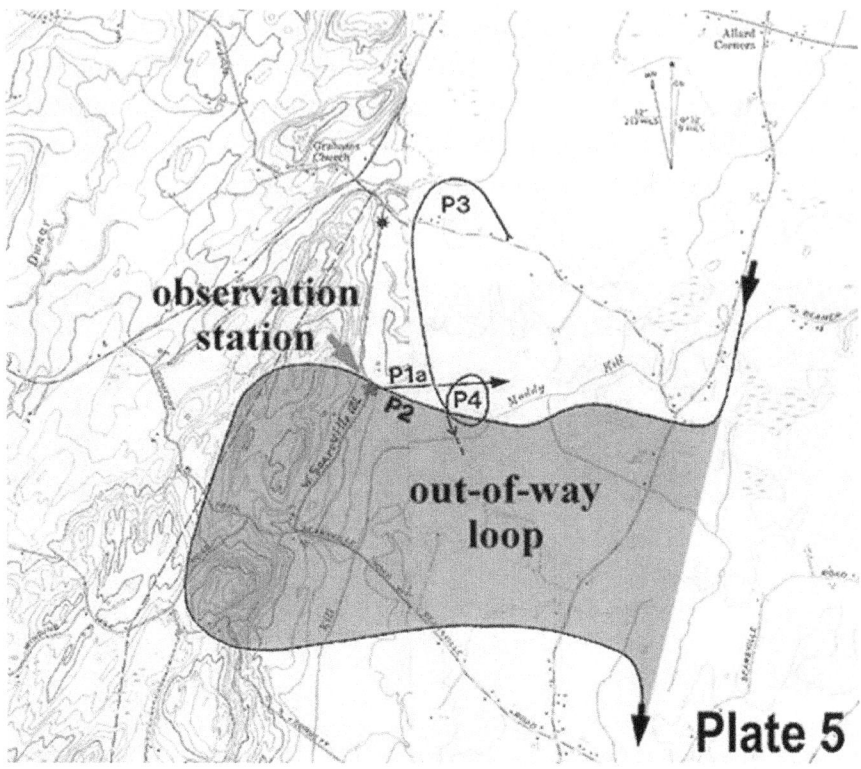

Plate 5

The Manta Ray was huge as it passed over me, its silhouette clearly visible. The craft then rapidly dove towards a low north-south oriented ridge, called Thompson's Ridge, and disappeared behind it. That ridge rose only 200 feet above the valley at that location, meaning that the Manta Ray's altitude was under 200 feet as it circled south, and then came through a gap in the ridge as it traveled east. I photographed it moving silently just above the tree tops as it returned to the very same area it had initially turned west to fly towards me. When it got back to that area, it turned south and proceeded down the valley on its original course.

Even if that had been a military aircraft, how did the pilot know I was there a mile away at night waiting with my camera? How was the pilot able to electronically jam both my camera and audio recorder, and produce different types of sounds or no sound at all? But when the sounds it produced were analyzed, even more surprises emerged: The gap in the audio recording is very obvious. The sound of the jet engines violates the Doppler Law, decreasing in pitch while increasing in loudness on approach. Frequency spectrographs reveal that no white noise is present. The sound is divided into several discrete bands, unlike that of recorded jet engine sounds. In addition, when played in reverse, the sound resembles a jet taking off from an airport, not one cruising overhead. Finally, the sound produced when the Manta Ray moved away from me was clearly synthetic.

Once again, an 'S' shaped pattern was produced, and the UAP disappeared traveling south. The jet-like sound may have been a recording, because its digital signature on my computer compares closely with that of recorded jet engine sounds. However, any high frequency white noise is missing, probably because the loudspeaker system used to replay the sound could not vibrate at those high frequencies of between two and five kilohertz. The clarinet-like melody that it produced after passing over me proves that's the sounds were synthetic or artificial, especially when it flew silently back east, completing its looping flight around me. Ellen Crystall is correct when she comments in her book that these UAPs can manufacture any type of sound. In my Chapter Five on sound, I will go into detail on the various types of sounds these craft can manufacture. Hynek et al. (1998) even describe two large circular globes on the belly of a large Manta Ray-like craft that could be the location of speakers, reported by one witness (reference: Night Siege, p. 166).

The Great Cycle of the Mayan calendar, also known as the Long Cycle, is 5,125 years long. The cycle we are currently in ended on December 21, 2012 (Arguelles, 2009). Until I recognized the possible significance of that cycle, the end of which was allegedly coincident with the alignment of planets and solar system with the Galactic Center (once every ~26,000 years), and the possible geologic changes that were predicted to occur with that alignment (which didn't occur), the performance of the Manta Ray on

September 29, 1994 did not have any special significance. There may be no correlation. Yet, why did the Manta Ray do a loop or circle (cycle) around me? Why did the pilot punctuate the sound he broadcast directly above me by electronically jamming my recording devices? Why was his flight path elevated the highest when he switched sounds, and why change from a familiar sound of human technology to a musical melody? Perhaps I am "reading" more into these events or aspects of the encounter than is warranted, but a possible reference to 2012 and the predicted changes anticipated for that year may not be out of the question (read The Mystery of 2012, by Sounds True, 2009).

The next two sightings and encounters with the Manta Ray on October 14 and 24, 1994. They involved five to ten sky watchers who frequently met along West Searsville Rd. near its intersection with Hill Avenue, between Walden and Pine Bush, Orange County, NY. One of the main reasons I did not associate with those who congregated along W. Searsville Rd. was the general lack of interest or commitment of anyone there to collect scientific data on sightings. That is not to say that individuals did not take time exposures and videos, or record possible sightings. Such data was not typically collected by those who frequently gathered there for social communication, storytelling, and peer support. On numerous occasions when I was in the field with Ellen crystal, people seeing us from the road would stop and want to chat with her. At that point any serious scientific observation would deteriorate into nonproductive conversation. On one occasion a craft was approaching from the east, low to the ground. The brightness of its light and low altitude were clearly anomalous. It was an opportunity to capture on film and video more data, but Crystall stopped videotaping to answer questions from a fan. A similar situation arose on October 14 with ten witnesses who preferred to debate the identification of UP/UFOS versus planes rather than collect data that could answer the question.

Vincent Polise reflects the sky watchers' viewpoint on personal experience in his book, The Pine Bush Phenomenon (2005: p. 26-27). "Dr. Bruce Cornet and the late Ellen Crystall performed extensive research on the area. Dr. Cornet is a geologist and had studied the Pine Bush area for years before I became involved. Dr. Cornet took a scientific approach to the area's phenomena and while I think this approach has its place, I think science tends to discount personal experience. Science and experience reveal two entirely different realities. I may hold no degree and be just a sideline experiencer but the UAP community has plenty of researchers with degrees who have NEVER experienced UAPs or paranormal activity. How do you study UAPs or the paranormal when you have never experienced it? Science has no place in an individual's experience and vise [sic] versa. I don't think science can offer any help or advice with what I've experienced over my fourteen-year period dealing with this area of activity. If a scientist entered the field and shared many experiences, his or her perspective on this topic would certainly change!"

I generally agree with Polise's viewpoint on scientists, who typically avoid the subject of UFOlogy and alien abduction because of assumed limitations on obtaining reproducible results, and because of peer criticism and job security. My scientific fieldwork and persistence did result in reproducible results that eluded other scientists relying only on anecdotal information. Exceptions are the late Dr. John Mack, a professional psychiatrist and hypnotherapist, and the late Budd Hopkins. Their work uncovered many aspects of alien abductions through hypnosis and scientific study that directly impact and helped those who have been abducted. But what fascinates me is the lack of interest among most UAP sky watchers in learning anything from scientists who have witnessed and experienced these phenomena. Appendix I, http://www.sunstar-solutions.com/AOP/Appendix I/APPENDIX I.htm) reflects the number of close encounters the author has experienced, along with some of his nine abduction experiences (also see Chapter Seven, The Awakening). Surely his more than 130 scientifically documented sightings in New York, New Jersey, and Virginia have the potential of adding valid information to our understanding of the UAP/UFO phenomenon that goes beyond personal experience and perception. It is the issue of perception that I address below and the reason for why scientific data can give answers and understanding that human perception and experience alone cannot.

Swamp Gas

http://www.sunstar-solutions.com/AOP/SOW/swampgas.htm

On October 14, 1994, at 9:25 pm, a pair of white lights with a red strobe in between them was spotted to the north – in a swampy area in a forest to the north of Hill Avenue, which ran east-west. Hynek's explanation for such lights in a swamp brought to mind, "swamp gas." I turned my camera towards the lights from the west side of W. Searsville Rd. and began a series of time exposures. The lights rose above the trees as they approached, and crossed Hill Avenue. At that point I recognized the UAP as the Manta Ray due to its size and shape. The craft flew over us on a southward path, rotated sideways to show us its belly without banking, and descended behind a low hill (Thompson's Ridge) to the immediate west of our location. It rose to its highest altitude almost due west of our location, and then began a slow descent behind the hill, forming a gentle arcing flight path. After disappearing for about three minutes behind the ridge, it re-emerged to the south as it came through a gap in the ridge where South Searsville Rd. climbs over the ridge.

Unconventional Aerial Phenomena

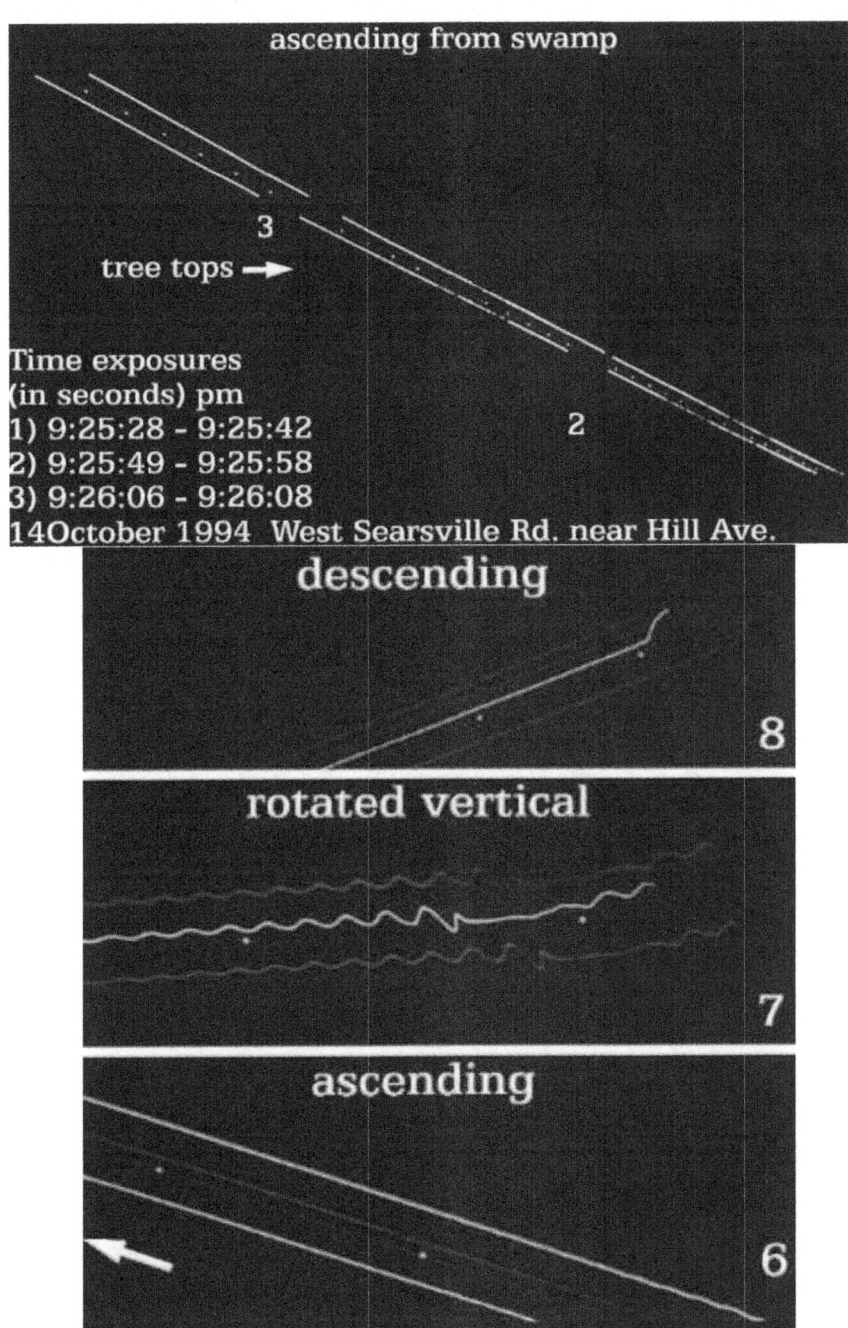

ascending from swamp

3

tree tops →

Time exposures
(in seconds) pm
1) 9:25:28 - 9:25:42
2) 9:25:49 - 9:25:58
3) 9:26:06 - 9:26:08
14October 1994 West Searsville Rd. near Hill Ave.

descending

rotated vertical

ascending

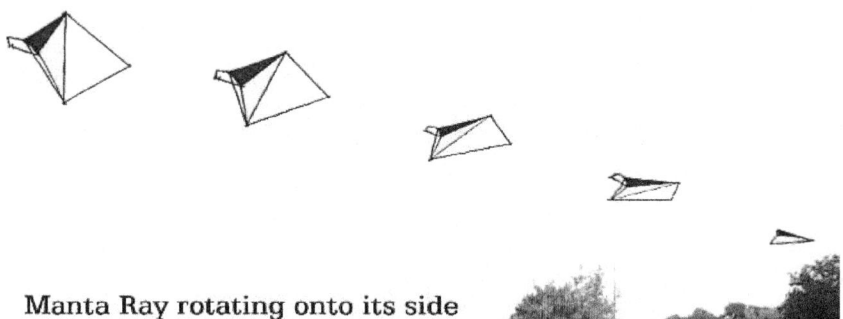

Manta Ray rotating onto its side

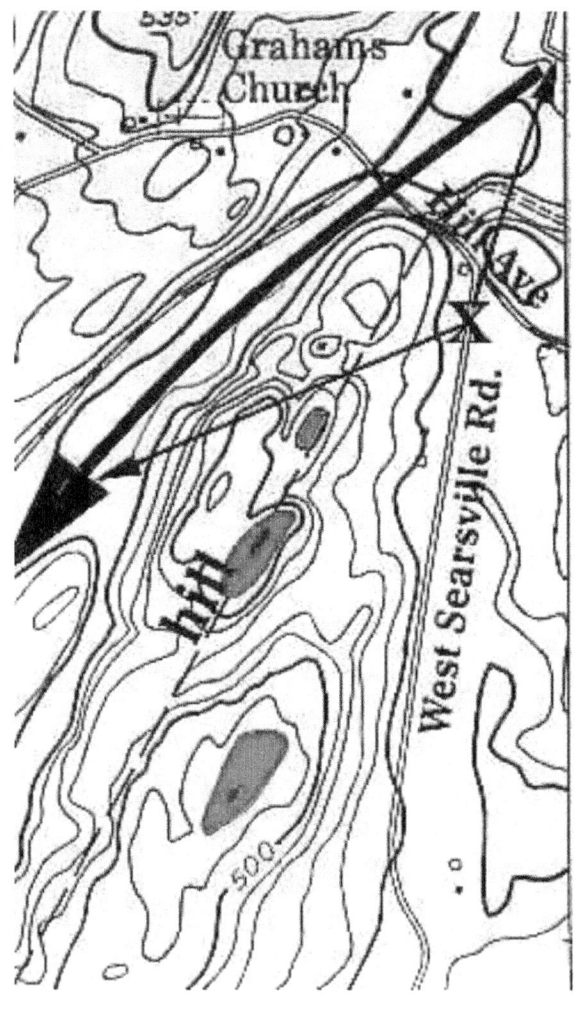

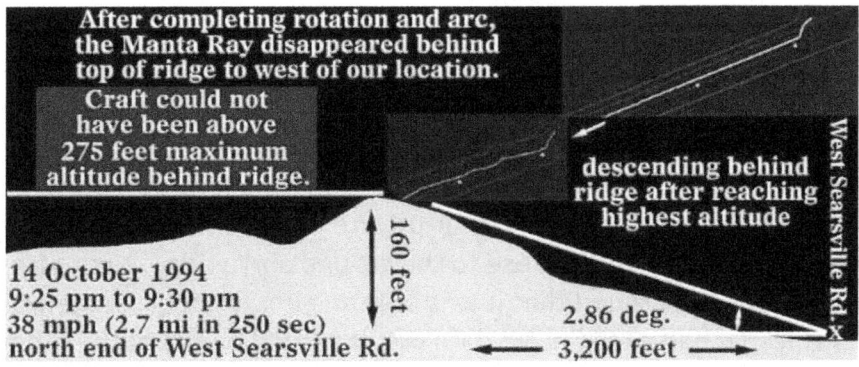

The craft rotated back to a horizontal orientation. The last time exposure (#10) was taken at 9:30-9:30:10 pm as the craft flew east over South Searsville Rd. It continued to descend until it disappeared behind tree tops to the southeast, at least eight miles short of the Steward Airport runway! At no point during the event did the craft reach minimum FAA-required altitude of 1,000 feet. It traveled 2.7 miles in 250 seconds, which computes to an average speed of 38 mph, well below stall speed for any fixed wing aircraft (with a vertical tail stabilizer). Without collecting scientific data, such analysis would have been impossible to derive only from recollections of the witnesses.

Keep in mind that at no time was this craft higher than 500 feet above us, and it had to drop below 160 feet altitude in order to disappear behind the Thompson's ridge just to our west. Greco and Gordon (1992) MUFON report that most the UFOs witnessed over and around Williamsport, PA, on 5 February 1992 were below 500 feet: 50-100 feet (46.1%); 100-150 feet (15.4%); 150-200 feet (23.1%); 200-500 feet (7.7%). This is important, because commercial and military aircraft (unless on a specific mission) are not going to fly below FAA minimum altitude (1,500 feet) unless on approach to an airport for landing, and there are no airports near this sky watching site. Orange County Airport is due South 3.75 miles away (longest runway 0.96 miles in length), while Stewart International Airport and Air Force Base is Southeast 7.8 miles away (125 degrees compass from due North).

Its shape was that of the diamond or kite. When it rotated on its side, its angular shape gave the impression of a tail stabilizer. One witness (Jim) said it was an airplane because he saw a tail fin through his binoculars. Analysis of the time exposures later proved otherwise, and no commercial or private plane would have pulled off that stunt at night so close to the ground and a ridge. No matter what I said, Jim argued that it was only an aircraft, because he was convinced it had a tail fin. No data was recorded by any of the other witnesses. With the time exposures, times for the exposures, map, distance data, and elevation of the ridge, I was able to determine that if the UAP had been a conventional aircraft, it would have crashed and burned on the other side of that ridge.

The Second Event

http://www.sunstar-solutions.com/AOP/SOW/swampgas.htm

Ten days later on 24 October 1994, an even more spectacular performance by the Manta Ray occurred before a similar skywatch group at the same location along West Searsville Rd. This time the Manta Ray took off from a farm field in front of us and flew in a loop South where it then spiraled down and disappeared into a forest. All of this was captured on time exposures. It is not just another reported sighting of a UAP, but provides clear evidence for this technology's size, capability, and existence as something far more capable than any known human technology.

George Filer, head of New Jersey MUFON, was present along with several other new sky watchers (Guerra, J.L., 2018). At 9:25 pm we witnessed a pair of lights turn on to the east, across an open field behind a row of mature trees. The lights were on the ground, and they quickly intensified to produce an orange-red glow that illuminated the ground and tree trunks in front of the craft. The craft began to lift up and travel forward at the same time its lights intensified; it climbed slowly and silently as it traveled north. The time exposures show the lights being blocked by the tree trunks and branches along the tree row until the Manta ray cleared the tops of the trees. Then its red and green navigation lights could be clearly seen (and recorded on film) as it continued to rise and begin a sharp U-turn to the east and south. Once the craft reached an altitude of several hundred feet it traveled south down to the general location of the dogleg bend in W. Searsville Rd. about 3,000 feet away.

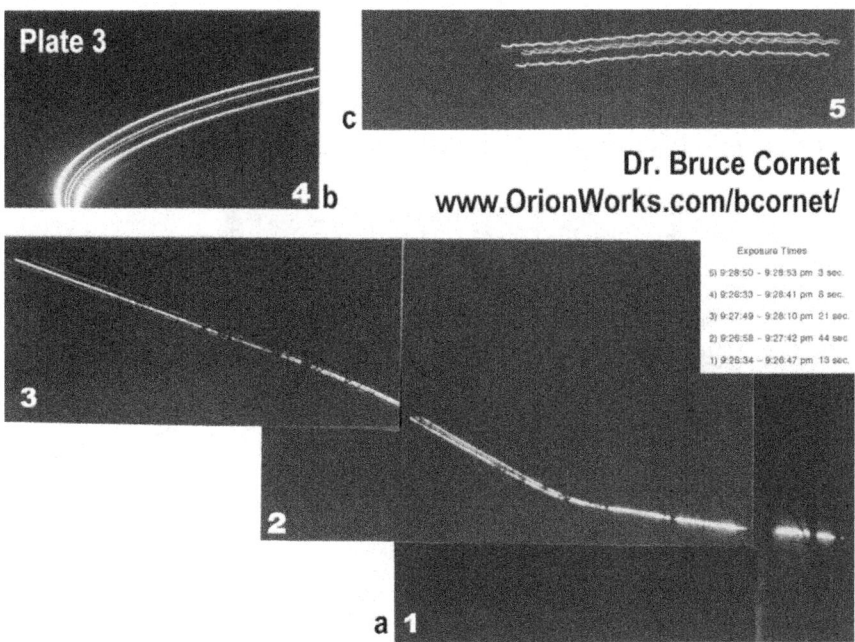

It was headed towards the Indian Mound when it began a spiral turn clockwise downwards, then reversed its spiral as it headed for a forest. To our amazement the Manta Ray continued to descend right into what appeared to be a dense forest, disappearing into the ground. The last time exposure I took clearly shows it passing between and through the trees as if they were not there. This could be an illusion because tree rows separating fields extend in different directions. However, trees in the foreground blocked its lights, and the lights can be seen descending to ground level *in front of trees* before completely disappearing. No one in the sky watch group could believe what they saw, and without an instant replay assumed the craft went down into an open field just behind the forest. The time exposures were long enough to image the trees clearly, so that they could be compared with daytime images to determine exactly what happened. The forest where it disappeared in an aerial view shows it to be 700-800 feet across (see image below), precluding an explanation that the image shows the craft landing in a field behind a tree row. The forest extends too far back from the road. Just as the lights enter the forest, they pass in front of some trees, then behind some trees which are closely spaced. But without instant confirmation, even George Filer was incredulous and barely mentioned the significance of his experience in his monthly "Filer's Files" newsletter.

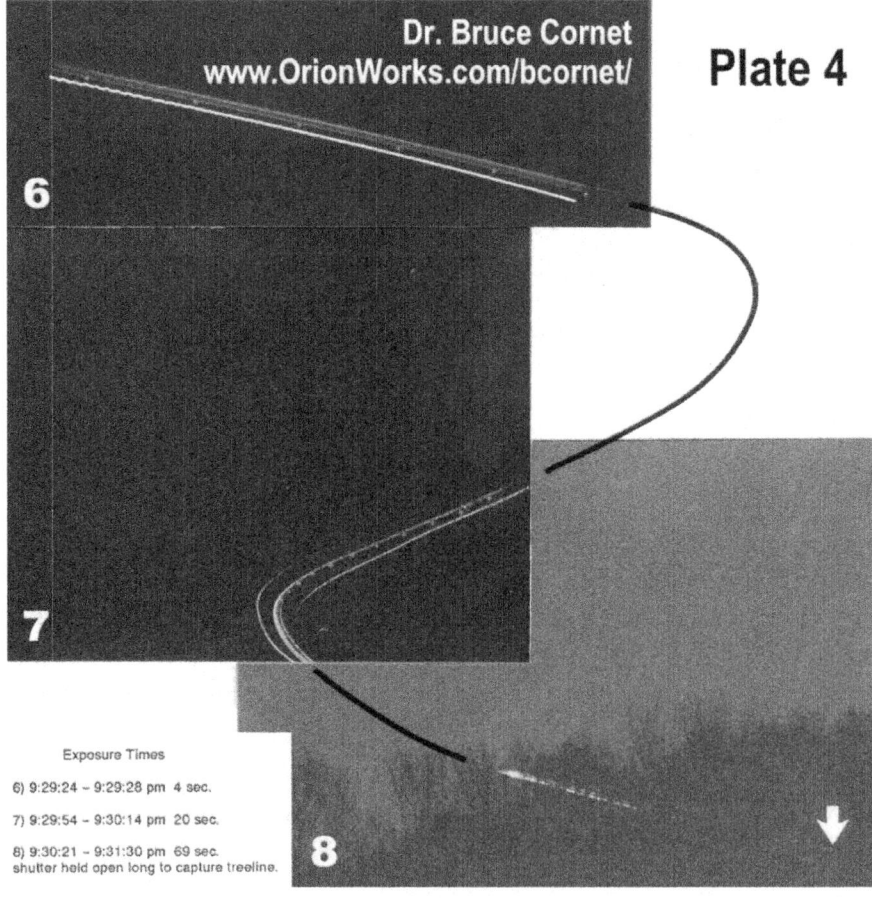

The arrow at the lower right corner of the last picture marks the last visible light of the craft just before it reached ground level inside that forest.

Vincent Polise, author of The Pine Bush Phenomenon (2005), has traveled to the Pine Bush area many times during the 1990s, and seen many of the same phenomena as Ellen Crystall and others who skywatched along West Searsville Rd. until local ordinance change and real estate development in 1997 prevented sky watchers from gathering along that road at night. Chris Burns wrote the Foreword to his book, stating: "Vinny's inaugural trips to Pine Bush to sky watch for reputed UAP activity was primarily to keep a friend company. Yet he was sufficiently impressed and fascinated by what he witnessed there to return to Pine Bush, night after night – eventually logging hundreds of hours – to further observe activity in the Crawford Township area that marks the border of Orange and Ulster Counties in New York. "*Silent Invasion*" by Ellen crystal documented the author's experiences chasing the UAP phenomenon in Pine Bush during the 1980s and the book's publication caused a surge of UAP buffs to line the streets in and around Pine Bush in hopes of witnessing activity." (Polise, 2005: p. 9).

The Manta Ray disappeared into a forest after it had put on a performance for sky watchers gathered along West Searsville Rd. on October 24, 1994. Activity within a forest is something that Ellen Crystall talks about in her book, and Vinny Polise corroborates such activity with his experiences in the same area where West Searsville Rd. joins Hill Avenue just south of the Beth Hillel Jewish Cemetery. In his book Polise describes his sighting of The Underground Craft: He writes, "On May 6th, 1995, I had the experience of a lifetime. It was just a normal Saturday afternoon. I looked up at the daytime sky and could hardly wait to get upstate and start sky watching. I had plans to meet some friends on the road....I pulled onto West Searsville Road at approximately 8:30 pm and noticed that there were a lot of people parked on the roadside. Honestly, there were more cars than I'd ever seen on the road at one time and while some of my friends were there, there were many new cars I had never seen before." (Polise, 2005: p. 72).

This is the same location described above for the October 14 and 24 group sightings at the north end of West Searsville Rd. Interestingly, the sighting Vinny and a new person had just north of that location on Heritage Rd., next to the forest and swamp the Manta Ray had risen from on October 14, helps corroborate the eye witness accounts of these craft defying conventional physics and/or human perception, in being able to pass through trees and Earth without interacting with matter in an expected way.

Polise continued: "As the night wore on with no activity, we began to swap stories of experiences and activity from the past...A group of us started talking about the crowded situation on West Searsville Road and considered going somewhere else to watch and avoid the crowds but in the end we decided to stick around. Around 10:00 pm when there was still no activity, many of the newcomers began heading home for the night. Just as I normally do, I continued taking pictures to see if any unseen phenomenon would show up on film. At 10:05 pm, I saw a sporadic, pulsing strobe light that seemed to highlight one section of the woods. Since the lights seemed close, I asked around to see if other people could see the lights too – everyone could. I decided to investigate the lights. I had seen these types of strobes on several occasions so they weren't new to me. In fact, everyone had seen similar strobes at one point or another, but we were never able to pinpoint their origin. The locals had seen them so often that no one really thought about them anymore. We just assumed the lights were probably weather-related or possibly someone setting off fireworks."

A little after 10:20 pm Vinny was able to break free from conversation with people at the gathering along West Searsville Rd. He wanted to check out the strobing lights in the woods, something Crystall talks about in her book, "Silent Invasion." (Crystall, 1991: Chapter 9, Aliens Underground; p. 86-93). He writes: "I drove to the end of West Searsville Road and made a right (turning onto Hill Ave.). I then made a quick left onto Heritage Lane. From the top of Heritage Lane, I was able to see the pulses of light starting to pick up speed. They were now pulsating at a rate of one flash per three seconds. They were becoming extremely rapid and as they pulsed quicker, I drove even faster to find them! I pulled the car over as fast as I could without causing damage – the strobes were coming from the dense forest at the end of this street! I was in such a hurry to get to the origin of the lights that in my haste, I left all my camera equipment in the front seat." (Polise, 2005: p. 74). I have made similar mistakes and regretted them, which is why I trained myself to focus on collecting data and taking time exposures or video. Emotional excitement can cause an observer to be unprepared, and having discipline and a protocol helps to insure that adequate data are collected for later analysis. It is usually those data which provide clues or even proof of what witnesses saw, whereas anecdotal stories are just that: stories. Multiple witnesses will help validate an experience, but not necessarily prove it beyond a reasonable doubt.

Polise continues: "The gentleman (Chris) I had just met parked right behind me and surely thought I was crazy because he pulled up in time to see me run into a neighbor's backyard to get a closer look at the flashes. Curious, he cautiously followed and ended up standing next to me.

"We watched the lights from that close distance for a few minutes before the pulsing got more rapid and more intense. They were now flashing at a speed of one flash per second – it was like a **laser light show**! Suddenly the treetops were enveloped in a white **mist** through which we could still see the strobes. This **mist** was chalky white and highly unusual! We saw an enormous half-moon-shaped object rise above the trees. We could see only half of the craft over the tree line but we could tell that the object was huge! When the **craft rose** above the trees it was **very bright**, but not so bright that we couldn't peer into it. It was brilliant and **full of colors** much like a **ferris wheel** on fire! The outer edges had balls of **sparks rotating** in a clockwise direction. These little balls were composed of beautiful colors: **blue, yellow, red and some green**. The craft's center was lit up in a brilliant **bright orange** with **shooting sparks** all around the outer edges. I couldn't see any metallic structure within the craft but such a light display had to be supported by one. The **craft floated** above the tree line for a full five seconds before it descended back *into the ground*. (emphasis added, as bold and italics type for comparison with scripture below).

"We watched the craft in utter amazement. What could we say to each other after such a sight – we were in such bewilderment! We had forgotten our cameras and had no way to film this craft. I wondered if anyone on West Searsville Road, no more then [sic] two blocks away, had seen the craft. We stayed there waiting and hoping to catch another glimpse of the craft but we never saw it again." (Polise, 2005: p. 74-75).

Polise's description reminds me of what Ezekiel describes in the Bible. From his perspective and experiences, he describes what he saw in encultured terminology. We might describe what Ezekiel saw differently, according to our knowledge of technology today. What strikes me is that people from 500 B.C.E. may have been having experiences similar to what we are having today, but were describing what they saw based on what was familiar to them then. In the Bible those experiences are correlated with the spirit world and divinity, not with alien visitors from a parallel dimension or another world. The human brain is wired to see faces and to identify unique features of every face. It is so wired that we can imagine seeing faces in clouds and in rock formations. Therefore it is not surprising that Ezekiel thinks he saw faces on someone's mechanical device or machine, since he had no prior experience with that technology. How might Ezekiel have described what Vinny Polise and Chris saw?

Ezekiel Chapter 1:
3 The word of the Lord came expressly to Ezekiel the priest, the son of Buzi, in the land of the Chaldeans by the River Chebar; and the hand of the Lord was upon him there.
4 Then I looked, and behold, a whirlwind was coming out of the north [**Crystall, 1991, describes such a wind coming up out of a field: p. 89**], a great cloud with raging fire engulfing itself [**chalky mist containing bright plasma lights?**]; and brightness was all

around it [**very bright?**] and radiating out of its midst like the color of amber, out of the midst of the fire.

5 Also from within it came the likeness of four living creatures. And this was their appearance: they had the likeness of a man.

6 Each one had four faces, and each one had four wings [**flight typically associated with wings, which is why angels are typically described as having wings**].

7 Their legs were straight, and the soles of their feet were like the soles of calves' feet. They sparkled like the color of burnished bronze [**bright orange?**].

8 The hands of a man were under their wings on their four sides [**craft floated?**]; and each of the four had faces and wings.

9 Their wings touched one another. The creatures did not turn when they went, but each one went straight forward.

10 As for the likeness of their faces, each had the face of a man; each of the four had the face of a lion on the right side, each of the four had the face of an ox on the left side, and each of the four had a face of an eagle.

11 Thus were their faces. Their wings stretched upwards two wings of each one touched one another, and two covered their bodies.

12 And each one went straight forward; they went wherever the spirit wanted to go, and they did not turn when they went [**craft rose?**].

13 As for the likeness of the living creatures, their appearance was like burning coals of fire [**plasma lights?**], like the appearance of torches going back and forth among the living creatures. The fire was bright [**very bright**], and out of the fire went lightning.

14 And the living creatures ran back and forth, in appearance like a flash of lightning [**strobes or sparks?**].

15 Now as I looked at the living creatures, behold, a wheel [**Ferris wheel?**] was on the earth beside each living creature with its four faces.

Ezekiel describes this object or wheels again in Chapter 10:1-22.

What I find particularly consistent between descriptions of various witnesses who see lights moving in the sky at night, is that when they dive into the ground or move through stands of trees or a forest, their lights are visible even though their solid structure is not. There are many descriptions of solid structure to these craft in Hynek et al.'s 1986/1998 book, "Night Siege," and in Crystall's 1991 book, "Silent Invasion." I have seen what appears to be solid structure reflecting light on these craft. The question becomes, how can such structure reflect light if it is not material, or how can the structure of these craft "dematerialize" yet their lights continue to be visible? Crystall does not believe these ships faded into a parallel dimension, but rather bent light around their hulls to become invisible. If that is the case, they should not have been able to pass through trees or disappear into the ground. Clearly there is technology here with which most humans are not familiar. In ancient times these phenomena would be called spiritual, but I have not yet heard of one explanation that adequately describes what spiritual is or of what spirit is made.

The sightings described above on 14 and 24 October 1994, and Polise's account from 6 May 1995, point out the difficulty humans have in documenting paranormal observations, sometimes looking at a sighting with camera in hand, but failing to take any pictures.

The controversy over Daniel Fry's publications and claims of ET contact between 1949 and 1956 developed a life of its own on the Internet, beginning in 2004 and continuing today on the "Daniel Fry Dot Com" blog at

https://danielfry.com/blog/2017/09/11/more-of-randy-morrisons-posts-from-the-paracast-forums/,

a controversy which seems to have begun after Timothy Good published his book, "Alien Bases" in 1998. The controversy between Randy Morrison, Ray Sandford, and Sean Donovan is extensive and detailed, and raises the question of whether Daniel

Fry faked his UAP pictures and movie film, as Sanford claims. Because this material is relevant to my discussions on the nature and aspects of the Contactee and eye witness accounts presented here, I will leave it up to the reader to pursue the pros and cons regarding the veracity of Fry's statements. By including this material from Good's book, I am not agreeing with either side of the controversy, but rather presenting this material for comparison with what happened to me and other sky watchers who stood on rural country roads many nights of the week during the early and mid-1990s.

Good describes a most interesting encounter between Daniel Fry, a rocket test technician at the White Sands Proving Grounds in New Mexico, an alien robotic probe or "sampling carrier," and a projected voice of Ahlahn (Alan for short, which some might think is pronounced, Allah), an ET aboard a mothership orbiting Earth 900 miles away. The date was 4 July 1949. Ahlahn's comments regarding the reason Dan was selected are relevant or germane.

"One of the purposes of this visit is to determine the basic adaptability of the Earth's peoples, particularly your ability to adjust your minds quickly to conditions and concepts completely foreign to your customary modes of thought. Previous expeditions by our ancestors, over a period of many centuries, met with almost total failure in this respect. This time there is hope that we may find minds somewhat more receptive so that we may assist you in the progress, or at least in the continued existence of your race...the fact that, in spite of being in circumstances completely unique in your experience, you are listening calmly to my voice and making logical replies is the best evidence that your mind is of the type we hoped to find." (Good, 1998: p. 60).

Regardless of whether you believe humanoid extraterrestrials have visited Earth, or consider the conversation reproduced above from the early Contactee Period in UAP History to be authentic, there are truths present in Ahlahn's discourse. Even though many of the sky watchers that turned out along West Searsville Rd. on any given night between 1991 and 1997 were open to the existence of UAPs, hoping to see one for themselves, others there were skeptics who came out purposely to disprove, discount, and ridicule any potential sighting or claim that a UAP had been seen. The skeptics were seemingly threatened, even though they alleged to have an open mind. That is why so many sky watchers were coerced into relegating good encounters with non-human technology to the trash bin of mistaken identification. My work and data have demonstrated that human observation alone is wholly inadequate in identifying non-human technology if the intelligence behind that technology wishes to remain unknown. The corollary to that reality is that when the right person with an open mind is selected to witness such technology, it will be presented in a form that can be recorded and analyzed. The key word, however, is "analysis." Without careful analysis of data collected by various instruments and recording devices, the intelligence providing the raw data is just pissing into the wind.

Whereas the Hudson Valley phenomenon east of the Hudson River lent itself to be witnessed by thousands of humans in groups or as individuals over a period of 14 years, and is continuing even today at a low level of activity, very few people documented their sightings with photographs and videotapes. Today that would be different with so many camera conveniently available in cell phones. The vast majority of sightings would never have made it into publication and a semi-permanent record without the early dedicated work of Dr. Alan Hynek, Philip Imbrogno, Bob Pratt, Marianna Horrigan, and Dr. Ellen Crystall.

The early work of Dr. Crystall, largely in Orange and Sullivan counties, New York (i.e. near the towns of Pine Bush, Montgomery, Walden, and Wallkill) yielded some of the most extraordinary photographs of non-human technology and beings to date, photos that were taken in 1980 before the Hudson Valley phenomenon (flap) began at the close of 1982, Additional revealing photos (other than Crystall's orbs and "Tesla" globes) were not taken by her until 1992 when I introduced time exposures to the recording methods.

Imbrogno and Horrigan (2000) introduced the Stone Chambers and their associated magnetic anomalies to the characterization of the phenomenon. Although these authors give credit to me for providing magnetic data, they never identify the three magnetic maps in their books as having been created by me back in 1992 after a field trip led by Imbrogno to the Stone Chambers. At that time I had already begun a 24 square mile magnetic survey involving 1,800 stations of magnetic measurements, which took me three years to complete. My fieldwork resulted in similar conclusions that Imbrogno and Horrigan came to for the Stone Chambers (see Chapter Six). Although I accompanied Crystall into the field for two years (June 1992-June 1994), and appeared on the Sightings and Encounters TV shows with her in 1993 and 1994, she would never go out in the field with me to help collect magnetic data or see what I was doing. Ultimately, my work would help validate her hypothesis that magnetic activity below ground was somehow connected to the aerial phenomenon above ground. Although no underground alien bases were confirmed, evidence for interdimensional portals, signals and beacons emanating from below ground and directed out into space, and possible ancient robotic mining probes entombed in Ordovician-age rock was revealed by my magnetic study.

The only other author to publish his research on the Pine Bush phenomenon to date is Vincent Polise (2005). Since then Linda Zimmermann (2013) published a book, "In The Night Sky," that relates stories of sightings in New York State north of New York City, and as far west as Pine Bush, but she gives little credit to Crystall, Polise, or Cornet. Her stories add to the wealth of anecdotal information for the Hudson Valley region, and together these studies, along with those of Hynek, Imbrogno, and Horrigan, firmly establish the existence of a UAP Corridor across lower New York, southwestern Connecticut, to Long Island.

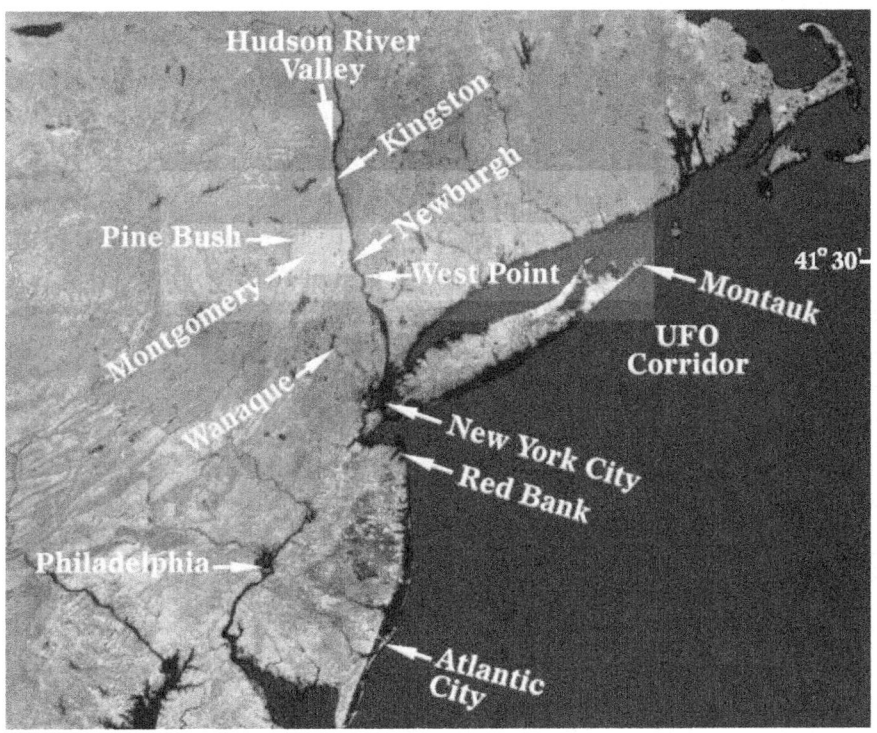

Chapter 3 The Trickster

The final account or encounter with the Manta Ray described in detail during the 1990s, occurred on June 10, 1995. Additional sightings and photographs are given in Appendix I. Pat Huff-Cornet, Sharon Cunningham, and I had spent the day and early part of the evening driving around the valley hoping to see something unusual. We stopped to visit with the sky watchers at the north end of West Searsville Rd. During our visit we saw nothing unusual in the night sky. Some people there said they felt a strange presence and heard strange noises near them but couldn't identify a source. Two men left to go into Pine Bush several miles away to get something to eat. They came back with a tale of being followed <u>on</u> the road by a light they thought belonged to a motorcycle, but then it suddenly took off into the sky and disappeared. As a group we sensed some trickster activity, but it wasn't until later that we discovered someone else's papers in our van. We had kept our van locked and had not been given those papers to read. We had no idea of how they could have found their way into our locked vehicle. We suspected an invisible entity had been at work, playing tricks on us. We later contacted the person who owned the papers, and she was amazed they had ended up in our car. She confirmed that she had not given us those papers or been in our car.

On the way home late that evening the mood was somber. After all our effort (Sharon had driven up from Maryland), she had not seen anything unusual in the night sky. As we drove through Walden on Rte. 52 towards Interstate 84 (I-84), we felt strange. Sharon said she felt nauseous. The Pat said she felt nauseous. Then I said I felt dizzy. We all thought that we were being psychically attacked, but didn't know why or by whom. After we passed through Walden, our negative feelings quickly disappeared. We took Rte. 208 to I-84 and got onto the highway driving east towards Newburgh, NY.

As we accelerated to highway speed, I looked in the rearview mirror to make sure I wasn't cutting someone off. I had the thought that I should accelerate fast when I spotted a pair of headlights approaching in the right lane a half mile behind us – usually more than enough room to merge with traffic. But a second glance in the mirror surprised me. It looked as though a large truck (based on the size of its headlights) was barreling down on us at over 200 miles per hour. I yelled at the women to look behind us. They both screamed in horror at what they saw, fully expecting a crash and being hurled from a demolished van. When they screamed the lights were only five or six car lengths behind us, closing very fast. All of a sudden the lights rose up into the air – reminiscent of the UAP coming up behind the pickup truck driven by Roy Neary (Richard Dreyfuss) while stopped at a train crossing in the movie, "Close encounters of the Third Kind," and then rising up above his truck.

I yelled at the women to get my VHS camcorder located behind them, and to turn it on. As they did this, the lights, now above us, moved to our right and low over the tree tops on the right side of the highway! As soon as Pat pointed the camcorder at the craft through the window, the Manta Ray dropped down out of sight – presumably into a clearing in the forest. Or it could have passed into the forest as it has been seen to do before (as described above), defying standard physics. In either case, its lights disappeared. The choice comments and language captured on the video says it all, as Sharon expressed her extreme dissatisfaction with the way she got her UAP sighting. A certain degree of mischief played on us seemed to be part of the encounter, as if they were saying, "We were told to give you a sighting, but we didn't want to. So here it is. Take it or leave it!" Sharon had had numerous sightings and some abduction experiences, so she was not a woman who would have been intimidated or disturbed by a UAP sighting. She acted as though she had been a choir member who had been scolded by the preacher for complaining too much.

Y Pine Bush

Pine Bush is one of four small towns located east of the Shawangunk Mountains on the west side of the eight-mile wide Wallkill River Valley. The Wallkill River is unique in being the only river in the Northeast that flows due North, entering the Hudson River near Kingston, New York. Pine Bush is the business section of Crawford Township, and is an unincorporated hamlet within that township. To the east next to or near the Wallkill River are the small towns of Wallkill to the north, then Walden and Montgomery to the south. About ten miles further south is a much larger town with interstate highway connections called Middletown. All but Wallkill are located in Orange County, NY, and that town is located in Sullivan county to the north.

Pine Bush is located at the intersections of Rte. 302 and Rte. 52. Rte. 52 runs west across and over the Shawangunk Mountain range to Ellenville, another small agricultural town nestled in a deep valley between mountains. Rte. 52 runs east for a couple miles from Pine Bush before splitting in two to form a Y. Also known as Hill Avenue, Rte. 52 travels southwest until it terminates at Albany Post Rd., which parallels the Wallkill River in that area. The northeastern branch of Rte. 52 continues east, crosses Albany Post Rd., and enters Walden, NY, on the eastern side of the river.

It is within the divided Y portion of Rte. 52 that the UAP mystery took off, literally, in 1980 when Crystall and her friend Harry Lebelson visited friends of his, Bruce and Wendy, on July 18th. Bruce and Wendy told them they had seen many craft in the area, and knew where they were landing. "Landing?" Crystall exclaimed, wondering why this area was different from so many other areas of sightings where ships rarely were seen on the ground.

The Y pattern of Rte. 52 is oriented east-west, with the opening or split facing east towards Walden, NY. On the northeast limb of the Y is a major magnetic anomaly, located under the Beth Hillel

Jewish Cemetery on the south side of the road. About a mile south of the southeast limb of the Y is the Indian Mound, located over another major magnetic anomaly. So many sightings were recorded over this mound, many originating at and/or ending near or around that mound that I have identified it as the Center of the UAP hotspot in this region.

I have recorded magnetic values just west of the mound that fluctuated or changed wildly [reference notes], and recorded a commercial American Airlines jetliner nearly crashing into the adjacent fields at night, presumably due to instrument failure from EMPs coming out of the ground (See Appendix I: July 2, 1992). On April 28, 1993 I even recorded a set of UAP lights manifesting a visual vortex over this hotspot center as it climbed in altitude traveling southward. That time exposure was the subject of a 30 minute documentary on the History Channel on March 26, 2008. The program was called, "UFO Vortexes."

http://www.sunstar-solutions.com/AOP/AAdescent/awesome.htm

Bruce and Wendy took Crystall and Lebelson to a field on the west side of Hill Avenue (southwest limb or branch of Rte. 52) before it reaches Flury Rd., which crosses Hill Avenue to the east. Crystall took me to this field, which has an east-west oriented low ridge or mound in the middle composed of gravel and clay sediment. That mound, which was hundreds of feet long before road construction destroyed a portion of its east end, has remnants of a ditch or moat around it, and it may represent an ancient burial mound rather than a glacial deposit. We hiked up onto that ridge, and had a commanding view of the surrounding area. It is there on that ridge that Crystall told me she saw Black Triangles taking off, which had landed on that ridge. Her account is perhaps one of the most startling sightings of multiple UAPs in the history of Ufology, in large part due to the closeness of the craft to the observers and their large number (read Silent Invasion, 1991). Were our visitors trying to point out the importance of that mound, which may be as old as the hundreds of Stone Chambers located on the eastern side of the Hudson River, dated by Imbrogno and Horrigan at about 4,000 years old (Celtic Mysteries, 2000)?

Crystall wrote, "About ten O'clock P.M., Harry and I got into my car with Bruce and Wendy, who directed us to drive east on Hill Avenue. The night was black but clear, with a thin crescent moon. A number of houses lined the road, although they were spread out with large parcels of farmland in between. They are definitely not off the beaten path. Within a mile, Bruce and Wendy told me to pull over next to a field with a slight rise toward the far side. I did.

"We got out of the car and looked up. Almost immediately we were surrounded by a dozen large triangular craft with amber-yellow lights in the form of a "plus sign" on their front. The plus-sign-shaped lighting panels divided four window panels that made up most of the front of the ships. I came to call those particular lights "starlights" because, as I was soon to learn, they could be turned up to a blinding degree of illumination and could also act as a headlight or spotlight. In fact, they could be used as a pair of headlights. When the intensity of the lights was raised to full bore, the entire sky lit up.

"The ships also had multicolored blinking lights all over them. I was staring at the triangular craft I had seen in California! But now, all the exterior sections I couldn't see in California were clearly visible.

"I was ecstatic. My searching seemed over. I was reunited with "my" ships after nine years. I hoped I cold uncover the great secrets they hold. I couldn't have been happier." (Crystall, 1991: p. 16).

The hook had been set and the bait taken. Crystall was now on a path of discovery for the rest of her life, going out into the field over and over again in anticipation of seeing more craft and aliens. But today these extraterrestrials are not very willing to expose themselves as much as they did to hook Crystall. I spent three years in the field with her (1992-1994) and she was just as committed and enthusiastic twelve years later as she was on that July day in 1980. Yet she would never get quite so close, quite so often, and capture images of these beings standing next to their craft as she did in the beginning. The craft she described on those first encounters would change and evolve over the years. That large four-pane window at the nose of the triangle would disappear as their craft became truly triangular (equilateral) in shape, and sometimes "extra stuff" as she called it would be attached in front and back of the triangles. It would take my research to understand that they were using us humans for quality control in developing their camouflage techniques, as they modified their craft to resemble (i.e. be mistaken for) conventional jets and single-engine propeller aircraft at night by mimicking their navigation lights and sounds.

The Ships Ellen Observed

Ellen Crystall, in her book "Silent Invasion," described the different types of craft, their shapes, and sizes over a ten year period of observation. Later I will add to and modify her list in order to show how the phenomenon changed or evolved during the 1990s.

- "Triangle" – about two hundred feet in perimeter. A large one is about four hundred feet in perimeter.
- "Diamond shape" – about four hundred feet in perimeter.
- "Rectangle" about three hundred feet in perimeter.
- "Boomerang" – the small one is about twenty-five feet from tip to tip, the large one appears to be about two hundred feet in perimeter [The "Westchester Wing" or "Westchester Boomerang" witnessed by thousands of people during the 1980s east of the Hudson River].
- "Walnut or turtle shape" – about four hundred feet or more in perimeter.
- "Manta Ray shape" – about four hundred feet in perimeter.
- "Jumbo or Big One" – about five hundred feet in perimeter; has two "noses" or "wings."
- "Small airplane" – about twenty-five feet wingtip to wingtip; has a black body, silver wings, no wheels or apparent engines." (Crystall, 1991: p.115).

Crystall told me (pers. Comm., 1992) that she had seen what looked like a Boeing 747 approach and land in a field in front of her. I am not sure if she was referring to the "Big One" above, but I can attest to the fact that some of the craft flown low over the farmlands and over highways do resemble conventional aircraft. They typically have no windows or markings on them; and have no visible engines attached to their wings or fuselages. They are either solid black or solid white (metallic) in color; can fly silently or produce jet and propeller engine sounds (that are synthetic in character when the sounds are recorded and analyzed); can stop and hover; can illuminates portions of their fuselage; and can even fold up their wings and tail stabilizers. I will describe more of these strange "aircraft" and their non-human technology in Chapter Seven. In 2000 I spoke with Betty Hill on the phone, who confirmed that she and others had witnessed these strange types of alien craft designed to look like human-engineered aircraft. Crystall describes one such small airplane (listed above) that she and a friend saw up close flying well below the stall speed of any fixed wing aircraft. She relates her story in her book: Incident in 1984.

"I continued driving south, and as we reached Route 17 in New Jersey, one of the lights followed us, paralleling the highway and matching the speed of my car. [I have had this done to me on several occasions: Appendix I. As I slowed to take the Oradell exit, the craft halted in the sky with a skidding motion, almost like a vehicle in a cartoon would stop. Then it descended much lower, moving towards the car even though we were still moving.

"About a block farther on, I started to pull off the road. The ship suddenly turned on many lights and headed toward the roof of the house next to us, skimmed over the top of it and headed directly toward us! I jammed on the brakes, stopped in the middle of the road, and we both jumped out, leaving the doors wide open. Cathy and I clicked off four photos of what we clearly saw was a small plane, about twenty feet across from wingtip to wingtip, with a black body and silver wings. No propellors [sic] or engines of any type were visible, and it was virtually noiseless.

"The "airplane" lifted one wing to pass over a small tree next to us. (We later measured the tree. It was about fifteen feet high, so the craft was barely fifteen feet above the ground.) The black metal of the fuselage was startling to look at. The metallic surface was quite smooth, although we could still see grooves and seams in the metal. The silver wings were also fairly smooth...

"At the plane's closest point to us – about ten feet away – we could barely hear a buzzing sound. The craft flew past us at about five miles per hour, continued past the nearest house, then stopped in midair." (Crystall, 1991 p. 117).

The Night Siege Ships

As stated in the Introduction, the types of craft seen east of the Hudson River tend to be larger than those seen near Pine Bush, NY. Hynek et al. (1998) summarized their findings in "An Analysis of the situation" in their second edition. Forty percent of the UAPs were 300 feet or wider in size. The most common size, however (about 60%), was estimated at 100 feet in width.

Crystall estimates the largest size of craft she saw or were described to her in the Pine Bush area as 300 to 400 feet in perimeter, which translates to width of only 100-150 feet. Whereas the Boomerang shape was the most common shape witnessed east of the Hudson River (33.6%), the Boomerangs seen near Pine Bush tended to be smaller at under 300 feet in width. Some very large craft were described over Pine Bush, but they were not as common. The enormous angular craft described by Russell Pardi that passed over Monticello, NY, to the north of Pine Bush is more the exception for craft observed to the west (see description under the Westchester Boomerang).

Triangular, boomerang, and V-shaped craft were described about 30% of the time, while disk (9.9%) and cigar shapes (1.8%) were the least common shapes witnessed (Hynek et al., 1998). Some of the larger ships appeared to be hybrid shapes between the boomerang and disc shape, based in part on the distribution of the lights on the underside and sides of the object. Not everyone was able to see the larger ships up close, but those that did sometimes saw structure behind the boomerang or V-shaped (chevron) front, that varied between an extended triangle shape, sometimes with a narrow trail, to cone-shaped that tapered quickly to a point, to rounded or even box-shaped. Some of this variation may be due to distance or perspective, but other observation imply that the shapes of these craft can be changed. We now have two videos of the Manta Ray shape morphing both its anterior and posterior as though the pilots could extend or contract portions of the ship to form a large triangle. Crystall reports large Triangles with a perimeter of 400 feet, which is about the size of the Manta Ray without extensions.

No small boomerang shapes are reported by Hynek et al. (1998), as are no UAPs shaped like conventional aircraft with wings. Whereas large ships that travel silently or produce only a faint hum are easily distinguished from conventional human technology, smaller craft that ae similar in size to conventional aircraft – at least in wingspan width – are much more easily misidentified at night as conventional aircraft if they are 1) outfitted with lights that are patterned after landing and navigation lights of conventional aircraft, and 2) if they produce sounds even remotely similar to loud engine sounds of jetliners and propeller-driven planes. I have found that most humans tend to react more to sound than to the lights in forming an opinion of what they saw in the night sky. But sounds can be just as deceiving (if not more so) if they are not properly (sufficiently) analyzed, and that can only happen after the event is over (no instant replay without a camcorder and computer in the field).

The Triangles

"Since the mid-1970s, people in more than one hundred American communities have reported seeing a very large triangular, diamond-shaped, or boomerang –shaped object flying slowly and usually fairly low in the sky. Little or no sound is heard. Just as in the Hudson Valley, the witnesses say the object is at least as big as a football field, and some say even larger. The object frequently has two very bright lights like headlights, along with a number of other lights. Sometimes it drifts about in one area for hours, as if it wanted to be seen by many people." (Hynek et al., 1998: p. 248).

Almost all of the triangular craft I have witnessed and recorded on film or videotape (some were videotaped by Crystall in my presence, while I operated a Minolta XG-M SLR on a tripod: Appendix I) are the small Black Triangles, about 60-70 feet on a side.

Some of the triangular craft reported by Crystall early in her Pine Bush investigation were not isolateral triangles, but were elongate with a square window at its nose. She illustrated them in figures A and C in her book (Crystall, 1991). Others she witnessed were more equilateral in shape (e.g. Figure F and illustration of Triangle over the Beth Hillel Jewish Cemetery). The equilateral triangles become the dominant versions flown during the 1990s and some of them were equipped with booms sticking out front where a window could have been located, and a whip-shaped extension out the back. Both boom and whip supported lights intended to mimic those on the DC-9 commercial jetliners that were the workhorse aircraft of American Airlines flying out of Stewart International Airport. Instead of wanting to be noticed as they did with Crystall and Lebelson in July 1980, they seemed to want to fade into the background as their purpose and activity became more clandestine.

A Shift in UAP Sightings to the East

Chopper device

, writer for the Syracuse Newtimes, and manager of the New York Skies, A UAP Blog online, provides sighting information for a 20 year window from 1995 to 2015, reporting 496 sightings in Orange and Suffolk counties of New York. Orange County is located in the western part of the UAP Corridor, while Suffolk is located in the eastern part. Although she underreports the number of sightings in

Orange County for the years 1995 to 1999, she documents a surge in UAP sightings in Suffolk County, or Long Island, NY, for the years 2000 to 2015, with the highest number of sightings between 2008 and 2015. UAP sightings in Orange and neighboring counties are not shown in her graph to decease abruptly at the end of 1997, but to gradually increase into the year 2000 before suddenly dropping. She claims 398 UAP sightings were made in Suffolk County while only 98 were made in Orange County. Costa reports 370 sightings in Suffolk County from 2000 to 2015, whereas only 89 sightings were made for the same period of time in Orange County. The importance of her data is not that Unidentified Flying Objects are more common over Long Island and its surrounding waters, but that the activity in and around Pine Bush seems to have dropped off intentionally in 1997, and moved elsewhere. Between 1992 and 1997 Crystall and I recorded on film and video well over 100 sightings in Orange, Sullivan, and Ulster counties, more than were recorded over Suffolk County for that period (Appendix I). The question becomes, why did the activity shift from the western part of the UAP Corridor to the eastern part? Did sky watching and public awareness become too focused on our visitors, or could there be another reason?

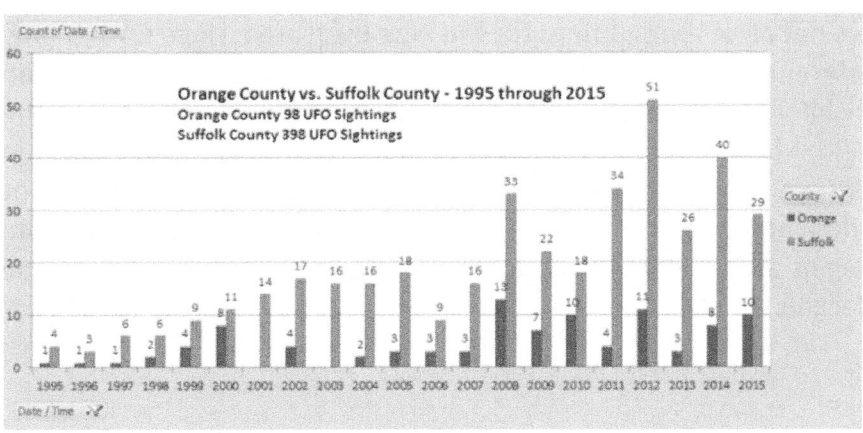

A barn on the west side of Rte. 17 near Middletown, NY, just before the Pine Bush exit, had been painted with a sign saying "ET Came Home to Pine Bush, So Can You. Put a Charge in Your Next

Real Estate Transaction." That sign disappeared in 1997 just as the sightings around Pine Bush waned. ET was no longer welcome in Pine Bush? See image below.

Rte 17 Middletown, NY
15 July 1994

Below is a list of sightings documented by film and video from 1992 to 2002 around the Pine Bush area. Cornet made only one sighting of the Manta Ray in 2003, but that was near Salt Point, NY, on the eastern side of the Hudson River (Dennett, 2008: p. 223-224). The tabulation below shows that most of the sightings made by Cornet and Crystall occurred in the years 1992 to 1997, with the highest numbers in 1992, 1993, and 1994 (see Appendix I). These are not all of the sightings for the Pine Bush area, since Cornet only references those encounters that could be verified as non-human technology through scientific analysis of the data collected in the field. Many other sightings were reported by other sky watchers in Orange, Ulster, and Sullivan counties during the 1990s, but not adequately documented for scientific analysis. Cheryl Costa's data begins after 1994, when sightings in Orange County were on the decline.

Unconventional Aerial Phenomena

9 June 1992
14 June 1992
18 June 1992a
18 June 1992b
18 June 1992c
23 June 1992a
23 June 1992b
1 July 1992
3 July 1992
7 July 1992
10 July 1992
13 July 1992
6 August 1992a
6 August 1992b
6 August 1992c
6 August 1992d
29 August 1992
4 September 1992
13 September 1992a
13 September 1992b
20 September 1992
24 September 1992a
24 September 1992b
2 October 1992a
7 October 1992b
7 October 1992
22 October 1992

8 April 1993
28 April 1993a
28 April 1993b
28 April 1993c
28 April 1993d
28 April 1993e
29 April 1993a
29 April 1993b
29 April 1993c
29 April 1993d
29 April 1993e
29 April 1993f
29 April 1993g
2 June 1993a
2 June 1993b
24 June 1993a
24 June 1993b
24 June 1993c
24 June 1993d
24 June 1993e
1 July 1993
7 August 1993
5 October 1993a
5 October 1993b
12 November 1993a
12 November 1993b
25 November 1993

26 March 1994a
26 March 1994b

10 May 1994a
10 May 1994b
10 May 1994c

23 June 1994

11 October 1994a
11 October 1994b
14 October 1994
16 October 1994
24 October 1994
24 November 1994

6 May 1995
31 May 1995
10 June 1995
2 July 1995

29 September 1995
28 October 1995

17 July 1996
9 August 1996a
9 August 1996b
10 August 1996a
10 August 1996b

22 January 1997
25 January 1997a
25 January 1997b
25 January 1997c
21 February 1997
24 April 1997
17 May 1997a
17 May 1997b
4 June 1997
10 August 1997
3 October 1997

25 July 1998

18 December 1999a
18 December 1999b
7 December 1999

15 January 2000
4 May 2000a
4 May 2000b

17 November 2001a
17 November 2001b

8 January 2002
12 January 2002
23 February 2002
7 March 2002

123

The last sighting I had was on 20 August 2003, with Billy McNamara at his home in Salt Point, NY, on the eastern side of the Hudson River, far from Pine Bush. That sighting was consistent with Cheryl Costa's and Linda Miller Costa's data published in their 2017 book, UFO Sightings Desk Reference, United States of America 2001-2015.

My first sighting of a Black Triangle occurred on July 13, 1992 (Appendix I). I had one more sighting during the mini-flap described on 6 August 1992, in Chapter **One**, with excellent data captured on video. In 1993 I recorded four more small Black Triangles, some with Crystall (three sightings) and one before the Sightings film crew on April 29, 1993. In 1994 I recorded three Black Triangles, two with Pat Huff-Cornet and one with George Filer (Head of MUFON, New Jersey) (Guerra, J.L., 2018). In 1995 I recorded two Black Triangles, one with Marc Whitford, who captured it doing double right angle turns on a time exposure, and the other with six witnesses at the Indian Mound, next to which the Triangle lifted off from a farm field, flared its paired headlights to produce an orange-red aura around the lights as it rose vertically, then dimmed its lights and arched across the night sky above us before descending behind trees on opposite side of the mound (Appendix I).

No sightings of Black Triangles were made by me in 1996, although six sightings of the Manta Ray were made, five on one spectacular night of 9-10 August 1996 (appendix I). In 1997 seven sightings of the Black Triangle were made, two on January 25 and two on May 17 before multiple witnesses. One of the best performances of a Black Triangle occurred on May 17, 1997; it lasted for nine minutes, and was recorded on two camcorders and one 35 mm SLR. The Triangle did flips, flew upside down and backwards (from its previous orientation), hovered, and produced a plasma plume discharge before disappearing into a farm field flying sideways (based on the orientation of its paired headlights).

http://www.bufod.co.uk/b_cornet/Vol_3/index.html?i=1

The last sighting I had of a Black Triangle was off the coast of New Jersey at Sea Bright (five miles from my residence then at Red Bank, NJ) on 10 August 1997. On videotape I recorded it flying south into the flight path of commercial jetliners (which flew by right after this performance) on course for Kennedy International Airport 20 miles to the North. As it flew past me it turned sideways about 1,000 feet above the ocean, presenting its belly to the camera, and produced 85 red and white flashes of its strobes in seven sequences or clusters of signals over 127 seconds. After careful analysis, observing the video frame by frame, a pattern emerged that indicated a deliberate coded message. The meaning focused on a five letter word, such as "Hello." Although that meaning would be significant based on who sent that message, it is by no means certain. Commercial and military aircraft are not engineered to produce such complicated signals with navigation lights. The pattern of signals is graphically displayed below:

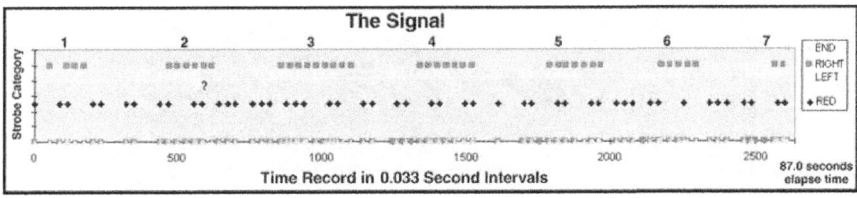

After finishing its signal, it rotated horizontal, traveled about a mile further south, then turned quickly out to sea (East) as it dropped in altitude. Shortly thereafter commercial jets passed by and in that same flight corridor, heading North. If the Triangle had continued on its course without turning, it would have headed directly into oncoming air traffic. (Appendix I).

I did not expect to see a Black Triangle while videotaping conventional aircraft at dark. I was doing research on patterns of navigation lights after recognizing that there is no standard for commercial aircraft other than for the presence and position of hazard lights. Other lights, such as tail lights, belly lights, additional wing lights, and lights on the fuselage can vary even for the same type of aircraft.

That was the last sighting I had of a Black Triangle to date, even though it was not in the Hudson Valley. Others have had sightings in Orange County near Pine Bush since then. For example, Sal Cirami of New York City witnessed a Black Triangle crossing Rte. 17 in Middletown, NY, as he drove north to visit his daughter. This sighting occurred on

2 January 2002 in the early morning (daylight), and was the first such sighting he ever had. The Triangle flew low over the highway in front of him, passed three religious crosses next to the highway, and stopped over a farm field next to a house along Midland Lake Rd. He saw it quickly drop to the ground before he lost sight of it.

He later went back to investigate, and even built a model to show what he saw. There was a blinking red light at its center, and bright white lights at the corners – similar to the Belgian Black Triangles witnessed and photographed in 1989. The Triangle made no sound, and Cirami felt it had purposely selected him for the sighting. He could find no other witnesses among residents along Midland Lake Rd.

Military aircraft, especially classified ones, do not fly low over residential communities and land in farm fields unless they are having mechanical problems. And yet I have heard many skeptics speculate that the Black Triangles are human hardware, and not alien spaceships. The fact that this Triangle chose to cross the highway in front of Cirami, and pass by three crosses next to a church when it did so, may have been a sign that we would both recognize. I say that, because Cirami soon found me on the internet and contacted me after his sighting. He was hooked by the event, which occurred in broad daylight just north of Middletown, NY. I wonder if that Triangle had passed across that highway several times in front of cars until the pilot could be sure one of the drivers would investigate the sighting. But how could the pilot know Cirami would contact me? We may never know. Yet the presence of three crosses as at Calvary, Jerusalem, did not go unnoticed by us or unrecognized as part of the intended message.

NIDS Involvement

For a period of four months (January – May, 2004) Cornet worked as Deputy Administrator at the National Institute for Discovery Science (NIDS) in Las Vegas, Nevada. NIDS was owned and run by the billionaire, Robert Bigalow. The Administrator, Dr. Colm Kelleher, was my boss. Just four months after I started working there, having moved all my belongings from Red Bank, NJ, Bigelow decided to shut down NIDS. My last day was at the end of May 2004. From there I moved to El Paso, Texas, and soon found a job teaching at El Paso Community College. In August 2004 the Administrator, Dr. Colm Kelleher, found another job in medicine, his primary profession, and the Institute was officially closed. Note uncanny resemblance of building to the 1952 Adamski Flying Saucer. Picture of Dr. Colm Kelleher standing in front of building in December 2003.

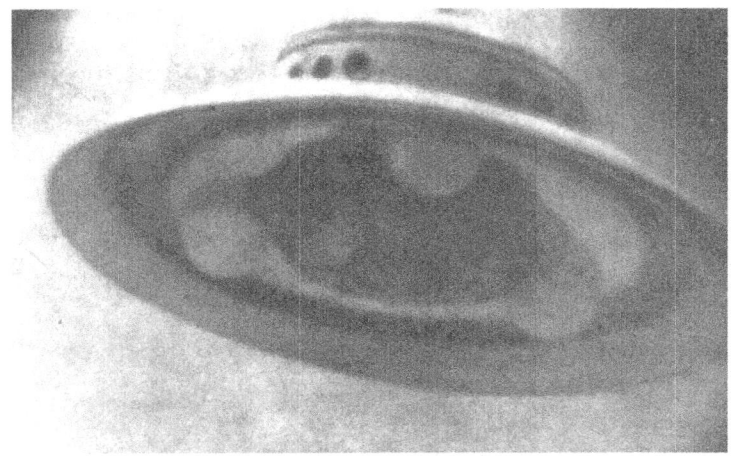

Also note the uncanny coincidence of the street number, 4975, on Polaris Avenue to the numerology of Bruce Cornet, where Bruce = 49 (2+18+21+3+5) and Cornet = 75 (3+15+18+14+5+20).

NIDS was involved in researching and studying UAP sightings and paranormal phenomena. Numerous projects and reports were generated between the years 1995 and 2004. Dr. Kelleher's personal interest was investigating the sightings and occurrences of Black Triangles across the contiguous United States. By the early 2000s he had developed a data base that suggested a significant number of triangle sightings were flying between U.S. Air force bases, implying that they were advanced human hardware not yet disclosed by the military, similar to sightings of the F117 Nighthawk (a triangular-shaped aircraft powered by General Electric turbofan engines) before the F117 was officially announced into public existence in 1988 (retired on 22 April 2008).

Dr. Colm Kelleher, Dr. Bruce Cornet, and Jacques Vallees at NIDS in May 2004.

Because the NIDS database was not as robust as Kelleher would like, and because other databases existed that were larger, he did not have "smoking gun" evidence for a military source of Black Triangles. Through negotiations with the owners of two other databases, he convinced Robert Bigelow to purchase copies of them. When those data were compiled with the NIDS database and plotted on the same map of the U.S.A., a different picture and pattern of Black Triangles emerged. To his surprise the military origin for Black Triangles was drowned out by a clear indication that sightings predominated over major cities, suburbs, and highways, not between Air Force and Navy bases. It appeared that whoever was flying these Triangles, they were interested in studying humans and our social infrastructure. The military would have no defense reason for doing so, especially using highly classified hardware.

When Cornet was working for NIDS, a caller to our UAP hotline was transferred to me. He told me of an incident he and a co-worker witnessed on a smoking break at his photography job in San Diego, California. It was later in the afternoon and the Sun was still high in the sky. Commercial aircraft traveling south to the San Diego International Airport would fly west of the airport and turn east after passing the airport. That airport is located just west of San Diego next to the bay. After making a U-turn, commercial aircraft would land on the East-West-oriented main runway. North Island Naval Air Station (Halsey Field), with its main runway oriented almost due North-South, and a second runway oriented northwest-southeast, is located just to the southwest of downtown San Diego, and due South of the San Diego International Airport. The two airports are separated by about 1.5 miles, mostly by water. The two men saw a large MD-11 jetliner (with an engine mounted in its tail stabilizer) approaching from the north on its way to the airport. To their surprise, they then saw a large black triangular craft rising up from the Halsey Field airstrip. It was on a collision course with the MD-11! The Triangle was as large (wide) as the MD-11 jetliner, which has a wingspan of 165.3 feet. They fully expected to see a collision, but the Black Triangle "stopped on a dime" just before reaching the passing jetliner, allowing it to pass and turn towards the airport. Then the Triangle accelerated very fast and disappeared beyond the building tops of the city. This sighting, direction of flight from the only other airstrip to the northwest, and rapid stop with hovering, followed by rapid acceleration, strongly indicate that the U.S. military has the Triangle hardware and technology.

But does that mean all Black Triangles are of human origin? One also needs to ask: Was I given this unconfirmed account in order to influence my opinion about the origin of Black Triangles? I cannot answer that question with certainty, buts the work of Dr. Colm Kelleher described above does bear directly on this question. After all, the CIA went to great efforts and expense to obfuscate and hide the truth about the craft seen over the Hudson Valley. Imbrogno and Horrigan (1997) in their chapter, "The Government Connection" (Contact of the 5th Kind: pp. 225-247), say that the CIA employed stunt pilots to fly out of Stewart Airport. The pilots flew O2As (built by Cessna), which were a plane with a single or double engine, and capable of long-range flights. The O-2 is used for surveillance missions and has very quiet, muffled engines. They were flown in formation with special lights in a failed attempt to mimic the lights and slow speeds of the Boomerangs and Triangles (Hynek et al., 1998; Imbrogno and Horrigan, 1997).

Dr. Kelleher developed a composite map based on Black Triangle sightings from 1990 to 2004, which consisted of four massive databases, two of which were purchased by Bigelow from MUFON and Larry Hatch. His composite map and frequency data are reproduced here with permission.

The MUFON Triangle Database
The Larry Hatch Triangle Database
NIDS Triangle Database
NUFORC Database

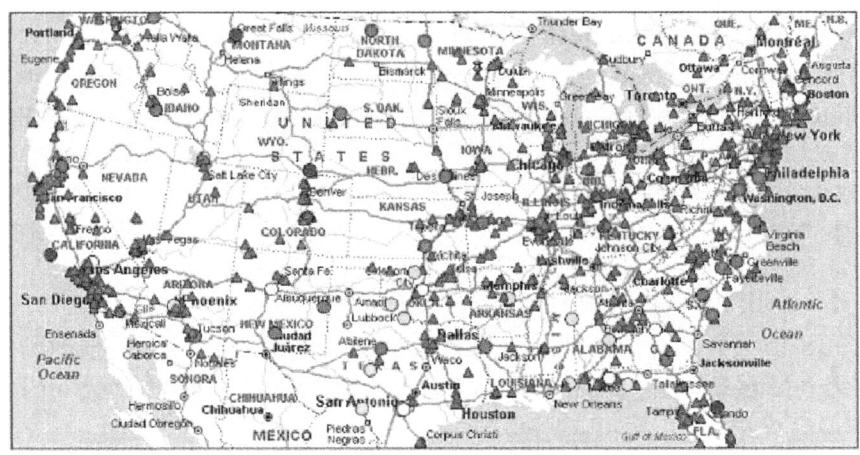

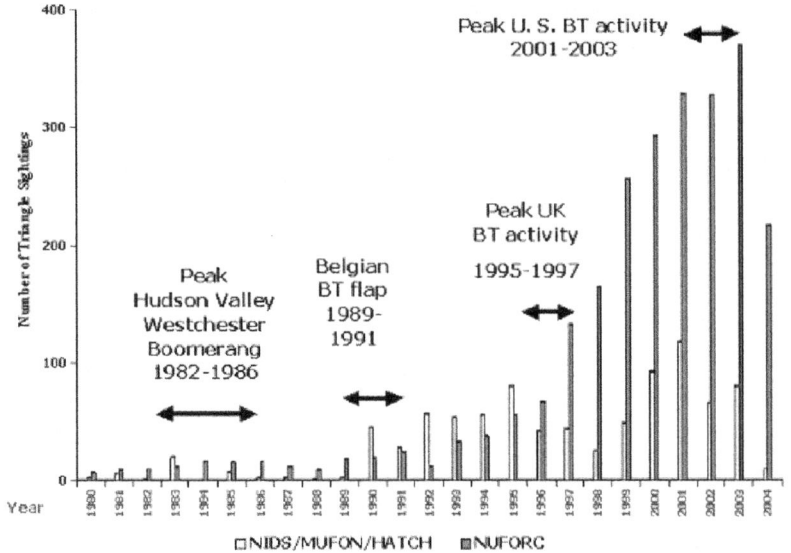

The NIDS report by Dr. Kelleher was finalized and published in August 2004, just before NIDS closed. NIDS last research is preserved on www.nidsci.org and www.cohenufo.org. The above two figures come from Kelleher's report, with permission.

Dr. Kelleher gave a talk at the 23rd Annual Society for Scientific Exploration Conference in Las Vegas on May 20-22, 2004, entitled:

NIDS Investigation into the Enigmatic Large Triangular Craft Phenomenon, while Dr. Cornet gave a talk entitled: Evidence for An Underground Magnetic Focusing System (large triangular alien mining probe entombing in Late Ordovician age black shales).

In that talk he noted the possible implications: A data-gathering effort not just to study our species and social behavior but a preparation for a worldwide invasion. He said this could be a "First Wave" to learn about our areas of strength and areas of vulnerability, because sightings covered so many areas – all related to human habitation and activity. If they were only interested in the geology and mineral resources of our planet, or in the natural species of plants and animals, they were focusing on the wrong places.

Dr. Ellen Crystall (1991: p. 79-80) recounts one of the closest encounters with a Black Triangle, sometime between 1980 and 1986. This incident occurred during the period of intense UAP sightings in the Hudson Valley to the east of the Hudson River (Hynek et al., 1998). This sighting is important because she got within 10 feet of the craft as it hovered just above the ground, and because of what she saw projecting from the sides of the craft. Unfortunately, Crystall was not always interested in giving dates and times, so exactly when this encounter occurred is not given in her book. On this particular night Crystall took Bob and his twenty-year-old daughter to see a ship, not realizing his daughter would get her wish to see a ship. Five of them went to the Beth Hillel Jewish Cemetery on Rte. 52 between Pine Bush and Walden, NY.

"When we arrived there, lights were flying all around the area. A large lit object came up from the trees towards us. Bob had brought his video camera, but it was in the car. [video cameras at the time were separate from the tape recording device, and were cumbersome and awkward to set up and carry.] The object cast a fabulous shadow on the house next to the cemetery as it approached us. This happened rather quickly. I noticed the object coming toward us and yelled, "Look at what's coming!"

"They all said, "Yeah, yeah...," but no one moved.

"I raced around the cemetery fence and cut through bushes to get into the field. The entire lighting system changed, and the hulk of a triangle ship came to the cemetery fence, entirely lit and completely silent. [It would have come across open fields to the south and east of the cemetery.] It was about sixty feet on each side, with antenna-like projections bobbing from it.

"I could clearly see every seem and detail. It came up to the far side of the cemetery, within twenty feet or so of where I stood, then stopped briefly and simply reversed direction, slowly heading back to where it came from. [The "far side" of the cemetery would be the south side, away from Rte. 52, where a large field would have given the craft plenty of room to maneuver, unlike the field on the east side that abutted the road.] The reversal without turning was staggering to watch. I could see every groove in the ship's metal exterior. A discharge or mist seemed to surround it, almost as if faint electrical currents were bouncing around its outside. Images below from Crystall, 1991.

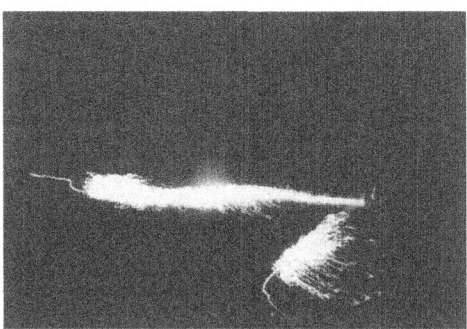 

21. *TRIANGLE CRAFT. Pine Bush, N.Y. July, 1980. What we saw was a clear-cut metallic triangle craft. The film registered what our eyes didn't see: shortwave radiation discharges coming off the craft out of vents. See tracing (21a) for clarity.*

TRIANGLE CRAFT. Pine Bush, N.Y. July, 1980. What we saw was a clear-cut metallic triangle craft. The film registered what our eyes didn't see: shortwave radiation discharges coming off the craft out of vents.

"As it moved away, I noticed a rustling sound. It was the others coming into the field to join me.

"I questioned them. "Why didn't you come into this field with me? Bob, why was your camera in the car instead of in your hand?"

"The conversation that followed raised some eyebrows. One of the four said he received a distinct mental message to stay exactly where we were. I, on the other hand, received a mental message to "go for it," which I did. My companion's message was quite clear: don't move. Mine was totally different in meaning. Someone or something wanted me to get close to the ship. Why, I asked myself." (Crystall, 1991: p. 79-80).

Crystall told me in 1992 that she wanted to reach out and touch the Triangle, but reading her mind the pilot suddenly reversed the ship's direction, moving away from her faster than she could run, which she did briefly before realizing that touching the ship could have caused her serious injury.

The Black Triangles and 'angular' craft appear to be more associated with human abductions, while circular or disk-shaped craft, which were more common during the first half of the 20th Century, were more asociated with contact between very human-like ETs and humans ("Alien Bases," 1998, by Timothy Good) [add more references] That is not to say disc-shaped craft were not involved in abductions, because they were (e.g. "Missing Time," 1981 and "Intruders," 1987, by Budd Hopkins).

Most of my nine abductions from earliest memory at age 9 to age 49 in 1994 involved disc-shaped craft. The agenda of the visitors clearly changed in the late 1960s as triangular and diamond-shaped craft sightings became more common. Crystall did not hae any contact experiences as people had earlier in the century, even though she was "allowed" to capture pictures of extraterrestrials inside or standing next to their craft (Crystall, 1991). In order to understand what is going on, and why Black Triangles are so much more commonly sighted than other types of craft today, we need to take a look at what contactees were told in the 1920s through the 1950s.

Chapter 4 Visitors Studying of Earth and Humans

Contactees have learned several things from the Visitors. Due to the warring and violent nature of humans, extraterrestrials may be legitimately concerned we will take our behavior and weapons into their space. Note how these visitors, as hybrids, can integrate into our societies without standing out, which gives them probable deniability. Excerpts from Alien Bases, by Timothy Good, 1998:

Zret and the NORCANS (Zret is short for Xretsim, or MisterX spelled backwards}. 16-year old Albert Coe saved his life after Zret fell into a rock crevasse and was injured and trapped. Zret and Coe became friends. Zret met Coe at the McAlpine Hotel, Ottowa, Canada, six months after their initial encounter in June 1920.

"Zret went on to say that his true identity and address and details of his personal life had to remain secret, though he did explain that he was one of the group which had come to monitor Earth's scientific advances. While on Earth, he 'doubled' as a student majoring in electronic engineering. Man's capacity for developing weapons ever-increasing power was the prime motive for the visits to Earth. 'Only recently,' he continued, 'many of the more "intelligent" and "cultural" nations of Earth concluded a long, bloody war, and during its progress several innovations, designed specifically for the mass slaughter of humanity, were introduced... As each new invention was applied to a military potential, its horizon broadened to the eventual horror, brutality and devastation that emerged as a "world war." This conversation of inventive genius from the brain of Earth's inhabitants, to ever greater devices of destruction, was the prime factor that motivated our mission...'

"Later, Coe was to learn that in 1904, Zret's people had paved the way for a hundred of them to infiltrate a very major nation of our planet – as small groups of technicians – to observe and evaluate every step of our scientific advancement. Their main concern was that we were on the verge of discovering secrets of the atom which could have disastrous consequences for our planet." (Good, 1998: p. 35).

Increasingly humans have been indoctrinated and conscripted to carry out an alien agenda as teachers, trying to educate and change human policies and behavior from within our societies and cultures.

The "Teacher"

Rose C., 24 years old, on 11 April 1952, near the French town of Nîmes (Gard), was told by the "normal man, a former teacher (who had been contacted in 1932 at the age of 25), who accompanied the three 7-7.5 foot "giants," acting as an interpreter:

"The teacher explained to Rose that the extraterrestrials had 'established' Earth for the use of its human inhabitants: this had been rather like a 'penal colony', consisting of banished individuals, from whom humans are descended. He went on to say that 11,357 years ago (9405 BC), because Man, there had been a cataclysm on Earth. (Many other Contactees had been told that Earthmen destroyed their civilization thousands of years ago.) [Compare this account to the Biblical war in heaven, and the banishment of Fallen Angels to Earth.]

"As to the purpose of their current visits, it was explained that the extraterrestrials wee here to take vegetation and soil samples to evaluate the consequences of our atomic explosions. The destructive, senseless behavior of humans, and disregard not only for our contemporaries but also for future generations on this beautiful planet, was commented on. Asked by Rose why the extraterrestrials did not intervene, the teacher replied that no good had ever come out of such attempts in the past. The visitors then told Rose that they had to go, owing to the fact that time spent on Earth exhausted the giants. Having warned Rose to stay clear of the craft and to hold the dogs, they boarded their craft and it took off, making a droning sound and creating a strong draught of warm wind." (Good, 1998: p. 98).

As to why extraterrestrials have not "landed on the White House lawn," the extraterrestrial Ahlahn (Alan) said to Daniel Fry in April 1954 at White Sands, NM, "...If we were to appear a members of a superior race, coming from 'above' to lead the people of your world, our arrival would seriously disrupt the ego balance of your society. Tens of millions of your people, in their desperate need to avoid being demoted to second place in the universe, would go to any lengths to disprove, or simply deny, our existence. If we took steps to force the acceptance of our reality upon their consciousness, about 30 per cent of the people would insist upon considering us as Gods, and would attempt to place upon us all responsibility for their welfare. This is a responsibility we would not be permitted to assume, even if we were able to discharge it... Most of the remaining 70 percent would adopt the belief that we were planning to enslave their world, and many would begin to seek means to destroy us. If any great and lasting good is to come from our efforts, they must be led by your own people, or at least by those who are accepted as such..."(Good, 1998: p. 72).

Ancient Hybrid Civilizations: Lemuria and Atlantis

"Alan explained that the symbol of the tree and the serpent was not unique to Earth. It is a natural one, 'perhaps because life is said to originate in the waters of a planet, and the undulations of a serpent are a convenient symbol for the waves of a sea. The tree is almost always the symbol of life, beginning in the sea, rising to the atmosphere, and finally into space.' But there was another factor, he added, that perhaps was significant.

"Your people, and some of mine, including myself, have, at least in part, a common ancestry (Read Dr. Susan B. Martinez's two 2013 books: The Lost History of the Little People, and The Mysterious

Origins of Hybrid Man). Tens of thousands of years ago, some of our ancestors lived upon this planet, Earth. There was, at that time, a small continent in a part of the now sea–covered area which you have named the Pacific Ocean (Martinez calls this lost continent "Pan"). Some of your ancient legends refer to this sunken land mass as the 'Lost Continent of Lemuria, or Mu.' (other ETs, above, give the date of 11,357 years ago, a time when sea level was rising at the end of the Wisconsinian Glaciation, the great meltdown. The submerged continent did not sink. It was flooded by sea level rise).

"Our ancestors had built a great empire and a mighty science upon this continent. At the same time, there was another rapidly developing race upon a land mass in the south-west portion of the present Atlantic Ocean. [Could this be the Falkland Islands, which are the highest mountains remaining of this land mass after sea level rose?] In your legends, this continent has been named Atlantis. There was rivalry between the two cultures, in their material and technological progress. It was friendly at first, but became bitter... In a few centuries their science had passed the point which your race has reached. Not content with releasing a few crumbs of the binding energy of the atom (a reference to quantum mechanics, which did not reach full development until atom-smashers and the 1980s for humans), as your science is now doing, they had learned to rotate entire masses upon the energy axis. Energies equal to 75 million of your kilowatt hours were released by the conversion of a bit of matter about the mass of one of your copper pennies.

"With the increasing bitterness between the two races...it was inevitable that they would eventually destroy each other. The energies released in that destruction were beyond all human imagination. [This may be the event that caused the extinction of the large mammals around the globe at the very time in history, and not due to some meteor exploding in Earth's atmosphere – as the Tunguska object did.] They were sufficient to cause major shifts in the surface configuration of the planet, and the resulting nuclear radiation was so intense and so widespread that the entire surface became virtually unfit for [human] habitation, for a number of generations." (Good, 1998: p. 68-69).

Whether you can believe these stories or not, much of what was said rings true, if we are willing to be honest with ourselves. Thus, as we approach the ability to control gravity, and travel to the far reaches of our own galaxy, we are the ones who could become the threat to extraterrestrial civilizations within reach in a few more centuries. Our technology is exploding exponentially as we reached and passed the end date of the Mayan calendar on 21 December 2012 (The Mystery of 2012). Even though nothing apocalyptic or catastrophic occurred on that date, there is strong evidence that our species is reaching a singularity in time, when we will be faced with the need for profound change in our behavior (collectively) and attitudes, or face extinction (Russell, 2009: p. 17-33). The increase in outside observation of our societies, as Colm Kelleher's Black Triangle map indicates, may not be an indication of imminent invasion, but a last ditch attempt to find a way to prevent our destruction, and with it save our planet. And if you can believe what Contactees have been told about past high-tech human civilizations, and the consequences of extraterrestrial collaboration and involvement, our species may be in for a very rough future. Can these extraterrestrials wake us up without causing worldwide panic and social collapse – in time, or will the following poem (reproduced here in part) by an alleged Biaviian named O-Qua Tangin Wann (i.e. "Tan") become our epitaph?

A Biaviian Poem

As recited to Riley Martin by Tan on 26 September 1990 (Martin and Wann, The Coming of Tan, 1995).

Stanza 3: Children of the living waters, possessors of the staff of reason:

Beings of celestial promise, doomed to perish in mid-season.
From the stone axe to the heavens, lo' the vision did not fade:
Still that pulsing seed of hatred, lay thy fate beneath the spade.

Stanza 4: Before me the planet lay in shambles, eco-destruction beyond repair:
By the greed of false controllers, acids now permeate the air.
Pristine waters from the mountains, die en-route down to the sea:
Thus to hasten soon the horror, the sapiens race may cease to be.

Stanza 5: Oh ye marvels of creation, why have you not sought the light:
Must you fade into the shadows, of that still, cold, azure night.
Unto you my sign is written, there upon the living fields (crop circles).
Lo' the circle is eternal, though you perish, you yet shall live.

Stanza 6: I have not returned to conquer, nor to alter the flow of fate:
But simply to gather a certain number, soon before it is too late.
Too unstable to embrace, yet far too noble to cast away:
Beautiful life form though unsuccessful, might succeed another day:
There upon the flowering meadows. Oh, shining precious Biaveh.

Within these selected stanzas from Wann's eight stanza poem could be a prediction of the fate of humanity within this 21st Century. To change our fate, we must take control of our destiny on a worldwide level. We cannot depend on outsiders to save us.

Although there is much misinformation, disinformation, and deception (secrecy) regarding our visitors, with most governments unwilling to admit to an ET presence, certain patterns and activities of the intelligence behind UAP sightings reveal purposes and agendas behind their activities. But before we can get to their purposes or agendas, more data must be presented on their interaction with humans in our environments (as opposed to their environments aboard their craft: i.e. abductions).

Do we see the Light?

One of the first things people see in the night sky are the craft's lights. The fact that their ships are frequently invisible or transparent to the human eye only makes their lights stand out that much more. The number, arrangement, and colors of the lights appear to be important, and are the primary means of communication with humans. Even though the ships can produce any number and types of sounds, all of which when recorded and analyzed appear to be synthetic (see Chapter 5), the absence of sound (42%) or a faint humming sound (55%) indicate that sound is not their preferred form of communication, as it is with us (Hynek et al, 1998). When sounds are produced, they appear to be selected to produce a desired reaction in humans. For example, synthetic jet-like sounds will elicit a familiar recognition in humans, and strongly influence human interpretation of what they are looking at in the sky.

When my wife, Pat, was abducted from our van parked at the top of Massanutten Mountain in Virginia on 25 November 1993, she said they ran a test using various colored lights, measuring her brain responses. They were apparently interested in gauging our responses as humans to their use of colored lights as a means of communication or modifying our thoughts.

http://www.sunstar-solutions.com/AOP/Massanutten/Massanutten.htm

The main craft hovered above us, illuminating the ground below as a dome-shaped shuttle craft came down and wrinkle-brow aliens abducted my wife, Pat. The aliens knocked me out. Drawings below of aliens from "Taken: Inside the Alien-Human Abduction Agenda," by the late Dr. Karla Turner (1994).

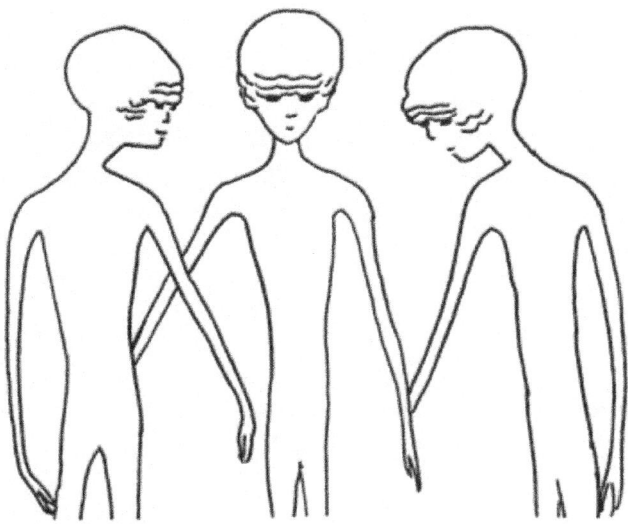
from Taken by Dr. Karla Turner
Polly's 'snapshot' memory of wrinkled-brow aliens.

Hynek et al. (1998) provide data on the number of lights and color of lights for the 7,046 entries in their Hudson Valley UFO sightings database. The minimum number of lights reported was four (24.6%). Six lights were not as common (11.6%), while eight lights were more common (27.9%). The majority of people reported ten lights or more (35.8%). Some witnesses reported more than twenty lights. In the reporting (determination) of the number of lights, Hynek et al. (1998) indicate even numbers, but whether this is an accurate reflection of what was seen, or involves witness bias, and/or analyst bias cannot be determined. If accurate, even numbers may be significant.

The color of lights for Hudson Valley sightings was also analyzed. White lights were most frequently reported (16.8%). Most witnesses said they saw many colors in a display (70.6%), while some reported only one or two colors. Some witnesses said the lights alternated in color, or changed in color during the sighting. Red was the second-most common color reported (6.0%), with green following (2.8%). It is not known if red and green have any particular significance to the visitors, or they selected those colors because of our use of red and green colors for our aircraft navigation lights, traffic stop and go lights, and symbolism for holiday celebration (e.g. Christmas). The least common colors used were blue lights (2.4%) and yellow lights (1.4%). Blue lights are not used on commercial or private aircraft (against FAA regulations due to depth perception problems).

In the study by Greco and Gordon (MUFON UFO Journal, No. 290, June 1992) for the Williamsport, PA, wave on 5 February 1992, amber with white lights were reported only 7.7% of the time, and green with white lights were reported only 7.7% of the time. However, only white lights were reported 84.6% of the time. Greco and Gordon's data are confusing, because they report nine different combinations of white and green lights, which ranged from one (green?) to 14 green and white lights. They add that the total (maximum?) number of green lights seen was two, and that the total number of white lights varied. The variation in numbers and colors was believed by them to be caused by the position of the witnesses in relation to the object and by the object possibly turning the lights on and off. Blue lights were not reported seen by any of the Williamsport witnesses.

However, blue lights are sometimes used for military aircraft for that very purpose, of confusing depth or distance. I have witnessed blue lights on angular craft, and they usually appear on the top of the craft, which is not often visible if the witness is too close to the UAP. Sometimes the blue lights are in multiples (clusters), unlike what military aircraft display, and they are mobile – videotaped moving over the back and top of a craft. The function or meaning of blue lights is more cryptic, while red and green lights could be used simply for mimicry and deception (camouflage). Military aircraft would probably not use blue lights when flying below minimum FAA altitude or putting on performances for sky watchers on the ground, unless the desire is to deflect source or cause from foreign invaders. And the CIA has already been involved in mimicking UAP lights at night in order to allay fears among the public (Imbrogno and Horrigan, 1997).

The use of multicolored or alternating patterns of colored lights was most commonly reported, and sometimes the lights changed color in sequence (order) to give the illusion that they were moving, for example, in a ring around the craft. Other times all the lights would appear as the same color, commonly red. Greco and Gordon (1992) report 23.1% of the Williamsport witnesses saw red lights or red and green lights. The number of such colored lights varied from five to 50 or more. These lights were observed to move individually, as in the case of the red lights only, or in unison, as in the case of the red and green lights. They were not discharged from the object or craft, nor were they seen as being on or part of the object. Interestingly, they were seen in movement at a distance near the craft and appeared to be either escorting or chasing it. I have witnessed a red orb being discharged from a bright white light (a structured craft was not visible) and doing controlled loop motions away from the white light on 2 June 1993 (see page 9 above; Appendix I), and a red light escorting close behind a bright white light on 2 July 1992.

Almost all UAPs observed at night are because of the lights on them. Reports exist of dark silhouettes – triangles and ovals – being observed at night if there is enough contrast between a lighter night sky (e.g. during full moon phase) and a darker object. A Black Triangle passed over me late at night when I was parked at a rest stop on highway I-287 near Bernardsville and Basking Ridge, NJ, in late 1992. It was large and silent, and was traveling southwest below cloud level. Against FAA regulations, it had no lights on, but could be clearly seen due to contrast with the clouds above it.

Only on occasion have witnesses observed interaction between two UAPs by the use of lights. One particular case is described by Hynek et al. (1998: p. 1113). This sighting was made by Mike Cobelli and his mother in Ridgefield, Connecticut.

"It was about 10:45 when we first saw it, and we watched it for fifteen to twenty minutes," Cobelli told us. "It was a round circle of lights, a little oblong because of the angle we were looking at it, a little to the north. It was all white, but the lights changed to blue and then kept changing back and forth from white to blue.

"Then another object came from the east. It was just a red light. At first we thought it was a plane. It kept getting closer to the thing. The circle was moving northeast, like on a collision course. As soon as they got close enough, all the lights in the circle turned red, to match the red one. Then the red one passed it, and the circle stopped and started going the other way, the same way as the red light." (Hynek et al., 1998: p. 113).

Was this form of interaction a recognition or a hail of one object to the other, or was it an acknowledgement that a message at some invisible frequency or method was received and understood, because the circular lights immediately stopped and changed course as if following the red light? Could the Cabellis have witnessed craft to craft communication using lights, just as naval ships do with lights and flags?

In rare instances the witnesses would flash their car headlights or a flashlight at the UAP, and the UAP would respond in kind. This type of very basic communication or acknowledgement was the most common, while any patterns or purposeful shift in light color and position was rarely mentioned, perhaps due to a lack of recognition or realization of a pattern (i.e. its significance). Typically, such higher levels of communication require a recording device and later analysis. Camcorders were rarely used (or available) back in the 1980s. It was not until the 1990s that affordable hand-held camcorders became readily available to the public, and frequently captured UAP sightings for later analysis. My use of time exposures has been used by only a few other researchers, and most of that type of recording occurred during the

1990s, after the Hudson Valley flap (1982-1986) was over. Videotape and time exposure recordings have yielded higher levels of communication, primarily in the form of visual symbols.

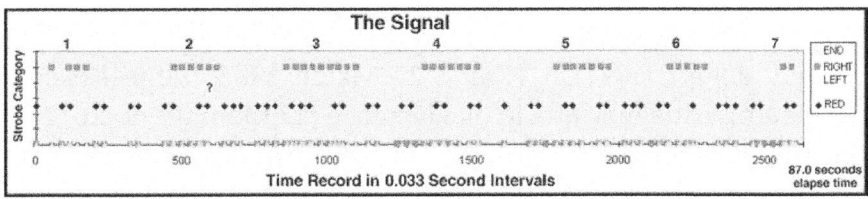

http://www.sunstar-solutions.com/AOP/SOW/signal.htm

Only once was a pattern of alternating red and white lights recorded, and that was for a Black Triangle on 10 October 1997, at the very end of the 1990s UAP flap in the Pine Bush area (see Appendix I). UAP activity continued, however, over Long Island, NY, in the eastern part of the UAP Corridor. For the most part, these craft were there just to be observed and realized, nothing more. They were attracting our attention so that we humans could be observed and studied, and experimented on.

Chapter 5 The Movement of Lights

On a number of occasions Hudson Valley witnesses saw bright beams of light projected down towards the ground. These lights frequently illuminated the tree tops and ground, and vehicles stopped along roads. On 17 and 24 March 1983, witnesses saw such beams, usually white. In one instance (17 March) a doctor and his family described a massive six story structure that moved to the horizon and back in seconds. They described a red object appearing within a white beam, moving down the beam, and then shooting off towards the horizon (Hynek et al., 1998). On 2 June 1993 Crystall and Cornet witnessed a white light ascending slowly from the tree line and perform a lazy 'S' sky glyph on time exposures (for pictures see Appendix I: 2 June 1993). At the top of the first time exposure, a red orb emerged, rapidly moved towards the ground, and formed a crude figure eight in the sky just above the trees (see Appendix I). Witnesses who observed large craft hovering less than 300 feet above them, described lights as recesses or wells within the bottom of the structure with flashes or movement of light deep within the wells (Hynek et al., 1998). None of these examples brings to mind conventional types of lights that use lamps or bulbs.

Changing patterns of lights can easily be explained by turning off of some lights and the turning on of other lights in different positions. If the lights being turned off and on are close to one another, the illusion of movement can be created. With the witness descriptions of banks of pipes and tubing across the bottoms of some larger craft, light movement by the switching of lights seems plausible, but it is not the method that is of interest here.

My time exposures have captured events or evidence that no single picture or videotape could capture, due largely to the potential failure of those recording media to pick up and record every light and very low levels of illumination, and for video frames (30 per second) or single pictures to miss lights. On several occasions I have photographed craft that were hovering and brightening their lights to the point that the air around the lights glowed an orange-red color. Sometimes this glow appeared to drift laterally away from the source like smoke or fumes on the time exposures. When enlarged from analog negatives, the lights sometimes appeared as strings of orbs generated from a single light source (for pictures see Appendix I: 4 September 1992). When orbs or bubbles are distinct from the craft they resemble plasma bubbles, or balls of highly ionized hot gas. Infrared film of craft lights almost always reveals lights much hotter than any conventional lamp or bulb light on a plane would produce (for pictures see Appendix I: 22 October 1992; 5 October 1993). This includes Black Triangles and diamond-shaped craft that try to mimic conventional aircraft lights.

On 2 October 1992 Evelyn and Fred Brock, and Bob Wisch accompanied Cornet out into the field at night. We first went to Muddy Kill Lane on the east side of the low ridge (Thompson's Ridge) that parallels West Searsville Rd. further north. Muddy Kill Lane is elevated about half way up the side of the ridge, and homes are built along each side of it, some much higher on the side of the ridge. It is a good vantage point to view most of the Wallkill River Valley, and observe air traffic coming and going to the two airports in the valley. It was on that night we witnessed a set of lights approach Walden from the east – not the usual flight path for airplanes flying to and from those airports. The time was 10:45 pm. The lights were too low to the ground, and appeared lower than the 1,500 foot minimum altitude allowed by FAA regulations (for pictures see Appendix I: 2 October 1992).

We observed and I photographed this set of lights turn south when they reached Walden and the Wallkill River. At first they flew level to the ground, but increased altitude as they approached two communication tower lights in the distance on top of the Ramapo Mountains (east side of the valley). After passing the tower lights they began to descend at a seven degree angle as if on landing approach to an airport, but no airport was on the craft's glide path.

When they reached three vertical red aviation beacon lights on a hill near Scotts Corner and Montgomery, NY, they stopped, changed configuration (new lights appeared), and then they began to descend towards the ground. The lights went all the way down below tree top level before becoming visually blocked by trees. The area was the edge of a golf course that had been built upon an old WWII practice landing strip, now with large trees growing on the old runway. The time of disappearance was about 10:50 pm.

We drove over to that area, getting as close as we could on a road on the west side of the Wallkill River. Evelyn said she briefly saw some lights hovering above ground behind the trees on the opposite side of the river where the UAP had disappeared. Disappointed that we could not see where it landed or even if it landed. On the next day I would photograph a military C5-Galaxy transport circling the area for hours, right over the field where we saw the UAP disappear. When I took magnetic measurements in that field, I discovered a magnetic anomaly that underlain the entire field.

We were excited by what we saw from Muddy Kill Lane, and my time exposures would show exactly where that craft reached the ground. Because we wanted to see more, I told Brock to drive over to West Searsville Rd. We stopped at the dogleg-bend and parked. We watched for activity to the east where the "Indian Mound" is located. At about 11:05 pm a bright light turned on just above the ground behind the tree row that defines the eastern margin of the field in front of us. I turned my camera (on a tripod) towards the light, focused, and began a time exposure at 11:06 pm. I had infrared sensitive film in the camera (see pictures in Appendix I: 2 July 1992).

The light rose slowly, vertically, and silently. When I started my second time exposure, the single light broke up into about five (possibly six) smaller lights that spread apart, perhaps defining the size of the craft.

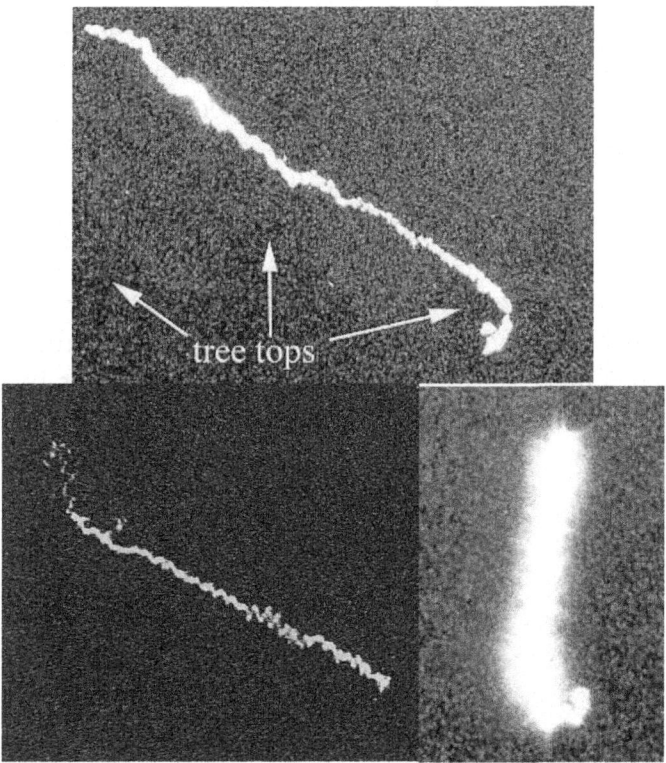

But then the lights rejoined into one light, and continued to climb vertically, moving first to the north (our left), then vertically again. As the single light began to climb, it brightened to the intensity of the Sun. It was so bright that we could not look directly at it. My infrared film recorded caterpillar-shaped hot plasma with bristling rays sticking out all over it like hairs. The width of the light increased at least ten times! This bright light rose until it reached an altitude of about 200 feet, then dimmed suddenly to a small single light. It then began moving south and a red strobe became evident behind the headlight. The overall duration of the sighting was about one minute. Most of the ascent was captured on film. No conventional aircraft or helicopter could duplicate this sighting and its distinctive infrared signature on IR film. The resemblance of the red and white lights to those on the craft Brock and I saw pacing his car on 2 July 1992 is notable, implying they may have been the same craft. Because its lights on film were so different from those of aircraft lamps, and showed the intense heat on the IR film, the use of plasma lights is an hypothesis that continues to be supported by data. Descriptions of UAP lights that behave like plasmas are described in the literature, but evidence I collected was accumulating that would scientifically test and ultimately support that interpretation, elevating an hypothesis to a theory. On 4 September of that same year I had captured on film one light on the Manta Ray breaking up into distinctive plasma bubbles, providing the best evidence to date for the unconventional lights of these craft (see pictures in Appendix I: 4 September 1992).

Fire in the Sky

The orange-red glow produced when the unconventional lights of a craft or UAP intensify their lights can now be explained as energized plasmas that ionize the atmosphere to the point of causing a chemical reaction between nitrogen and oxygen, the two most abundant gases in the atmosphere. Similar orange to red glows or auras around the surfaces of discs or spherical UAPs have been described, and can be explained the same way. As the nitrogen and oxygen molecules are broken up into ions in the intense electromagnetic field of a plasma, they recombine to form reddish-brown nitrous oxide gas as the ions move out and away from the light source and cool. This is the same color seen generated by fireworks and enhanced in flares as NO_2 gases are produced chemically, but it does not mean that the lights of these UAPs are flares. They come in different colors that are described as remarkably pure (Hynek et al., 1998), and do not generate nitrous oxide fumes unless intensified greatly. The location of the light sources deep within wells of the craft may be necessary to control the ionization process and prevent too much atmospheric nitrogen from contaminating the purity of the plasma in those wells.

In a number of sightings I have videotaped light that moved relative to one another, and not by any illusion. In some cases such movement appears to be due to physical change of the ship's structure. In other cases the lights appear to move across the bottom or top of the craft as if not attached to any mechanical source or device. On 6 August 1992 a Black Triangle was videotaped as it passed us to the east (see pictures in Appendix I: 6 August 1992). The four large white lights on its bottom changed position relative to one another like billiard balls on a table. Frame by frame analysis reveals no change in orientation of the Triangle that could explain this shift in light position.

In another case on 24 April 1997 the Manta Ray flew low and slowly over my neighbor's property in Red Bank, NJ (see pictures in Appendix I: 24 April 1997). I was alerted telepathically to go outside to my back yard. When I did I saw a set of white lights approaching low over the trees to the south, traveling north. I went inside and got my camcorder. I recorded it passing by my house over my neighbor's yard. After replaying the video and analyzing the craft's movements, I realized that most of its forward and perimeter lights did not change position, indicating that the craft had to have rotated clockwise as it passed me. Even though its lights frequently were blocked by trees between the craft and my camera, a consequence of the Manta Ray being so low to the trees and ground, one can make out the number of lights on its front and near side. Then I had the realization that the craft departed flying nearly backwards in order for me to see its forward-facing lights as it departed. In addition, a pair of powder blue lights (which appear greenish blue in some frames) on its back or top moved around independently of all the other seemingly fixed lights. This could be an illusion created by a craft that was rotating as it flew in one direction.

The ability of craft to produce lights on its outer surface and move them around is not something that humans have invented or would want to create on an aircraft unless for specific display purposes for other humans. The amount of energy to produce plasma lights would be prohibitively expensive and inefficient unless we had an unlimited and/or very cheap source of energy. Putting plasma lights onto any surface of a metallic plane would be tantamount to lighting a fuse over a fuel tank. Even if a plasma-type of light could be created by current human technology, the use of plasma exclusively by these unidentified craft argues very strongly against human technology, if it doesn't provide extraordinary evidence obviating human technology.

On 28 April 1993 Crystall videotaped a triangular craft that had wing-like lateral extensions. It could have been one of the "planes" she describes in her book. The craft in her video can be eliminated as a misidentified aircraft, because its right "wing" had an enormous bright plasma light extending over part of its anterior right surface (see below). How it could have been generated by lamps is puzzling, since the light changed shape and size. Why it would have been created on a conventional aircraft is even more puzzling. A plasma light illuminating that much surface of a conventional aircraft wing filled with aviation fuel would be an explosion waiting to happen.

Movement of lights along the bottom surface of a stunt plane could be designed by aeronautical engineers. Grooves or slots for mobile lamps, or rails mounted along the belly could direct lamps and their lights in different positions. The undersides of some Hudson Valley UAPs have been described as looking like the back of a refrigerator. Could such tubing has been used for mobile plasmas or lamps? Even if they were, the reason for such heavy and non-aerodynamic structures eludes me. Then one must explain how craft 300 feet or more in width could hover. IHS Jane's Defense Weekly, 17 June 2015, reports on Lockheed Martin's heavy-lift hybrid airship revealed at Paris Air show (Gareth Jennings, London). Other hybrid air vehicles have been designed, built, and tested during the last decade. Their surfaces, however, are not weighted down by piping or tubing, and plasma lights are not used on their surfaces. There was even a Mechanics Illustrated version of a gigantic triangular-shaped blimps or lifting body. All these lighter-than-air blimps are slow moving, whereas the giant ships witnessed during the 1980s in Night Siege (Hynek et al., 1998) were described as being able to move from one location to another with lightning speed.

Evidence for structural changes (morphing) that aid in the movement of lights on the Manta Ray comes from two videos. The first was recorded by Crystall at the Beth Hillel Jewish Cemetery along Rte. 52 between Walden and Pine Bush, Orange County, New York. It was recorded on 4 September 1992 (see pictures in Appendix I) with Brock, Wisch, and Cornet as witnesses. As the diamond-shaped craft flew overhead (at its closest approach to those on the ground), the two central lights at its chevron-shaped nose, one on top of the other, began to change position. The top light moved to the side and then descended downwards until it was next to the bottom light. Conventional aircraft lights cannot do this. As the top light descended, it began to stick out further in front and became brighter, as if to call attention to itself. When it reached the bottom light, it did something unexpected: It began to move backwards along the bottom of the Manta Ray. As the Manta Ray departed, the video shows this light well past the bottom nose light on the craft's belly. The structure of the craft should have blocked both nose lights as the ship moved away from us, but they were clearly visible in altered positions from front to bottom as the huge diamond shape flew away (see collage below).

The last time I saw the Manta Ray was on 20 August 2003 (reported by Dennett, 2008: p. 223-224), but not over Orange County.

http://www.sunstar-solutions.com/AOP/Millbrook/Millbrook.htm

I was taken by Billy McNamara to a field on a horse farm across from his relative's house in Salt Point, New York, on the east side of the Hudson River. McNamara is a Hollywood actor, who began to see strange angular craft flying around his residence. He was directed to my web pages, and recognized the Manta Ray as one of the craft flying over his house. He called me on the phone, and eventually convinced me to take a long drive from Red Bank, NJ, to Millbrook and Salt Point, NY. As soon as it started getting dark, he

Collage showing one of two nose lights moving forward of the other one.

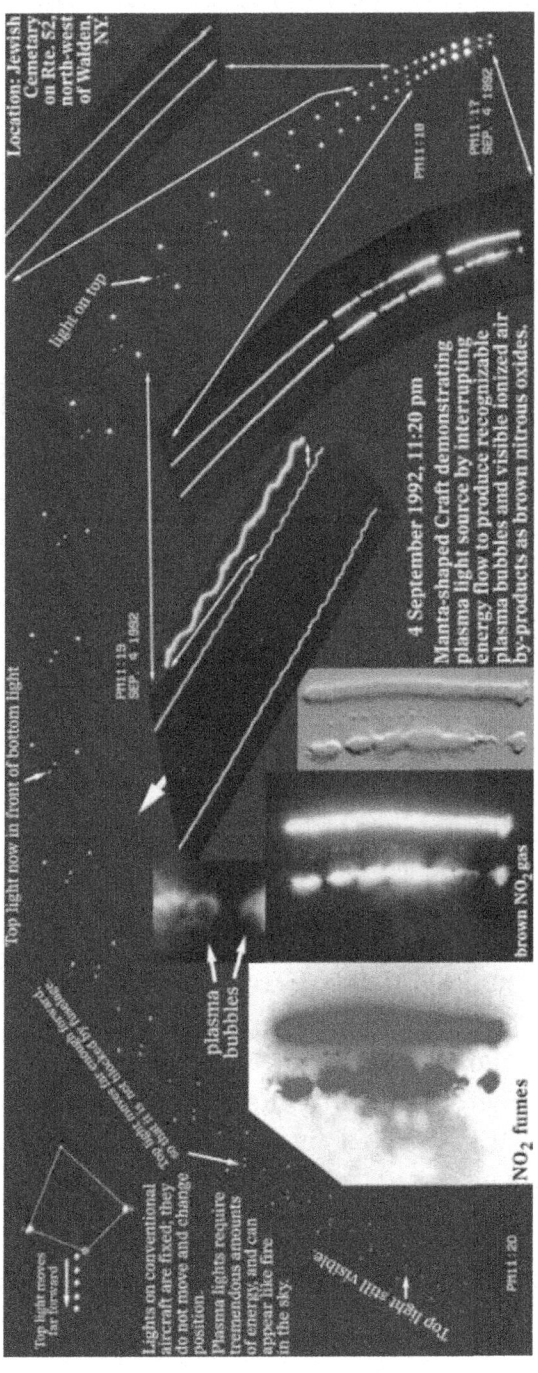

wanted to take me to see these craft. I went with him, a little skeptical that he could "produce" a sighting, as if on demand. I would soon be surprised that he knew what was there and knew we would have a sighting.

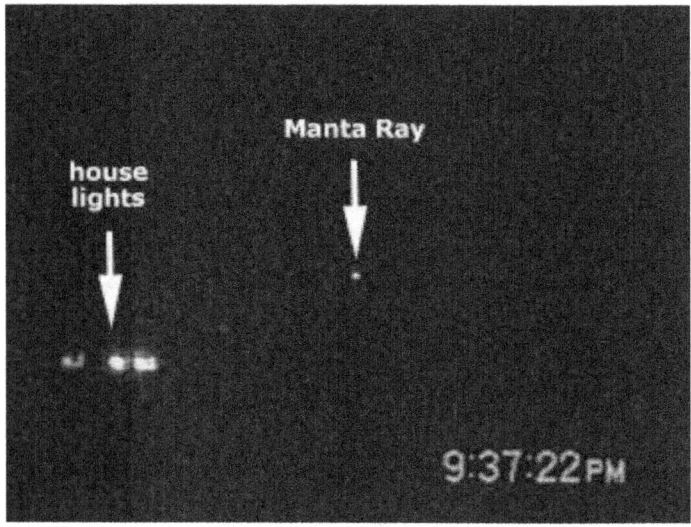

No sooner had we climbed a wooden fence and ventured out into a small field surrounded by wooden fences, than he spotted some lights past an adjacent field to the north. I turned on my camcorder and zoomed in on the lights. I didn't have time to set up my tripod. I recall seeing a light move from left to right over the house lights in the foreground, then stop directly opposite us about a quarter mile away. In hindsight I suspected that this craft had emerged from an underground base, where McNamara had said he saw cargo trucks disappear down a private road (cone C). This hidden location is shown on the map below at cone B. He thinks this activity is being controlled by humans, due to the amount of strange activity he has witnessed by his neighbors, such as a man climbing out of a hatched entrance to an underground tunnel, located in the middle of a farm field across Clinton Avenue from his property (cone A in left map). The prompt arrow on the right map points to the location of this mysterious road.

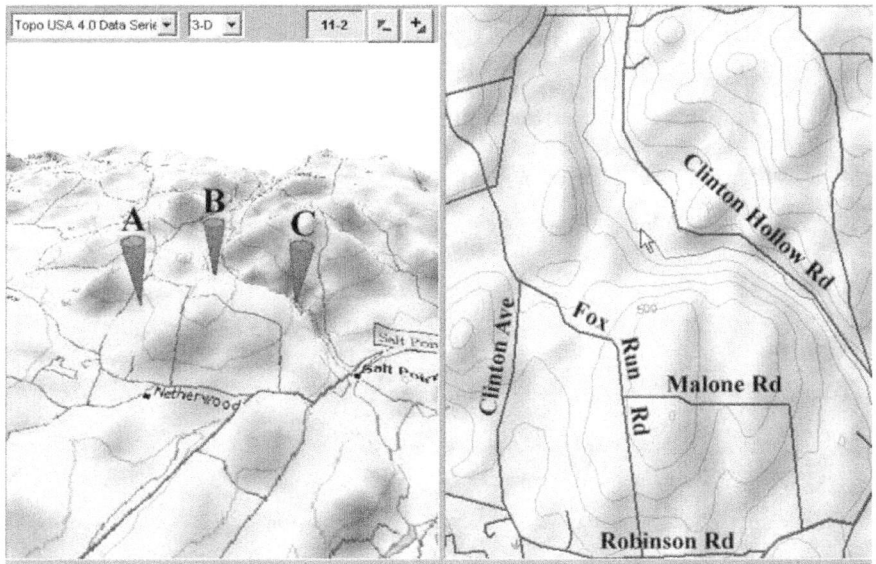

McNamara and I witnessed a pair of lights with a red pulsing strobe in between. I immediately exclaimed that it was a ship, thinking of how Ellen Crystall had responded when she recognized a familiar pattern of lights near Pine Bush.

As soon as I said, "This is a ship," the lights brightened. McNamara said, "See, I told you. I told you," excited that his "faith" was fulfilled, confirming his expectations. The craft had been stationary until I began videotaping. Then from less than a mile away, the bright "headlights" began to slowly move towards us, no more than a couple hundred feet above the trees. Frequently performances by the Manta Ray and Black Triangles near Pine Bush began with a flaring of their headlights just before they began a deliberate run at us. The lights were too far away to know what type of craft this was. We had to wait about three and a half minutes for it to reach us. Its speed was no more than 22 mph (2,000 feet in 3.5 minutes), well below stall speed for any fixed wing aircraft other than a jump jet.

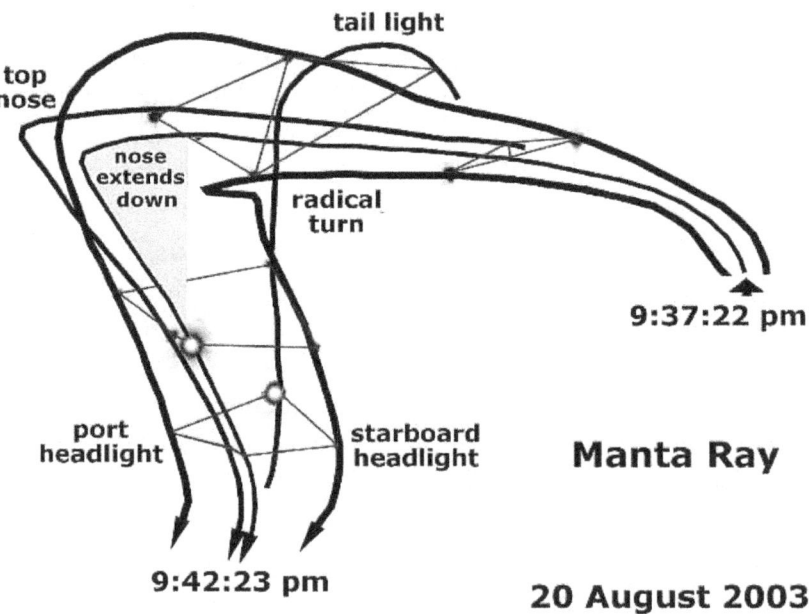

Manta Ray

20 August 2003

As the craft got closer, we realized that it was huge, reducing the possibility it was a small Black Triangle (60-100 feet in width). As it came up over tall trees beyond the wooden fence, where we could see it clearly, it began to turn west. It did not bank. It kept on gliding almost silently towards us as it rotated. Then I saw its trail light at the end of a narrow tail, and I called out to McNamara that this was the Manta Ray.

The craft slowed as it turned, like a skater turning sharply o0n the ice in order to go back in the opposite direction. We could hear a low drone of a turbine-like sound, so quiet that it was barely audible. The Manta Ray briefly stopped and then started to move away to the northwest. It had previously been traveling south towards us.

As the Manta Ray cleared the trees in front of us (previously it had been partially blocked by branches in the tree canopy), still pointing mostly at us, we could see a pair of lights at its nose, one on top of the other. Like the performance on 3 September 1992 at the Jewish cemetery, the top light began to move forward and down, until it was alongside of the bottom light (to its right). The top light (now on the right side of the craft) brightened considerably – more so than it did over the cemetery. It also could be seen extending further out front than it did on the cemetery video. Once it reached the lower light, it began to move back along the belly of the craft. By this time the Manta Ray had turned or rotated clockwise 230 degrees from its previous orientation, and was beginning to move away from us. The tail light, which was not very bright, then began to flash or strobe brightly as it moved away.

Here we had two nearly identical performances by the Manta Ray for comparison, separated by nearly 11 years. As the Salt Point Manta Ray turned, its "headlights," located at the lateral angular margins of the diamond, illuminated some of its structure. The reflections accented the chevron shape of the front of the craft, ruling out any mistaken identification of a conventional airframe.

The final altitude of the Manta Ray over the trees was no more than 300 feet above us, and it was no more than 150 feet away from us.

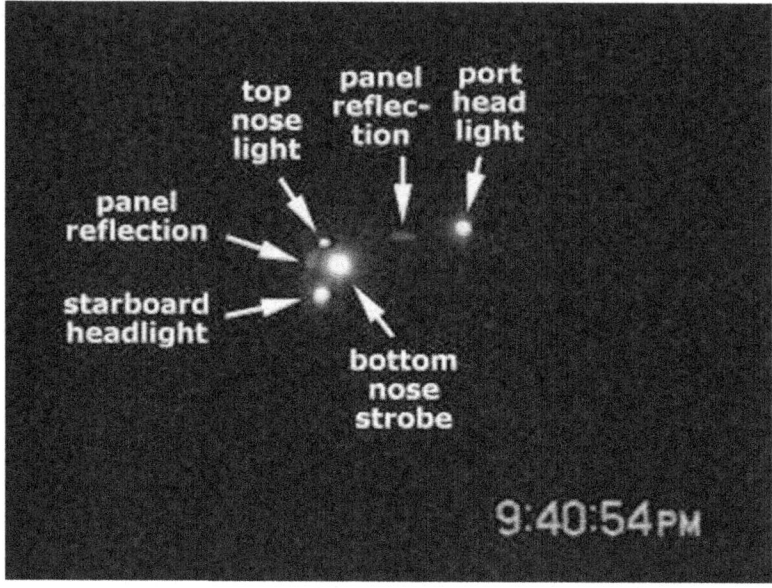

Light Control and Telepathy

In order to adequately describe the movements of UAP lights off of the physical structure of their ships, I need to talk about mental telepathy as a triggering event that resulted in some of the most spectacular time exposures in my collection of photographs.

In "Alien Bases" Timothy Good (1998) provides information exchanges between several Contactees and alleged extraterrestrials. In the case of Daniel Fry, whose initial contact with Ahlahn (Alan) began on 4 July 1949 at the White Sands Proving Grounds in New Mexico, a discussion on mental telepathy, or 'extra-sensory perception' is given" Alan said, "In the first place, it isn't extra-sensory at all. It is just as much a part of the body's normal perception equipment as any of the others, except that during one phase of the development of the race it falls into disuse because it is a rather public form of communication, and during this phase of development the individual requires a considerable degree of privacy in his words and thoughts [this leads to keeping secrets and to lying]. Most of your animals use the sense to a greater degree than your people, and for some of your insects, it is the only form of communication..." (Good, 1998: p. 67).

Good (1998) also says: "Adamski was the first to proclaim that some people from other planets were actually living and working among us – illegal aliens, as it were – and stated that his contacts often took place in the anonymous surroundings of restaurants and hotel lobbies. Having spoken with a number of other, witnesses who had related similar encounters," Good decided to make an attempt at telepathic communication with alleged "aliens" among us. On two separate occasions, one in a roadside restaurant near the Arizona/California border on 13 November 1963, and another in the lobby of the Park-Sheraton Hotel in New York City in February 1967, he attempted to initiate contact with individuals who caught his interest. In both cases he had startlingly successful results, but the contact consisted of only acknowledgement that they had picked up his mental thoughts. Just as highly advanced avatars and masters said to be living on Earth can allegedly levitate, render themselves invisible, project their images, and walk through walls, extraterrestrials living on Earth can supposedly do these feats and more. So, how does one distinguish between human avatars and non-human visitors?

My first exposure to the possible use of telepathy occurred on 2 July 1992, less than a month from when I went out into the field with Ellen Crystall for the first time (on 9 June). Crystall had introduced me to Fred Brock, who at the time was an editor for the Wall Street Journal. Crystall had taken Harry Lebelson, editor of Omni Magazine, with her into the field during the early 1980s, and she was inviting other reporters and editors to join her in the early 1990s. Fred Brock, an editor with the Wall Street Journal, picked me up at my Middletown condominium, and I gave him directions to meet Crystall on West Searsville Rd. where it meets Hill Avenue (the favorite meeting place for sky watchers). I took him the long way so we could have time to get acquainted that evening. He was very interested that a professional scientist had become involved. By that time I had been into the field with Crystall and associates three times (Appendix I). We took Rte. 208 to Walden, NY, from Middletown, and then turned west on Rte. 52, heading towards Pine Bush. Brock was driving (see Appendix I: 2 July 1992).

No sooner had we crossed the bridge over the Wallkill River and followed Rte. 2 as it turned north to parallel the river, than a bright white light with a small red light behind it flew low over Brock's car, and began pacing us just ahead of the car and over the road. I noted to him that this was the same type of UAP that had paced me home on Rte. 17 the very night before. Brock was shocked that we would make contact so soon. Because the same type of UAP lights, with no visible indication of structure between the lights, had met me at the Harriman toll plaza and exit (New York Thruway exit for Rte. 17) on the previous night, I thought it would match our movements as it had done before. I told Brock to expect the craft, which by now was clearly not a helicopter – no noise or prop wash, to stop with us at the traffic light ahead (intersection of Rte. 52 and Albany Post Rd.). And indeed the light stopped with us at the red light, hovering just above the traffic lights.

There we were, waiting for the light to turn green, and a UAP was hovering directly in front of us (about 20 feet away) and about 30 feet above the road! In our excitement and shock, I almost forgot to tell Brock to turn left. When I said, "We are going to turn left," Brock turned on his left turn signal. But before he did that, the craft turned "on a dime" so its red light (tail light) was to our right. The light changed to green, and we turned onto Albany Post Rd. heading south. The UAP took a position over the trees on the right side of the road, and paced our car just enough ahead of us that we could watch it.

We were speechless, but I had to warn Brock to slow down so we didn't miss our turn onto Hill Avenue, which was 1.68 miles ahead of us on the right, and concealed by a bend in the road. As soon as I told him to slow down, and before I finished my sentence, the UAP shot ahead silently, and disappeared out of sight, blocked by trees. When we rounded the gentle turn to the right, there the craft was, hovering above the intersection of Albany Post Rd. and Hill Avenue, waiting for us to turn!

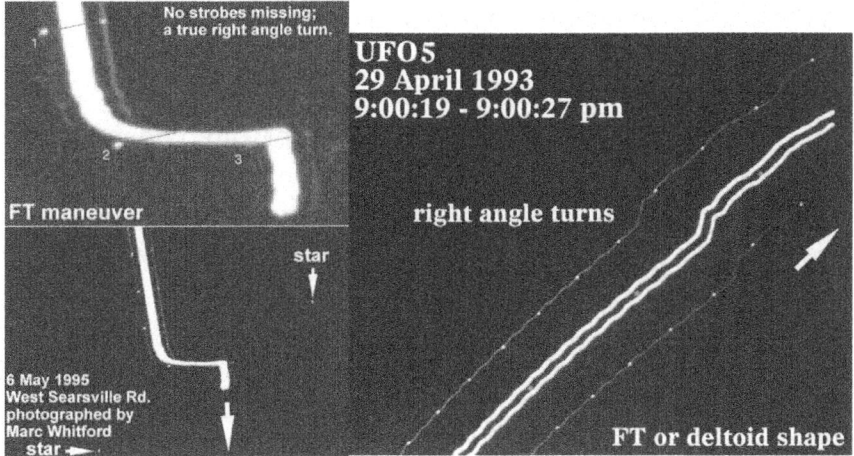

Two Black Triangles performing right angle turns on 29 April 1993 and on 6 May 1995, demonstrating control of gravity, apparent mass, and inertia.

We proceeded on Hill Avenue, with the UAP still in sight, now over the trees on the left side of the road. When we got to the turn for West Searsville Rd., the UAP took off. We pulled over and waited for Ellen Crystall to show up. At no time did we see a physical structure connecting the larger white light and the small rid light, which led me to think this was the same craft I had seen the night before.

One can always argue that the pilot had high-tech listening devices aboard, and that is how he could respond to our conversation, and seemingly anticipate our moves. That would also imply that's the pilot knew English. But the swiftness of his responses left me wondering whether he could also pick up on our (my) thoughts. I would soon collect data that confirmed my suspicions that mental telepathy was indeed being used.

I have many photographs that show some of the lights on a craft moving freely off of the craft, but in a very controlled manner. Plasmas, because they are electromagnetic, can be controlled and moved in an electromagnetic field that can be intentionally shaped and distorted (see pictures in Appendix I: 28 April 1993).

alien craft moving lights in sync with camera shots Sunset 28 April 1993

Nikolas Tesla demonstrated how he could illuminate bulbs or lamps in an electromagnetic field, even though they were not connected to any wires (Nikola Tesla: Imagination and the Man That Invented the 20th Century, by Sean Patrick, 2013). The same is true for plasmas, which can be maintained within an electromagnetic force field. The photographs showing this phenomenon in action were in response to thoughts I was having when operating my camera. As I thought, "Open shutter," and "Close shutter," the operator of the craft would make deliberate movements of some lights (usually the brightest) on or off of the craft. Those movements were recorded on my time exposures and audio tape recorder just after I made my self-talk in my head.

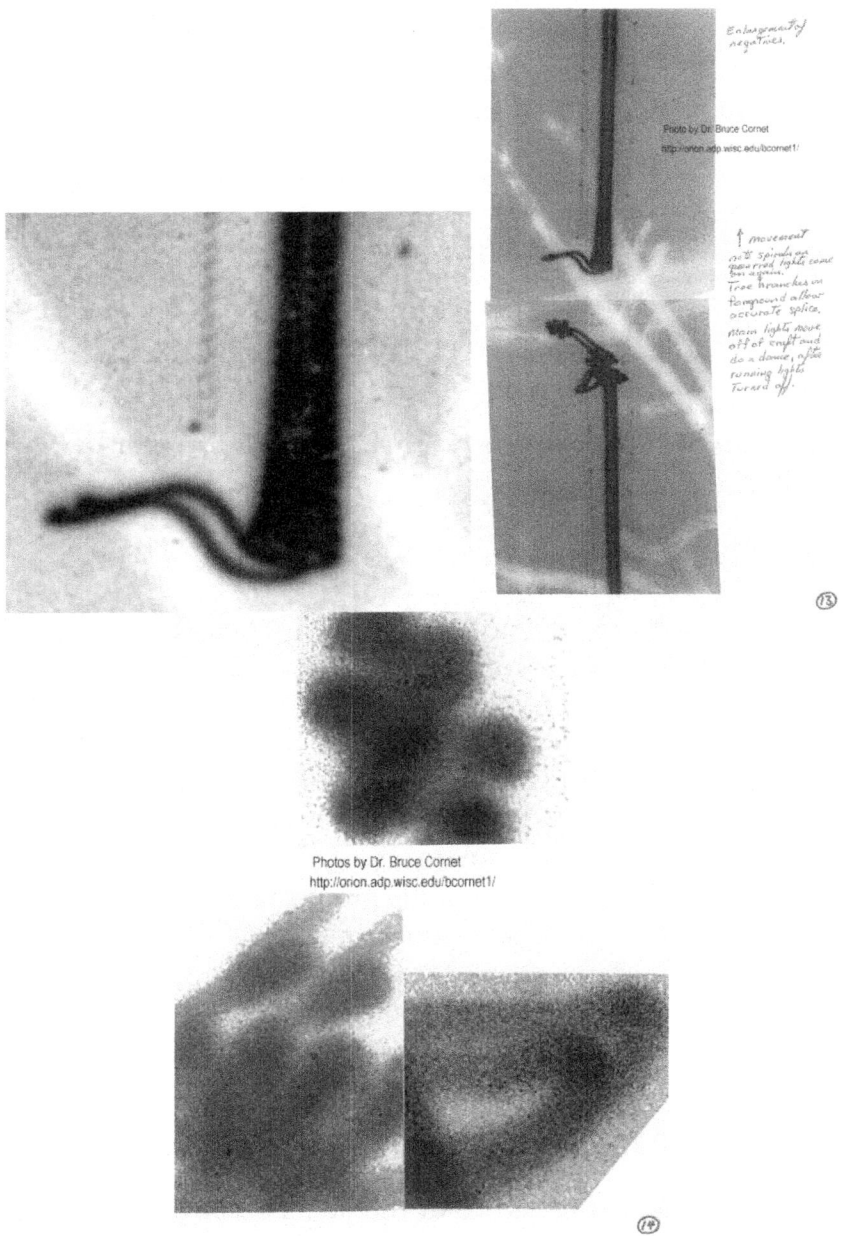

In left image above, note outer left navigation light coming back on, then spiraling.

In the examples given below taken on 6 January 1993 and on 28 April 1993, the up and down oscillation of a single light occurred either just after the camera shutter opened, or in some cases, just before the shutter closed, indicating that the pilot "heard" me think, "Close shutter."

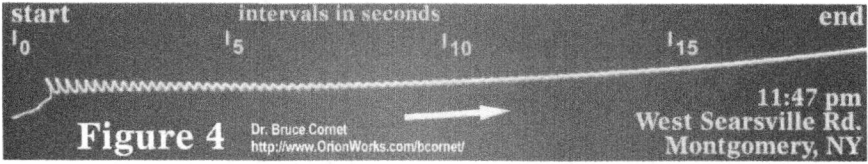

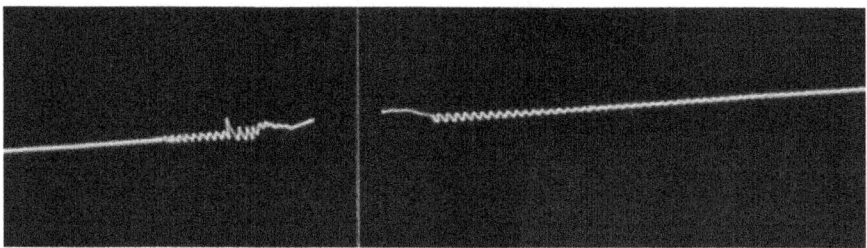

Once I recognized with the above evidence that the pilot was "listening to my thoughts," I tried what Ellen Crystall said she did: Act without thinking. When I stopped my mental self-talk, and opened and closed the camera shutter without thinking, the oscillations stopped. I had scientific proof that my thoughts were being read by the pilot. That experiment would be confirmed again with other types of contact, and I would get direct mental voices in my head coming from the pilot. The Garden State Parkway escort on 11 October 1994 would include conversation with two different pilots in two ships that paced me for as much as 72 miles along that highway (see pictures in Appendix I).

The most frequent response of the lights occurred when I first opened the shutter for each (or any) time exposure. The movement of the UAP lights would typically be either what I call a "tail, or a strong oscillation of the light that dampened down slowly. The "tail" looks like a tail, when the light or lights shoot up a short distance above the traces of other lights in the picture. The tail is typically at an angle or slant, and begins just as the shutter opens. In other cases the UAP lights oscillate to produce a sine wave pattern just as the shutter opens. Note the movement of the paired bright white lights in the negative image below, while the other lights do not change course.

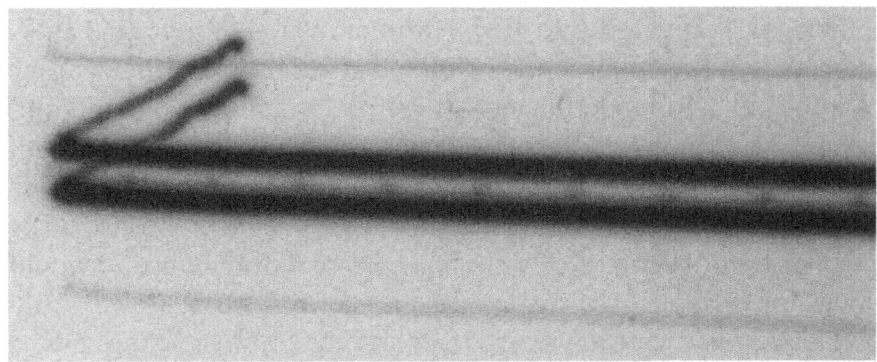

In the images below (taken on 29 April 1993), note differences in tails for different craft in the same pictures, as well as oscillations and spirals, which conventional aircraft cannot make. Angles and lengths of the tails are all different, implicating source. At first I thought these anomalies were due to some accidental movement of my camera. I turned to using an electronic shutter release so that no mechanical movement would occur when I pushed the button at the end of an electrical cable. I did many experiments. I even took numerous pictures of conventional aircraft flying over or near airports for control. I determined that the most vibration the camera system would produce on the film by the movement of the reflex shutter was a ½ second oscillation that was short in length on the film, and was low in amplitude. When I was photographing a suspected UAP set of lights, the oscillations were much more pronounced, and lasted for up to 10 seconds!

I suspected some sort of interference when oscillations occurred just before the camera shutter closed. If the pilot was sending shutter movement, how was he able to predict when I would close the camera shutter? Even if the vibrations were caused by some sort of energy beam hitting my camera, predicting just when I would close the shutter was like trying to photograph a bullet being fired from a rifle without knowing when the rifle was going to be fired.

I then began paying more attention to what I was doing and thinking. I discovered that I was forming words in my mind (self-talk), as an exclamation of what I was doing with the shutter button. So, I stopped "talking" in my mind, and just opened and closed the shutter. Results: No oscillations. Then I began experimenting. I would delay my mental words for a few seconds after opening the shutter, or think "Close shutter" consciously seconds before actually closing the shutter. In other words, I would delay my response to my conscious self-talk to see when the oscillations would occur. To my delight, I got 100% correspondence and confirmation that my conscious words were being read (not my unconscious thoughts), and they were the trigger for the oscillations or tails on the film. Telepathy or a mind link with the UAP pilot was confirmed.

In some series of time exposures I would get anomalous light movements just after I opened the shutter and again just before I closed the shutter for each exposure when I deliberately (consciously) said in my mind, open or close shutter. On 28 April 1993, one a Black Triangle approached the camera at dusk or twilight. Looking West, a set of six time exposures shows a pair of "headlights" moving at the beginning and end of each of the first three time exposures. The craft literally jumped or moved slightly across exposures. But between the second and third exposure the craft did something different. The pilot turned off the outboard red and green lights and embedded strobes, and caused the pair of "headlights" to move well off to the side of the Triangle and do a dance, before returning to their original position on the front of the craft!

There is no way any human pilot could know what I am thinking and doing with such split second accuracy three times in a row. An advanced ET might be able to do this who has telepathic ability, but even he might need technology to enhance his ability. It is highly unlikely that the military has such technology, or that government psychics and remote viewers could do this without enhancement. Then we need to ask the question: Why would our military or government be displaying and revealing technology and ability which would be classified above top secret? Why would our military put on a public display over rural farm fields and housing developments, and let our enemies know our secrets? That explanation simply makes no sense.

Now let's review what they could make their lights do, given that you now know the circumstances involved when they demonstrated such extraordinary technology (with extraordinary photographic proof) to Ellen Crystall and Bruce Cornet. Extraordinary claims demand extraordinary evidence: Now you've got it!

Electromagnetic Force Fields

In addition to sine-wave-like oscillations, "tail" at the beginning of time exposures, and precisely-controlled movements of plasma lights in the atmosphere around a craft, I have photographic evidence for single-light spirals, double-light spirals that resemble the DNA helix, vortex-like spirals, and alternating clockwise-counterclockwise loops. I also have photos that show irregular loops and swirls of plasma lights. All these motions and patterns produced by plasma lights are independent of the craft's position and movement. In other words, the ships are not spiraling or looping or oscillating up and down. Only their lights are doing this motion. Such motion requires a containment electromagnetic field that extends out beyond the physical craft structure some distance. The shape and strength of that field has to be controlled precisely, and would require special field generators aboard the craft. Humans do not have such technology yet.

I have time exposures of several craft doing light movements where all but the moving lights on the craft are turned off until the performance is over or complete. Then the accessory lights are turned back on. When they come back on, they show a spiraling motion at first that tightens up to what looks like solid lines on the time exposure. If accessory lights are not turned off, all the lights on the craft move in unison, indicating the presence of an all-encompassing force field that must surround the ships.

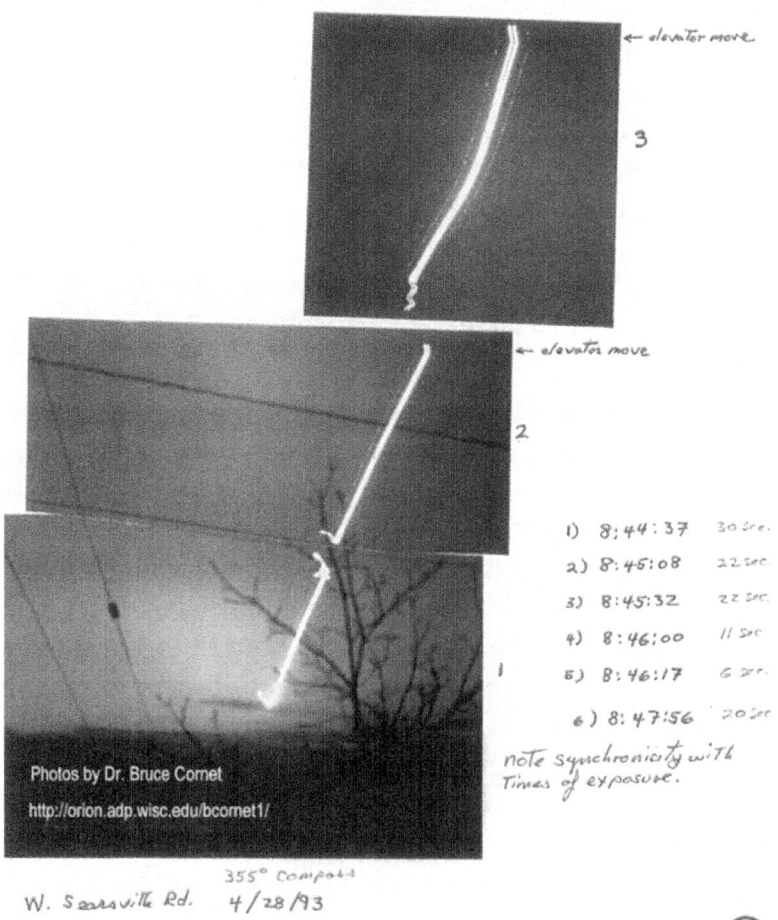

Photos by Dr. Bruce Cornet
http://orion.adp.wisc.edu/bcornet1/

W. Searsville Rd. 4/28/93 355° Compass

1) 8:44:37 30 sec.
2) 8:45:08 22 sec.
3) 8:45:32 22 sec.
4) 8:46:00 11 sec.
5) 8:46:17 6 sec.
6) 8:47:56 20 sec.

note synchronicity with Times of exposure.

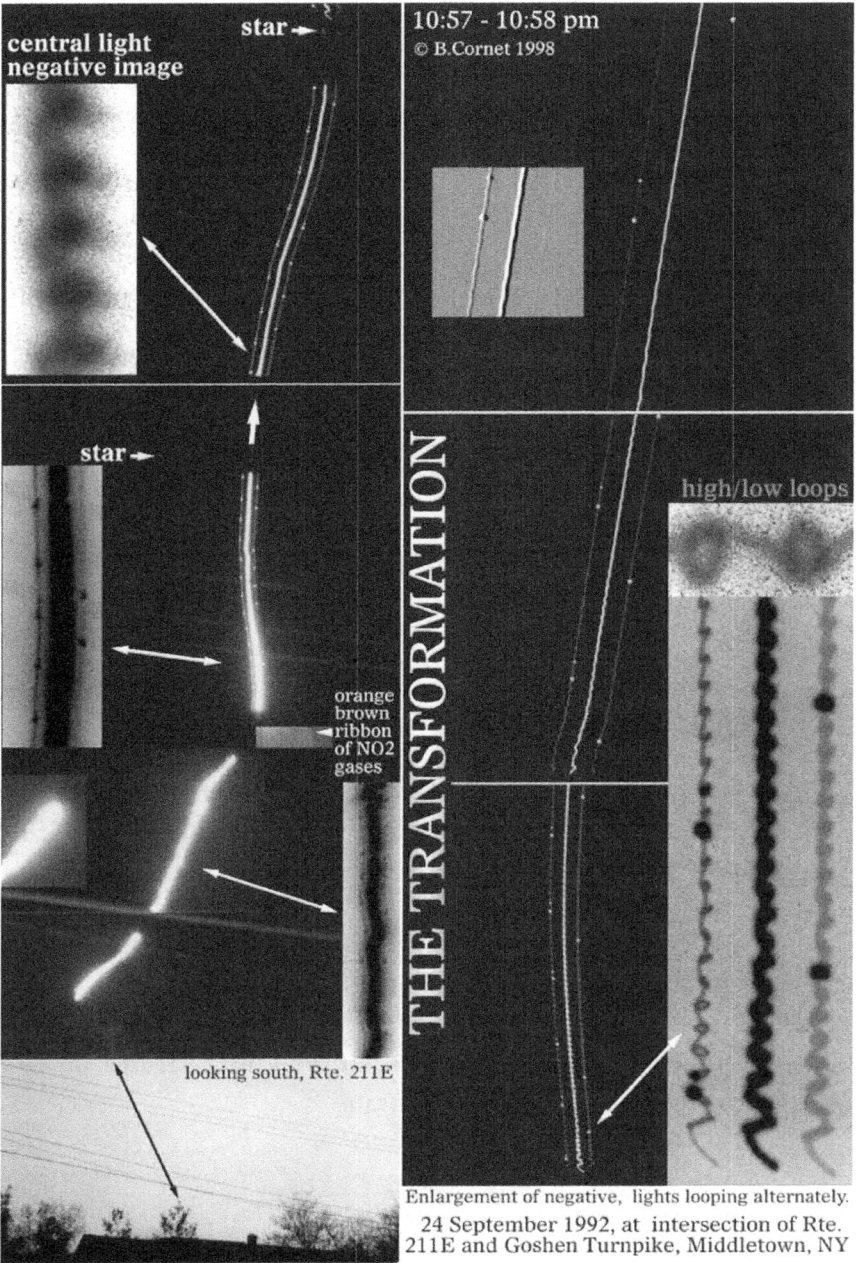

Enlargement of negative, lights looping alternately.
24 September 1992, at intersection of Rte. 211E and Goshen Turnpike, Middletown, NY

Unconventional Aerial Phenomena

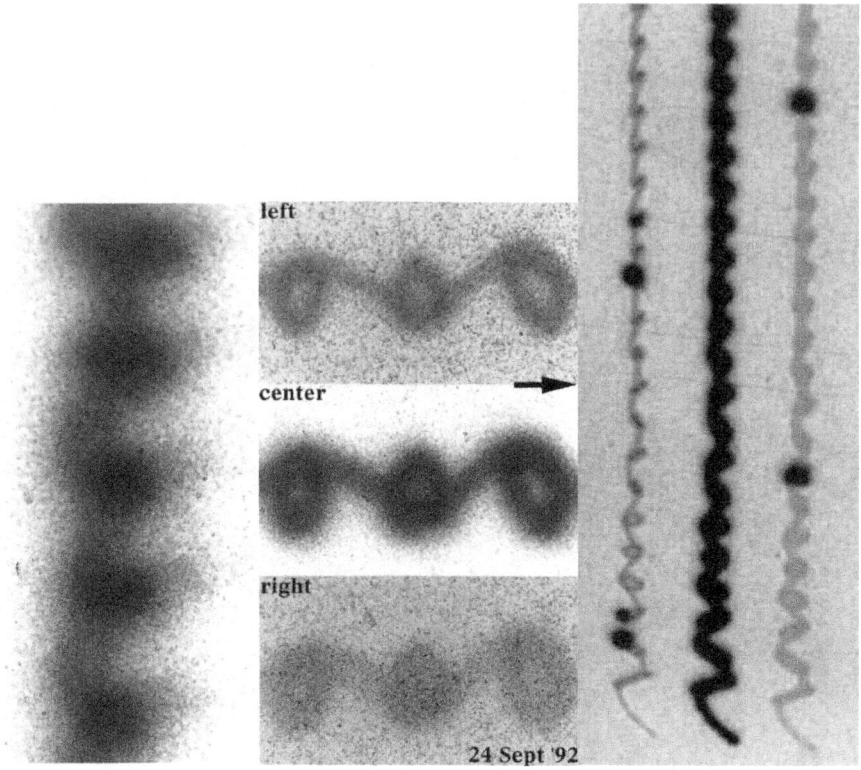

Above: Note the spiraling and looping of the lights on this UAP, which flew directly over Cornet at 10:57 pm on 24 September 1992 (see pictures in Appendix I).

The Transformation: a ball of bright plasma progressively transformed into what seemed like the navigation lights of a small propeller-driven aircraft. However, detailed analysis of the time exposures proved it was not a conventional aircraft, just an ETV pretending to be a conventional airplane.

http://www.sunstar-solutions.com/AOP/SOW/transfor.htm

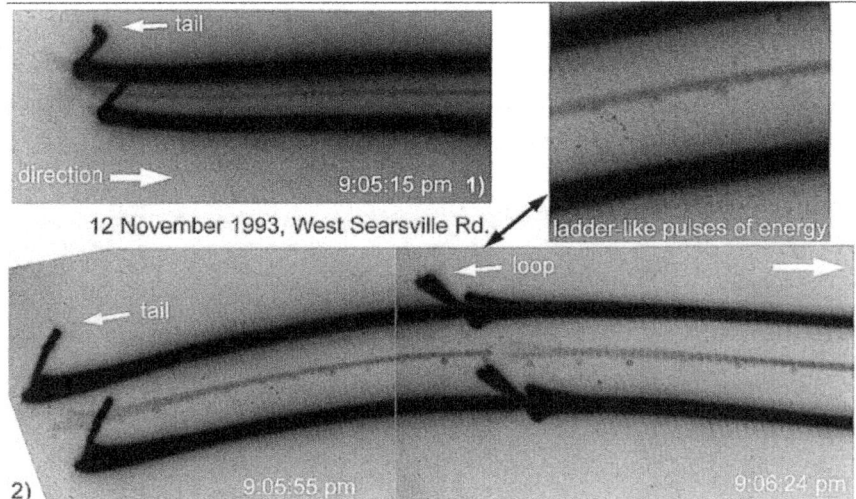

Flew over us, then banked hard and dove towards the ground; disappeared below tree line 1/2 mile to the east.

Above: Flyover at 9:05 pm on 12 November 1993 at dogleg bend in West Searsville Rd. Negative enlargements show anomalous light movements in the form of tails and loops, and ladder-like illuminations between the two main lights. These are not characteristics of conventional human-engineered aircraft (see pictures in Appendix I).

Evidence for the redirection of power from a single power source or a limited number of power sources comes from the above descriptions of lights being turned off before other lights are put into motion. Another source of information on power capacity comes from lights that are made to loop repeatedly, but have non-uniform levels of plasma illumination, where the light has "brown" or lower intensity streaks within it. Such examples were captured on film twice: 7 July 1992 and 29 August 1992 (see pictures in Appendix I). Two of those pictures are shown below.

On 29 April 1993 a group of UAPs showed up when the Sightings TV camera crew was there, shooting a program with Crystall and Cornet for their show (see pictures in Appendix I). We had no idea so many craft would show up or that some of them would put on performances for the TV cameras that came close to matching scenes in "Close Encounters of the Third Kind." Before the main "show," as many as six UAP lights appeared to the east, in the direction of Stewart International Airport. Crystall

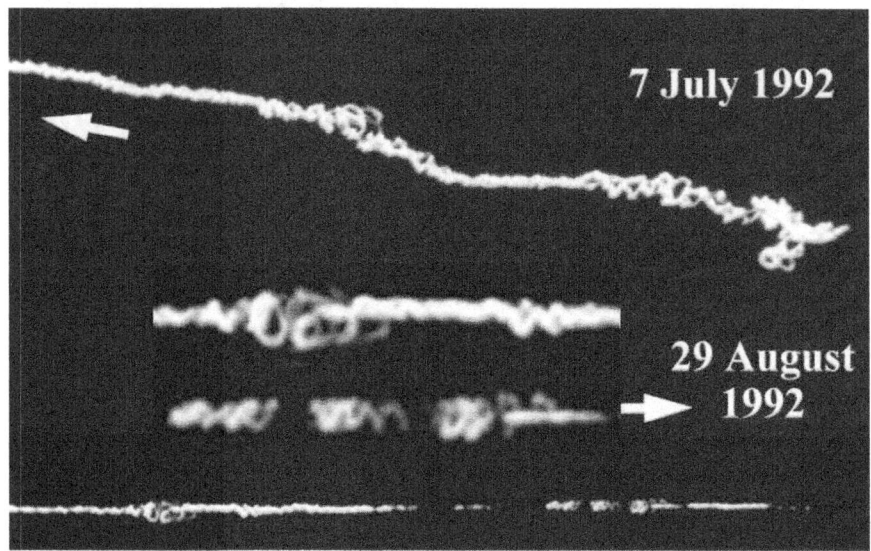

commented, "What is going on here?" as I captured four of the lights in the same camera field of view on one time exposure. When I got my prints back from the photo shop, I was surprised to see all four lights had "tails" at the beginning of their traces on the film. Close examination revealed that all four "tails" were oriented at distinctly different angles relative to the picture frame. Had these "tails" been caused by camera movement, they would all have been oriented at the very same angle. Not only do the "tails" indicate that the objects were not conventional aircraft flying to and from the distant airport near Newburgh, NY, but they indicate communion between the pilots and what I was doing with my camera. All four pilots had to know exactly when I opened my camera shutter!

Different angles for tails at beginning of light traces indicate intentionally created at source, not camera movement.

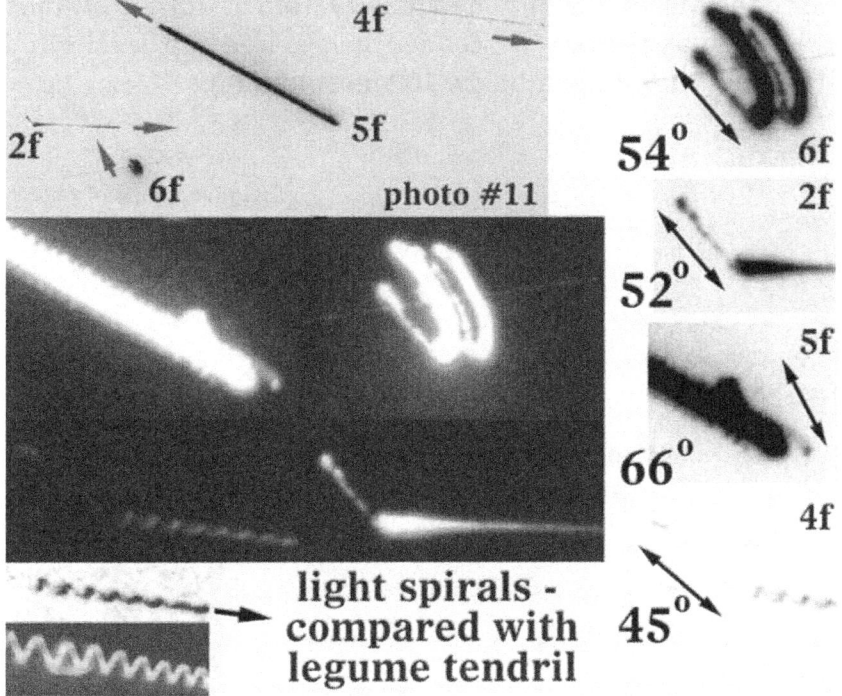

photo #11

light spirals - compared with legume tendril

In a spectacular example of multiple plasma generators being used, the Manta Ray flew past and above the Sightings TV camera crew. With Crystall recording the event on her camcorder, I took a series of time exposures as this craft, after hovering about a hundred feet above a distant farm field and forest for eight minutes, made its run towards us at 8:49 pm. It made a wide turn to our left, and passed in front of us as it climbed in altitude, and turned its belly towards us. At that point the Manta Ray was travelling only about 30 mph. If it had been a fixed-wing aircraft, like a large jetliner or C-5 Galaxy transport aircraft, with its wings pointing towards the sky and ground for more than a minute, it would have dropped out of the sky like a rock. As it flew past us a battery of parallel lights turned on along its belly (midline). Simultaneous with the turning on of additional lights, we could hear (recorded on the videotape and on my audio recorder) what sounded like multiple turbine engines winding up their rpms, as if needed to generate the additional power for those lights and for lift. Ten the Manta Ray turned east away from us, turned off those extra lights, and slowly descended below tree top level into a distant field, out of sight (below 100 feet altitude).

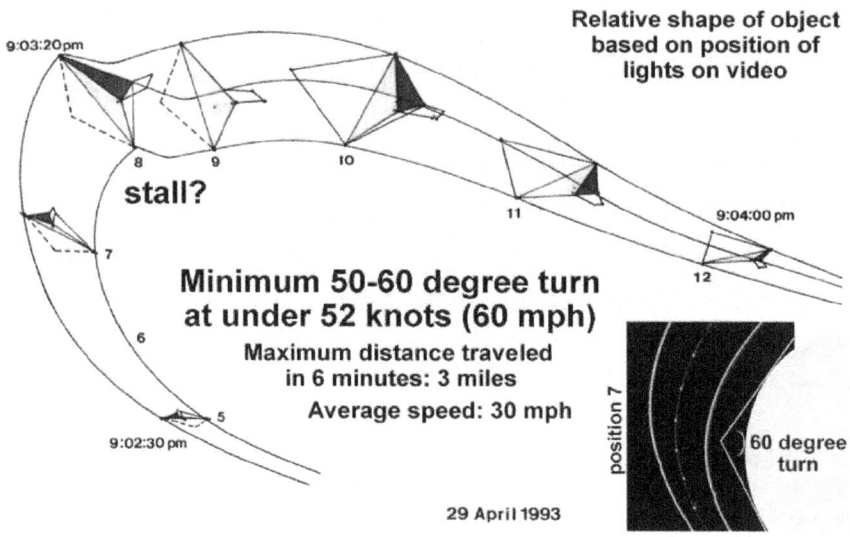

The "tails" at the beginning of light traces on time exposures sometimes involve the peripheral red and green lights and sometimes do not. For the force field to control only the central lights means that the pilot can focus or adjust the field size to affect just the central lights. Presumably the peripheral lights can be controlled independently also, accounting for the mobile pair of blue lights videotaped changing position on top of the Manta Ray (see pictures in Appendix I: 27 April 1997). Controlling energy patterns and intensity within and around these craft must be a technological marvel.

On 29 April 1993 I captured a picture of a set of lights on the ground behind the performance for the Sightings camera crew as a conventional airliner took a familiar flight path, turning at a red aviation beacon on top of Thompson ridge to our west. That picture shows the lights on the slope of that ridge just to the west of West Searsville Rd. If you look carefully, you can see what appears to be a humanoid figure sitting inside, who is visible only because of the lights on the outside of the craft. But his visibility means that the hull of the craft is transparent for the light to pass through it, reflect off of the pilot, then go through the hull again to my camera.

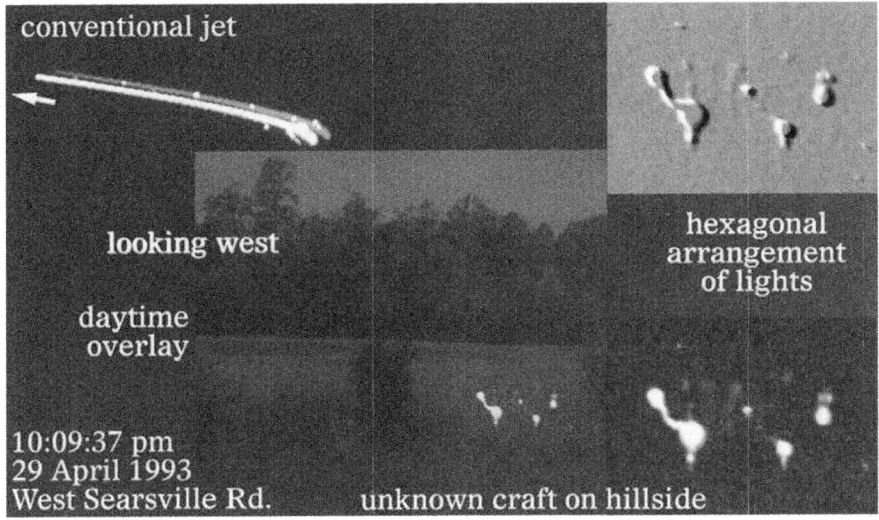

Crystall describes the underside of a Black Triangle that seemed to have a sheeting of glass around the outside of a metal hull. "On another occasion Harry [Lebelson] and I were at the bottom of Hill Avenue around one O'clock in the morning, en route to another site, when we saw the dual headlights of a craft coming towards us. I slowed down and it suddenly turned on a flat sheet of yellow lights on the underside, displaying its triangular shape clearly as it cruised by us, passing over the roofs of some nearby houses.

"Our jaws dropped open. The underside of the triangle craft was a solid plate of yellow lights, but we could see seams in the metal under what looked like glass sheeting, indicating different-sized plates had been connected to form it. On numerous other occasions I've seen the underside of the triangle craft. With each encounter I've observed more and more details of the craft. The underside has looked the same each time but, interestingly, the right and left halves of the underside are not identical.. Their seams are somewhat different, as I confirmed on later occasions." (Crystall, 1991: p. 55-56).

What is interesting about Crystall's description, is that the lights seemed to be associated with the glass sheeting, which was separate from the hull, which appeared metallic. In addition, the illumination of the glass sheeting appeared to be comprised of solid "lights," not "light." In other words, the light had individual point sources on the glass sheeting, giving the appearance of multiple lights joined together ("solid). The separation of the glass sheet from the hull is understandable if the illumination is formed by plasmas. The hull needs to be insulated from the lights, or it would ground and short circuit the high energy electromagnetic charges of the lights. The appearance of many continuous light sources, giving the appearance of continuous illumination, implies the control of energy to specific points on the sheet, which may have acted like a plasma television screen.

The craft came much closer to Crystall and Lebelson than they ever did for me and witnesses with me, including Crystall when we were in the field together. The craft would keep a certain minimum distance and altitude from us, climbing in altitude as they approached us. Perhaps this was a form of security after Crystall published her book. We would be prepared to look for details that Crystall and Lebelson had overlooked because they were unfamiliar with the alien technology. By the 1990s our technological progress might have given close observers the knowledge of what to look for in order to figure out both how the visitors were able to control their lights and how their stealth capability (i.e. invisibility or transparency) worked.

Compiling descriptions of all lighting on UAPs from the Hudson Valley and Pine Bush regions, certain observations and interpretations can be made:

Their lights are not lamps, but high energy plasmas that can be controlled by electromagnetic force fields and/or by sheet-like screens insulated or separated from the metal structures of their hulls.

The plasmas are generated or formed in specific structures sometimes embedded within wells or cavities in the hull, or sometimes positioned outside the hull.

The large craft over 100 feet in width appear to employ lights that originate within sunken cavities or wells, while the craft 100 feet or narrower in width appear to generate their plasmas external to their hulls.

The maize of tubes and pipes seen on the undersides of some of the larger ships may operate in cooling the hull from the excess heat generated by plasma lights, and in removing excess nitrogen from our atmosphere prior to ionization, or in removing nitrous oxides generated in the plasma wells to keep the color of the lights pure and under control.

The appearance of light movement around the hulls is created in two ways: 1) mechanical or structural morphing with the physical movement of plasma generators, and 2) external to the hull by the manipulation and control of electromagnetic fields.

There are multiple power sources for plasma lights on large craft, while energy from a limited number of power sources on smaller craft has to be proportioned and redirected, causing a reduction in performance.

Although the primary form of communication appears to be mental telepathy, lights are the primary method of communication with humans. Abductees tell of tests being conducted aboard ships where the intellectual (left brain) and emotional (right brain) reactions of humans to various colors of lights is measured. Responses to UAP lights by humans using light as a form of communication usually has positive results, even though the quality or value of information being transmitted is minimal.

The types of light motions created by the pilots of these craft vary considerably, from
   A. simple oscillations;
   B. "tails" at the beginning of performances;
   C. movements of lights to different positions around the hull;
   D. loops as the craft moves (non-structured);
   E. alternating loops of different rotation, clockwise or anticlockwise (structured);
   F. figure eight pattern, spiral pattern, DNA pattern, and vortex pattern, to light motions created by the entire motion of the craft: The 'S' sky glyph, for example.

Some of the light patterns created may represent signs or symbols, while others have unknown meaning, if any, or are related to the actions of observers on the ground.

Mechanical Sounds and Synthetic Sounds

Ellen Crystall told me back in 1992 that the ships could create different types of sounds and that she was sure that they were synthetic and not real, i.e. created by engines or generators on the ships. The phenomenon had changed or evolved from the time she began her field studies in 1980. She was accustomed to seeing ships or triangular craft up close, and many times saw two or more ships hovering over fields at the same time. Her popularity grew and with the spread of her stories, the size of groups accompanying her into the Pine Bush area grew. "We found ourselves quite literally running through the fields, chasing the ships." (Crystall, 1991: p. 71). She also noticed something was changing. By 1982 the ships' behavior was detectably different towards her and those with her than it was in 1980. Her photographs also changed, and no longer revealed high energy emissions, or alien silhouettes standing near their landed craft. She also changed her photographic method, by using a flash to capture the foreground in her pictures for orientation and reference. By the 1990s she had a Canon camcorder, and was recording video with sound. During the 1980s no one thought of taking an audio recorder into the field. The ships flew and hovered silently, so there was no need to record sound. That would all change once audio was being recorded, and was part of camcorder video.

During the night siege to the east of the Hudson River, Hynek et al. (1998), who analyzed 7,046 reports in their study, reported that 42.8% of the witnesses reported that the objects they saw in the night sky produced no sound. Their reports extend sightings past 1986 to 1995, and therefore do not represent just the night siege part of the study from 1982 to 1986. Witnesses in their study report that 55.1% produced a humming sound, compared to a finely-tuned electrical motor. The hum was very faint. The vast majority of witness accounts described in their book report no sound. Only five reports of the 62 sightings in their book for the years 1982 to 1886 describe a sound: In one case it sounded like that of a single engine airplane but faint (24 March 1984). Another witness on that same day also reported hearing a clearly audible but not loud engine-like sound. Only three of those 62 detailed reports indicate a hum or faint buzz. Even Crystall, when describing a UAP that resembled a small airplane with no visible engine or landing gear, reported only a faint buzzing sound (Crystall, 1991: p. 117). This would all change in the 1990s when camcorders with audio were routinely used by the sky watch groups, and when I would almost always have a separate audio recorder operating during a sighting (1992-1997).

Hynek et al. (1998) do report engine sounds heard by 2.2% of the witnesses in their reports. Witnesses compared the sounds to aircraft engines – mostly to the reciprocating sound of a propeller-driven single-engine aircraft, but much reduced in loudness. Almost no witnesses reported sounds like that made by turbofan jet engines. But this was all to change on 13 July 1992 when Ellen Crystall, Ralph, Steven and I witnessed a Black Triangle lift off from a field behind a tree row to our east, and ascend towards us, flying over us as we stood along the side of West Searsville Rd. just north of the dogleg bend. Crystall had her Canon camcorder, while I was taking time exposures. The lift-off and fly-over was well recorded (see pictures in Appendix I: 13 July 1992).

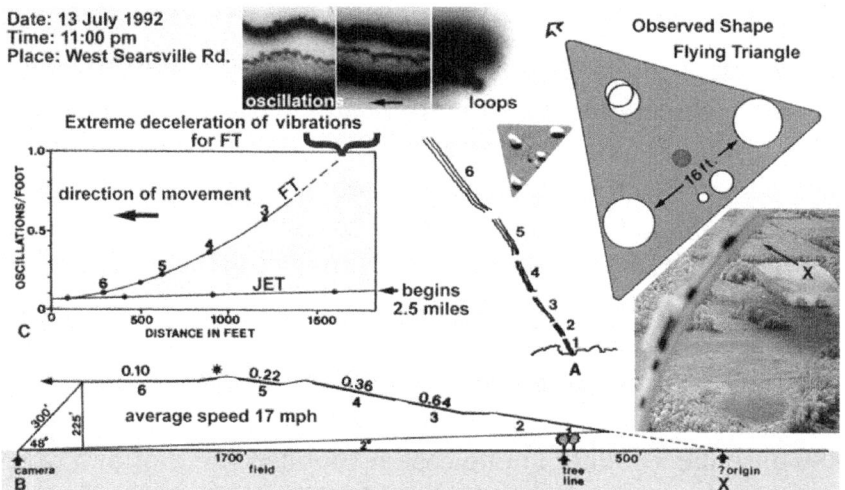

Between 1992 and 2003 I recorded 133 sightings, many with two or more witnesses (Appendix I). Of those the vast majority had no sound (75%). Only 33 (25%) produced sound ranging from

A harmonic clarinet-like sound;
A power turbine generator sound;
A turbofan jet engine sound;
A single-engine propeller-driven sound.

A statistically significant correlation was observed between an engine sound and the use of a camcorder to record the sighting. Twenty five of the thirty three sounds (76%) were recorded by a camcorder, while of the 59 audio recordings without a camcorder present, 49 did not record the UAP producing any sound. The incidence of sound being produced decreases to 17% when a camcorder is **not** used. It becomes clear that the human observers were being observed and the type of performance by a UAP depended on what equipment the visitors could detect being used by the sky watchers.

The use of a camcorder increased significantly after the 1990's Pine Bush flap was over in 1997. Activity apparently shifted east to the Long Island area and to east of the Hudson River. There are only ten video recordings of sound between 1992 and 1997 for about 128 sightings, whereas almost all of the post-1997 sightings had sound that was recorded due to the use of a camcorder. In addition, digital cameras were becoming more available and popular after 1997, and I was taking few time exposures once I had my own camcorder available.

In 1997 alone, 13 videos of the Manta Ray and Black Triangles were recorded, 10 of which captured sound coming from them. The three that did not were because the craft were more than two miles away; they were the same craft that had been recorded earlier that evening of 25 January, as they flew over the camcorder producing engine-like sounds (jet-like). They then circled across the valley and landed in the area of Lake Osiris just north of Walden, NY.

Thus, we can discern a trend as the method of recording the sightings shifted from time exposures, with some use of a camcorder, to almost the exclusive use of camcorders: sounds were produced more frequently when camcorders were used, but not when only tape recorders were used. When I used a small pocket-sized Radio Shack audio recorder, it was usually concealed in my shirt pocket, with a wire extending to a lapel (lavalier) microphone. When a camcorder was present, it was clearly more visible, and could have invited the UAP pilots to produce a sound.

In summary, as the methods of recording information changed, so did the type of information being given by the craft. This can be confirmed as time exposures gave way to videotape. Sky glyphs were commonly performed by slow-moving UAP lights when the film canvass of a time exposure was available. Once camcorders replaced time exposures, sky glyphs were no longer created. Instead, the craft began to mimic conventional aircraft more and more, duplicating their patterns of navigation lights at night and their sounds.

Because I had both audio recordings and video-recordings of sounds going back to 1992, I could analyze the sounds produced over that period of time to determine whether the physical characteristics of the sound produced by the Manta Ray and Black Triangle changed. It was almost impossible to mistaken either craft for a conventional aircraft if its shape or silhouette at night could be made out. Video recordings, however, did not have the sensitivity to record low levels of light at night, especially when they first hit the consumer market in the 1980s. By the early 1990s not much improvement occurred. By the late 1990s my Sony camcorder could image the silhouettes of craft if the background sky was light enough. By the early 21st Century "night shot" was available, and faint detail on black craft could be pulled out using computer enhancement techniques. These data are the result of the scientific method and technological progress, in which most sky watchers did not partake.

The video of the Manta Ray Crystall videotaped on 4 September 1992 (Appendix I) only shows the lights well, but not any structure behind the lights. The video of the Manta Ray I videotaped on 20 August 2003 with the latest model of Sony camcorder captured details of some of the structure behind the lights of the Manta Ray.

This was a clear improvement of human technology, but it did not stop many UAP enthusiasts who gathered on west Searsville Rd. at night to watch for UAPs from becoming skeptical of video records, when what they saw and heard looked and sounded too much like a conventional aircraft. It was misdirection in its most basic magical form.

Reversed Doppler

Suspecting that a sound is synthetic or fabricated will not convince skeptics, and I had many people question my identifications without even asking to hear or see the evidence or wanting to see the analyses of the sound that proved they were synthetic. I felt very much alone in my investigation, because I had very few other scientists and experts to talk to and ask for opinions that were objective and not biased by fear of ridicule. However, from the initial spectrographic frequency analyses of these sounds something very anomalous showed up. All the sounds of suspected craft – suspected based on other evidence – turned out to have distinctive and anomalous characteristics. None had any evidence for high frequency white noise, as is always present on recordings of turbofan jet engine sounds. None of the frequency spectrographs had a continuous spread of frequencies. Instead, all were divided up into about a dozen separate and "pure" frequencies bundled together as with sound synthesizers. The human ear would interpret those bundles as one sound, just as the human eye cannot see the individual frames of a motion picture film. But most unusual of all, most of those sound recordings appeared to violate the Doppler Law. One theoretical physicist, Dr. Jack Sarfatti, calls this evidence "the smoking gun" for non-human gravity-controlling propulsion technology (Pers. Comm. 2003; 2019).

Everyone should be familiar with the Doppler effect. If a sound source is moving, the frequency heard by a stationary observer/audio recorder, or one that is moving slower than the sound source will be different than the frequency of the sound source. In other words, a car horn will appear to a stationary listener to increase in pitch if the car is moving towards the listener, and decrease in pitch if the car is moving away from the listener. This is an absolute principle of physics. Recordings of jet engine sounds of an approaching jet liner will increase in all embedded frequencies, and decrease in pitch as the jetliner flies away. The sound spectrogram of a Boeing 737 below demonstrates the expected Doppler effect, as well as noise increase as the jetliner passes over the microphone. Note the embedded white frequency lines take a turn downward on right.

Reversed Doppler spectrograms below: 4 May 2000.

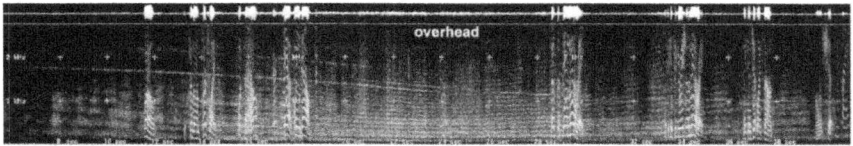

Frequencies uniformly drop in pitch and bandwidth.

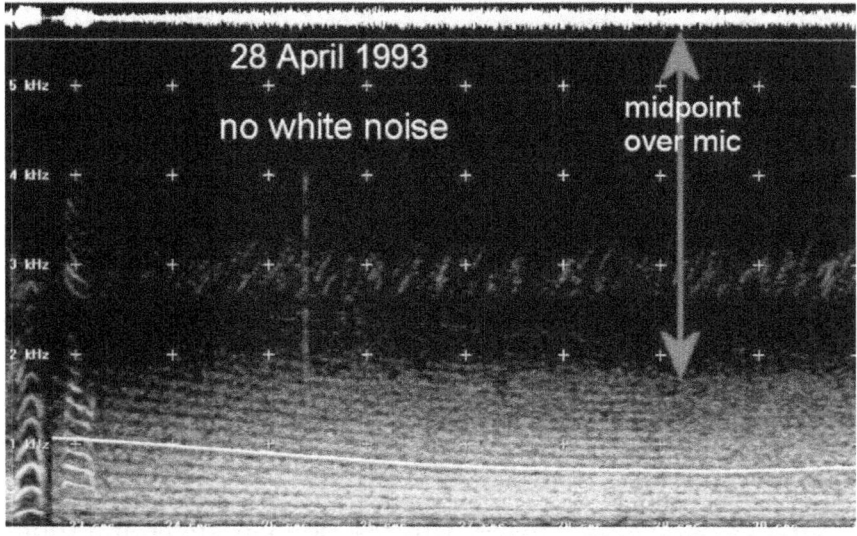

Unconventional Aerial Phenomena

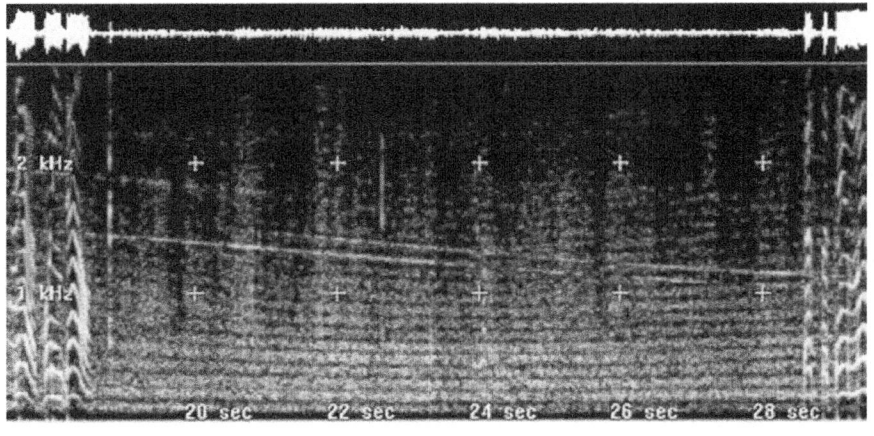

Another Reversed Doppler Spectrograph

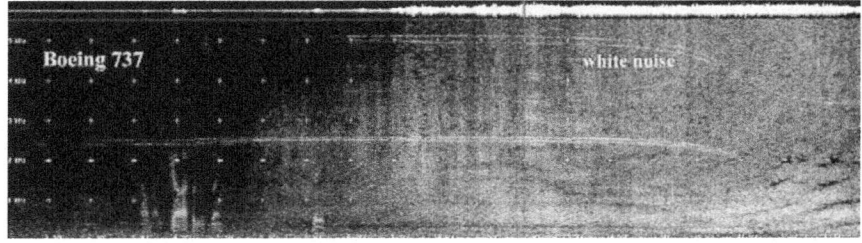

The frequency spectrogram of a Harrier jumpjet below also shows a distinctive white noise envelope around the primary noise of its engines. This spectrogram also shows a rise in frequency (accentuated by black lines) after the aircraft passed the microphone, which is due to hot compressed gases coming out of its engines, which locally reverse the normal Doppler effect. This is important, because it shows how and why the sounds coming from the alien craft cause the Reversed Doppler effect. Sound frequency decreases when the air/atmosphere is rarified or expanded, while sound frequency increases when the air/atmosphere is compressed and made more dense. That means that the propulsion system of UAPs is fundamentally different from that of our jet propulsion system, and involves the stretching of space/time in front of the craft, and the compression of space/time behind the craft, causing the vehicle to move from more dense space (behind the craft) to less dense space (in front of the craft). The manipulation of gravity can do this.

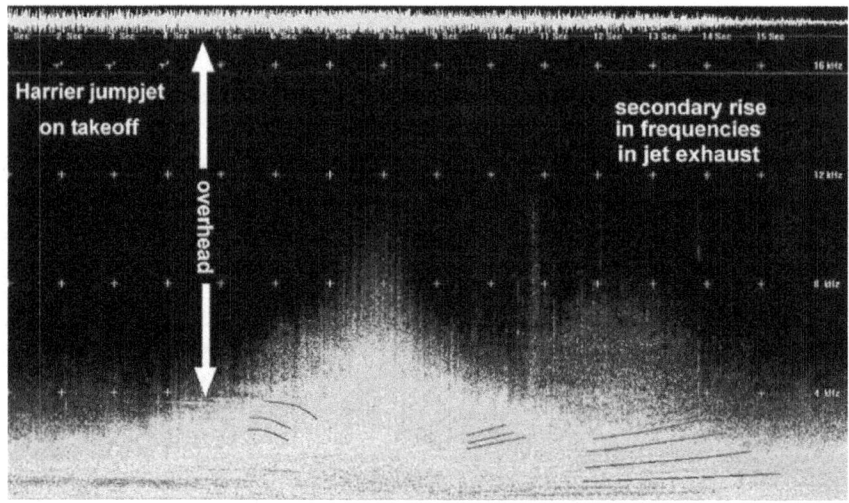

Normally a jet approaching cannot be heard until it is close to the listener, if it is flying directly towards that listener. That is because the sound coming from the engines is directed behind the jet in the form of expanding hot gases and air vibrations. Audio recordings of jet noise amplitude (volume) show this. Only when the jet is nearly on top of a listener on the ground does the sound volume coming from the jet increase rapidly. If the jet is passing the listener at a distance, the engine sounds can be heard sooner, and the full effect of Doppler is reduced.

As a jetliner passes over an observer, the peak sound of loudness is not heard until the back of its engines just pass the listener. See volume and frequency spectrograms below: Note the asymmetrical distribution of sound amplitude relative to the microphone ("overhead").

Below are spectrograms of a Boeing 747 flying low over coastal New Jersey on approach to Kennedy International Airport.

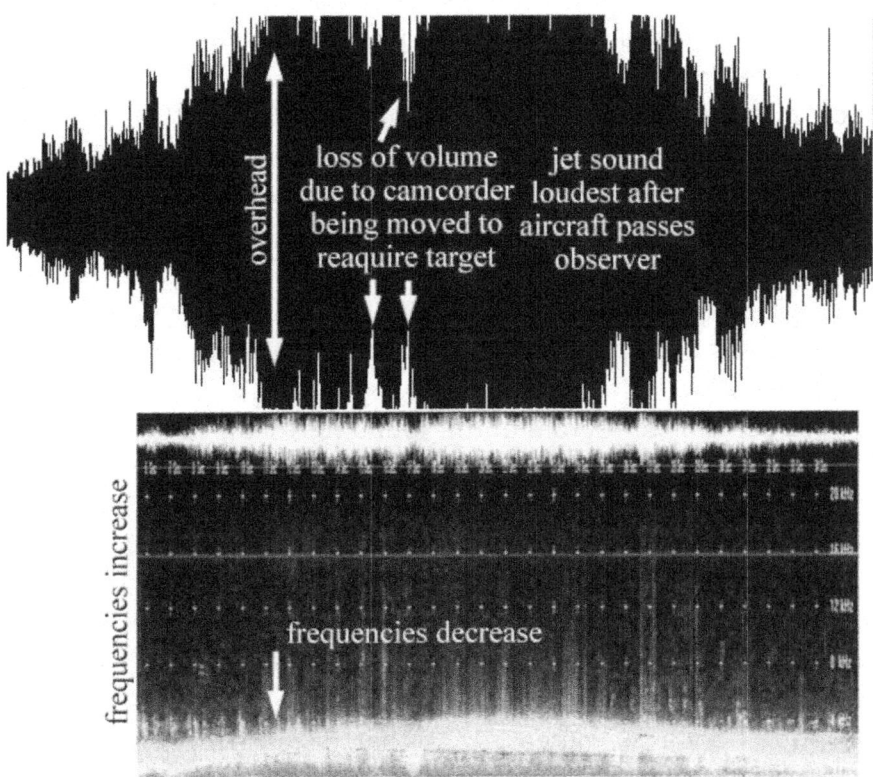

From that point onwards in time as the jetliner flies away, the amplitude of the sound steadily decreases. At an airport, for example, a listener positioned at the end of a runway or online with the runway will not hear much sound from a jetliner taking off until it is almost on top of the listener, even though the engines are at or near maximum thrust. The sound volume reaches maximum just after passing over the listener – not when the jetliner is directly above the observer. Again that distinction is due to the sound being projected directly behind the jetliner. The sound volume stays high on take-off until the jet reaches its initial flight altitude and engine thrust can be greatly reduced. Then a recording will show a steady drop in engine noise.

But there is a phenomenon that occurs just after a jetliner takes off and passes over a listener on the ground that is called "Reversed Doppler." Because the hot gasses coming out of the engine form a cone of compressed gas out the back of each engine, the sound of the engines passing through that expanded cone initially rise in pitch, higher than the sound radiating laterally from the engines. Thus, the listener on the ground will hear the roar of the engines above, which will suddenly increase in pitch or frequency as the cone of hot gas passes overhead. Once the jet is a quarter mile away from the listener, the cone of hot gases will have expanded and cooled enough that the sound will assume the normal Doppler characteristic of decreasing in pitch and volume with distance.

A frequency spectrogram of the sound for that jetliner taking off from an airport will show low noise, a sharp increase in noise or volume just before the listener/microphone is reached, along with a steady increase in pitch. That is because the distance the sound travels as a moving sound source approaches is decreasing. That steady decrease in distance causes the apparent pitch of the sound to increase. Sound pitch is a function of the distance between peaks and troughs of the sound wave. As that distance decreases, the pitch/frequency increases. As that distance between peaks increases, the pitch/frequency decreases. As the jetliner moves away from the listener, the distance between sound source and listener increases, and should result in a steady decrease in both pitch and volume. But if the jetliner is taking off with engines producing near-maximum thrust, there will be a short period when the pitch of the engine sound increases even more, until the pilot throttles back.

All conventional aircraft whose sounds I have recorded exhibit on frequency spectrograms normal Doppler. I have found no exception, taking into account exhaust reversed Doppler on takeoff. The reason for that reversal (increase in frequency as the jetliner is moving away) is due to atmospheric compression from hot engine gases. As the air and exhaust molecules get compressed – distance between gas molecules decreases – the peaks and troughs of the sound waves are forced closer together. It is the vibration of the air molecules that carry the sound, and the distance between them is inversely proportional to the frequency. In other words, as molecular distance increases, frequency decreases, and as molecular distance decreases, frequency increases.

In the example of Reversed Doppler for jetliner exhaust, physics is not being violated. Another explanation – compression of atmospheric gas – has to be introduced in order to explain what appears to be a violation. For the examples of Reversed Doppler given below, that Reversed Doppler is required in order to explain this phenomenon for unconventional flying objects or UAPs, and why it indicates foreign technology.

Almost all of the audio recordings of sounds coming from the Manta Ray or Black Tri9angle exhibit Reversed Doppler. Only those recordings where the path of the UAP was tangential and not over the audio microphone was Reversed Doppler not evident, as would be expected. I will describe in detail seven encounters where Reversed Doppler was recorded, two for the Black Triangle and four for the Manta Ray. In each case a loud sound was produced well before the craft reached the microphone, and in all six cases the craft deliberately flew towards me and the recording devices (i.e.

camcorder and/or audio recorder). These differences in Doppler are important in determining whether the UAP was the misidentification of a conventional aircraft or helicopter, or something that no one will find parked on a human airport tarmac.

Encounter with Reversed Doppler

On 13 July 1992 Crystall and I, and two other sky watchers (Ralph and Steve) were standing near the shoulders of West Searsville Rd. just north of the dogleg bend. We were in the edge of a farm field looking east towards a tree row separating fields. The tree row is about 1,000 feet away. A farm house, barn, and silo to the Wilde residence was to our right (to the south).

This account was given previously on page 11 under The Experiment, but needs to be described again in more detail with focus on the sounds this Black Triangle made.

It was 10:55 pm. We saw a pair of lights turn on just behind the distant tree row. They were too close to the ground to be distant lights of an aircraft. Beyond the tree row and fields there is a forest. No aircraft lights could be seen through that forest if this was a distant conventional aircraft. The lights began to rise and brighten. They brightened to the intensity of the Sun as they cleared the trees headed directly towards us. The lights soon dimmed down to the intensity of aircraft landing lights. Crystall was videotaping the approach, while I was taking a series of six time exposures. The audio was recorded on my pocket Radio Shack audio recorder, and on Crystall's video audio track.

I had infrared film in my 35 mm SLR Minolta XG7 camera that shows intense heat coming from the plasma lights. But during the event we didn't have the benefit of photographic analysis. Transcript of recorded conversation given below, starting just as the Black Triangle cleared the trees (see pictures in Appendix I: 13 July 1992):

Ellen: "What is that?"

Steve: "That's either a UFO or a plane getting ready to land in the field."

Ellen: "Yah, in front of us."

Ellen: "Tilting. Oh man!" [Craft was about 1,000 feet away.]

Steve: "This is going into Stewart."

Ellen: "This ain't goin'...this is comin' our way. Oh man. Where are we? Oh Jesus, I'm screwing this up, big time here." [Ellen loses sight of craft in her viewfinder.]

Steve: "Now without a flash, you would pick them up on the film, wouldn't you?"

Ellen: "Yah, without a flash you will pick them up if you aim it at the sky."

Ralph: "It ain't making any noise."

Ellen: "Not yet, but it is very big. Where are we? Well..."

Ralph: "That's got three lights..."

Steve: "That's like what we saw the other night, Ralph."

Ellen: "Now there's some noise." [Craft was about 500 feet away: faint noise...]

Steve: "That sounds like jet noise."

Ellen: "But this was hanging around too long. This was from fifteen minutes ago." [Average speed of craft later calculated at 17 mph.]

Ralph: "Car noise."

Steve: "Well why is he so low. That is my point."

Ellen: "Well....I don't know, but I've got video, and that's what I wanted to see."

Steve: "The one the other night was about half this closeness."

Sound gets louder and louder to the point that Ellen cries out loud: "Oy yoy yoy!"

Ellen: "The problem with looking through this [camcorder viewfinder] is that I can't see with my naked eye. That's the problem but..."

Steve: "Now do UFOs ever make simulations..." [jet-like sound begins to abate as Black Triangle flies away to our west –

not the direction of any conventional aircraft flight path in the area.]

Ellen: "Yah, they make all kinds of stuff."

The craft then reduced the intensity of its right light (aircraft lights are not designed to dim gradually: only low beam or high beam) to match intensity of its left light, and dropped slightly in altitude (obvious on my time exposures) and levelled off. The first sound we heard was when it was only about 500 feet away, and the sound was distinctly like that coming from a jet engine. The craft (its shape not visible yet) levelled off its climb and then brightened its right light gradually to an intensity that ionized the atmosphere.

As the craft flew over us, the sound grew then dropped proportionately to distance from the observers, unlike that of a conventional jetliner, where the loudest sound occurs just after the jet passes over the listeners. As I looked up, I saw a Black Triangle, not a cigar-shaped fuselage with wings and a tail assembly. The Triangle was about 500-600 feet above us. Ralph and Steve argued that it was just a plane, while Crystall was certain it was a Triangle. The sound had clearly biased and confused the less-experienced observers.

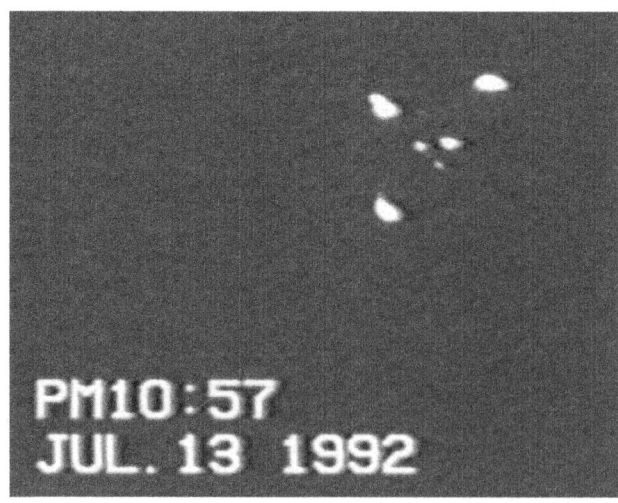

Unconventional Aerial Phenomena

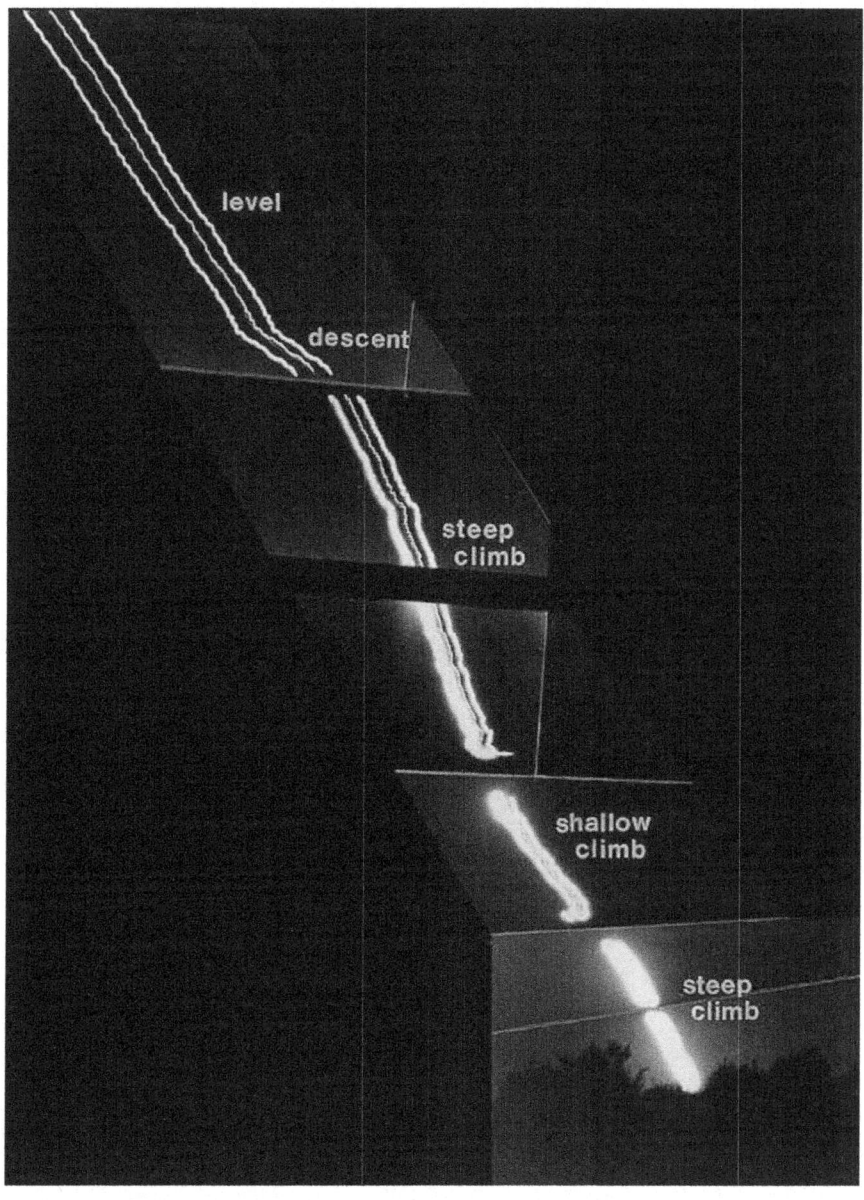

Later analysis of my time exposures showed the craft's lights actually spiraling through the air, something conventional aircraft lights do not do. That spiraling is probably connected to the operation of its propulsion system. An aspect of the sound that was strange was its decreasing frequency as it approached the microphone. It would not be until 1998 and later when Personal Computers were capable of running sophisticated sound software that I was able to analyze the sound track. What I discovered astonished me.

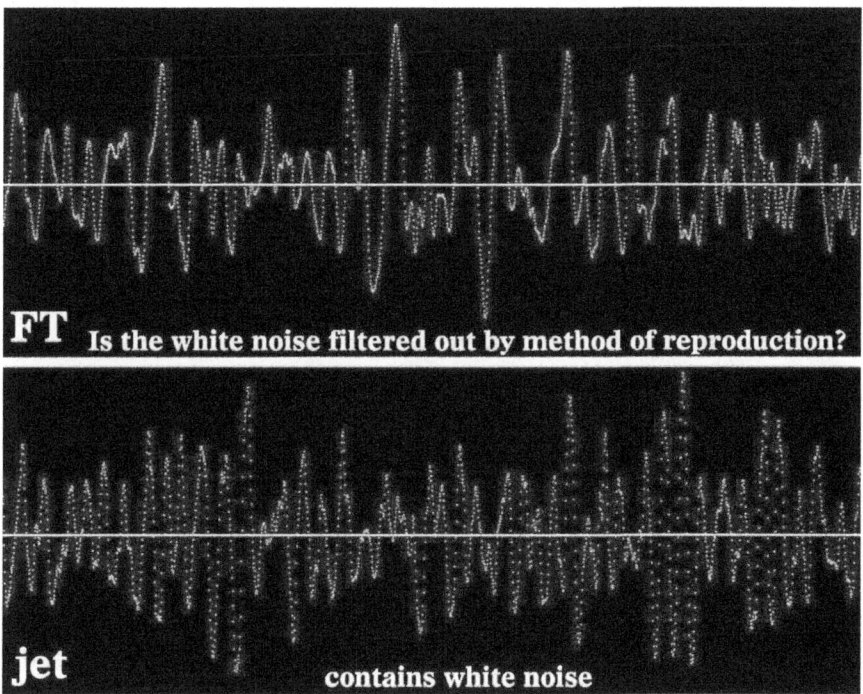

Frequency spectrographic analysis shows a lack of white noise, and the division of sound into discrete frequencies, both indicating a synthetic or artificial sound. White noise, if present, will be recorded on analog tape. It is machine-generated sound above 3 kHz, and ranges as high as 12 kHz. The UAP sounds are mostly below 2 kHz in range. If an audio tape of a conventional jet sound is replayed from speakers, white noise will be reduced or cut out, because most speakers will not be capable of reproducing it, and because there is no reason to reproduce frequencies above 8 kHz, since most humans are incapable of hearing those high frequencies. Dogs can hear them. Microphones are designed to record those high frequencies, because they appear on frequency spectrograms, but if they are played through a diaphram speaker system, much of the high frequency noise will be lost. Newer types of speakers can reproduce those high frequencies, but that does not explain why the sounds coming from UAPs lack them. The only explanation that makes sense is that the sounds coming from the Black Triangles and Manta Rays are synthetic, and not full range analogs of conventional jet sounds. One can see this when one compares sound files for unconventional and conventional aircraft sounds.

Above is given wave signatures for a normal conventional jet sound below that of the FT (Flying Triangle) describe above for 13 July 1992. Even though the two wave forms are similar (especially to the human ear), the jet wave has more white noise interference. The FT wave is less complicated with fewer oscillations. As I would discover, most sound coming from the alien ships was artificial, as Ellen Crystall believed.

The 13 July Black Triangle also turned on red and green outboard lights, trying to mimic a conventional jetliner. This was the first apparent attempt to camouflage a Black Triangle captured on videotape (i.e. Crystall's Canon camcorder). It would not be the last. The skywatchers may have been used by the visitors to guage their effectiveness with mimicry. Even though the camouflage and sounds produced by the Black Triangles and Manta Ray became concealed their identity to most people on the ground, they could not mask the characteristic frequency patterns of turbofan jet engines. There was always something about their spectrograms that gave them away, as is illustrated in the frequency spectrographs below, which show frequencies below 2 kHz and multiple parallel bands of frequencies, rather than blended overlapping frequencies, and no white noise.

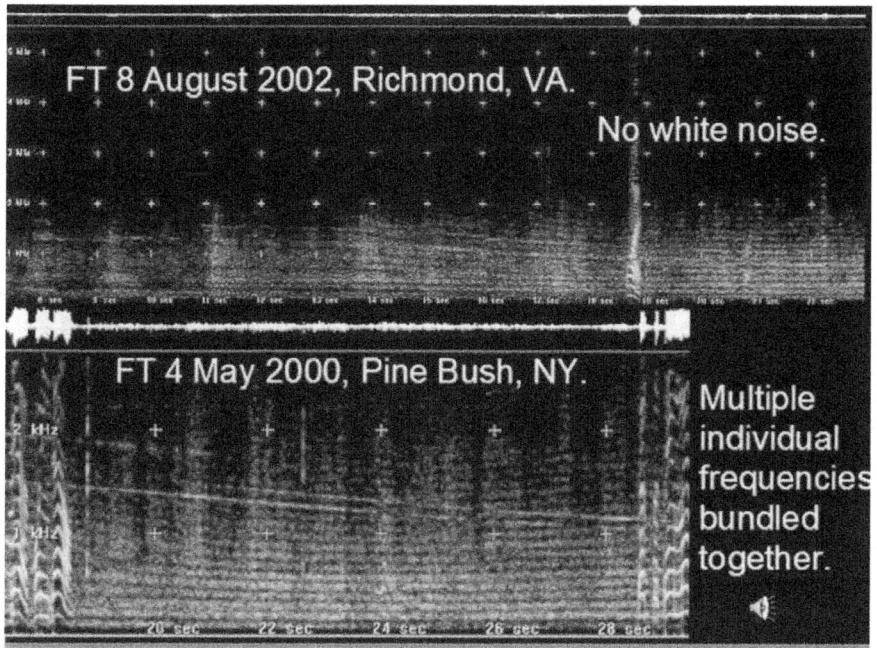

In addition to having synthetic characteristics, the sound also violated the Doppler Law. Although sound increased in volume as the craft approached, the frequencies all decreased, then reversed and increased as the craft departed. The graphic spectrograms below illustrates this for a fly-over by the Manta Ray on 10 August 1996. This is exactly the opposite of what the sound should have done, indicating that the atmosphere in front of the ship – a considerable distance away from the ship (as much as 600 feet) – was being stretched or rarified, while the atmosphere behind the ship was being compressed.

Sound of AOP captured on 10 August 1996, West Searsville Rd. on hillside at bend, Montgomery, NY

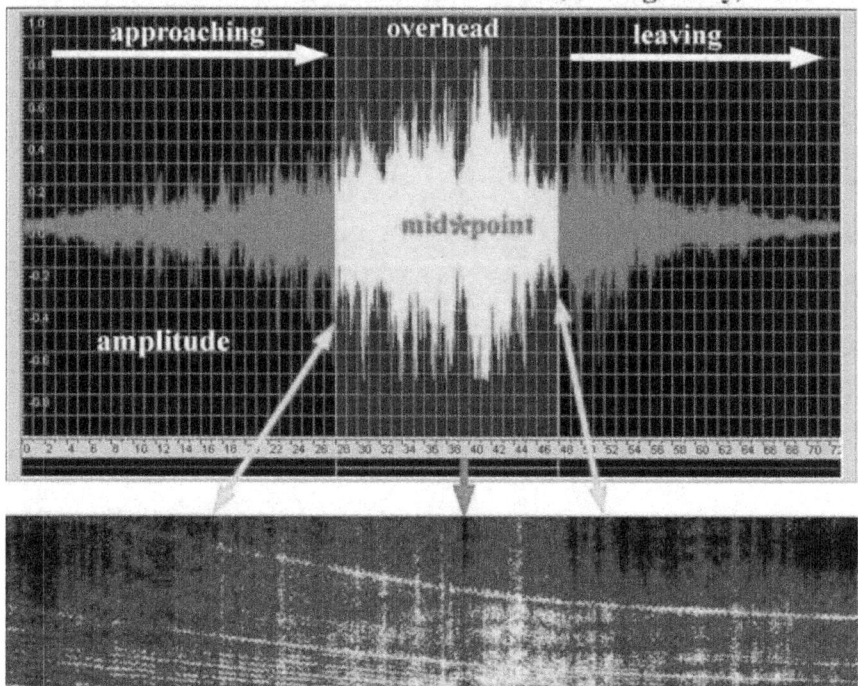

Spectrogram shows descending frequencies on approach - irrespective of distance to mic in violation of Doppler effect.

A turbofan engine will do something similar, but to a much lesser extent: The air directly in front of a running jet engine is being sucked into the engine, causing it to stretch and atmospheric pressure to drop. But this effect is so small, and the noise of the engine so loud, that a decrease is frequency is almost completely drowned out. Behind the jet engine the air is hot and compressed, reversing the Doppler effect. This is clearly audible only when a jet is taking off, accelerating, or has its afterburners on. The Reverse Doppler phenomenon around a turbofan engine involves the same principles we see with the sound envelopes around Black Triangles and the Manta Ray, but is on a much smaller scale. The Reverse Doppler effect for the ships can extend as much as half a mile in front and half a mile behind the craft. I will discuss the implications and cause of this phenomenon later in this chapter.

The second example of Reversed Doppler occurred on 28 April 1993 at 8:13 pm, just after sunset. The night sky to the west was still in twilight or not completely dark. The place was at the dogleg bend on West Searsville Rd. Crystall had her camcorder, while I had my Minolta SLR and pocket audio recorder. No sooner had we gotten there and I set up my tripod and camera, then Crystall said, "Here's something coming," pointing to the West. We saw a pair of headlights on the front of a black silhouette, and recognized it was a Black Triangle by its size. I began a series of six time exposures, three on approach, two above us, and one as the craft departed, descending rapidly below the tree line across the field to the east.

As it approached we didn't hear a sound until it was about a half mile away. The sound of a jet gradually increased in volume, something that wouldn't happen with a conventional jetliner coming directly towards an observer (see above explanation why). As it got closer we could see the craft had red and green outboard lights with white strobes behind the navigation lights. Clearly this craft was trying to mimic a conventional jet. Anyone expecting to see a conventional jetliner or military jet would not take further notice and would probably be convinced it was conventional before it even passed overhead. Crystall even comments on tape that, "I think this is a military something or other," based on its sound.

When it flew directly over us the sound volume had increased (gradually) to such an extent that the air vibrated around us. Crystall said, "Whoa it's loud," as I turned my camera around on the tripod to continue taking time exposures as it flew away. Then Crystall said in exclamation, "Wait a minute. That body is solid. That body is solid. Where the hell is it? Did you see the body on that?" I said, "No." (I was too busy operating my camera). Ellen continued, "That was a solid body, a solid triangle body with extra things in the front and back. Shit. I took the video camera away, because I wanted to get a better look at it." Careful analysis and enhancement of her video frames indeed show a triangular shape. I would later videotape Triangles on 25 January and 6 June 1997 (see: Appendix I) showing more clearly booms sticking out front and back that held lights on them – an obvious attempt to camouflage the Triangles by mimicking the lights of a DC-9, which was the American Airlines commercial workhorse in and out of Stewart International Airport on the other side of the valley.

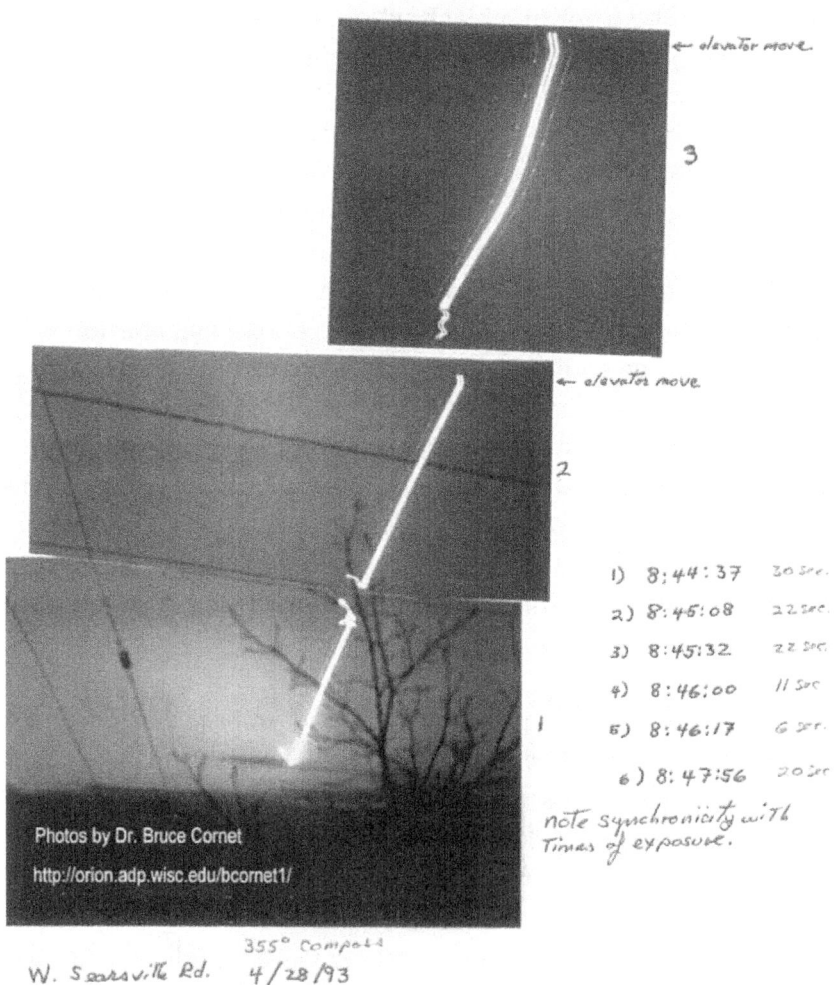

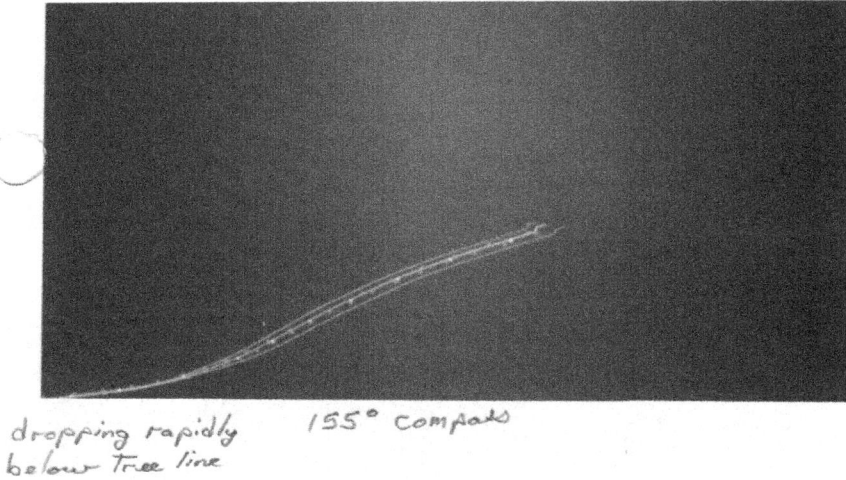

dropping rapidly below tree line   155° compass

The last time exposure shows the FT dropping rapidly below the tree line to the East (155 degree compass reading). Conventional and military aircraft would not do this at night over a forested area.

The time exposures I took are spectacular in that at the beginning and end of each exposure, as the Black Triangle approached, the lights would jump or move, indicating that the pilot knew when I opened and closed the shutter to my camera. No human pilot and technology could duplicate that precise timing and pattern. The 28 April 1993 Triangle showed no physical movement on Crystall's video, and one can align the sequential time exposures by tree branches visible in the pictures to show that the Triangle flew straight as an arrow. Only its paired central plasma lights moved off the flight path, and they did so for only fractions of a second as my shutter opened and closed. They did a dance next to the craft, indicating precise control of the electromagnetic fields morphing to produce this effect.

When the sound on Crystall's video was finally analyzed in the late 1990s, the jet-like sound dissolved into a cluster of 12 individual and parallel frequencies with no white noise, proving that this craft was not a conventional aircraft. The frequencies all decreased as the FT approached, then increased on departure, confirming a Reversed Doppler effect. The volume or amplitude gradually increased from about a half mile away, becoming loudest directly overhead – unlike that of a jet aircraft where the loudest volume for the listener is just after the aircraft passes over the hand-held microphone.

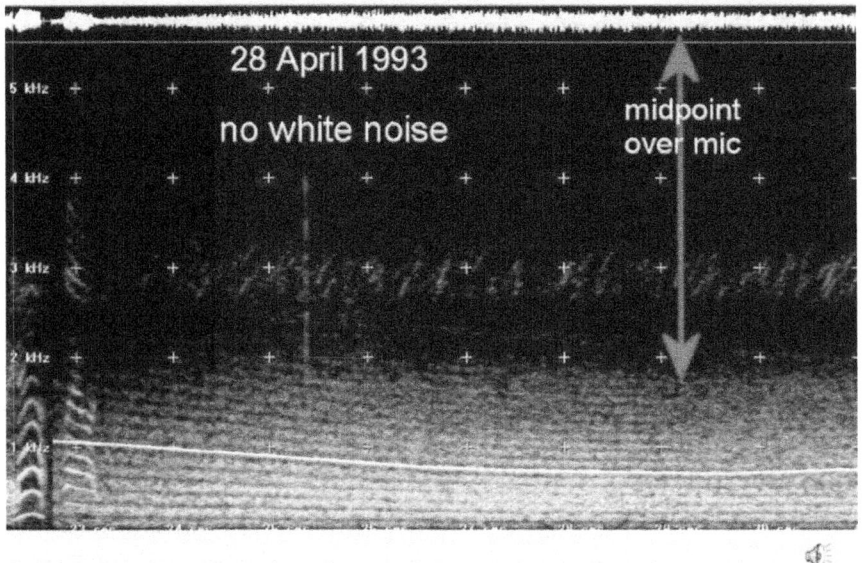

In addition, the spread of the ~12 frequencies decreased as the craft got closer to the microphone, with the spacing almost disappearing when the craft was overhead. Then the spacing on the spectrogram increased between frequencies as the Triangle departed. Not only did the frequencies shift, but the difference between frequencies changed, becoming the least when the sound source was directly overhead. Such a shift appears to be due to factors other than a change in the sound at the source. If an artificial gravity (or net-gravity) field existed around the craft, tis shape or volume could account for the clustering and dispersion of the frequencies, but not for the Reversed Doppler. That would be related to how the gravity field is modified to produce thrust or propulsion. More on this subject later.

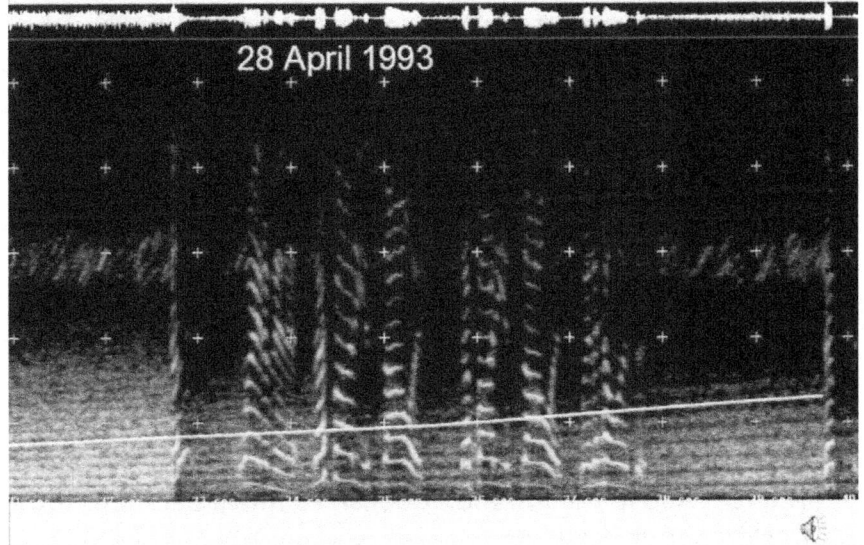

The third example of Reversed Doppler occurred on 29 September 1995 at 7:33 pm just after sunset at the dogleg bend on West Wearsville Rd. (see pictures in Appendix I). I was alone. David Ring from California was supposed to join me that evening, but he missed his flight. Instead of a Black Triangle performing the fly-over with jet-like sound, the Manta Ray was involved. But what this large ship did demonstrates convincingly that these visitors have technology to fool us, and that they also have a sense of humor.

After setting up my camera and tripod, I didn't have long to wait until I spotted what looked like a distant set of aircraft lights flying south, about a mile away to the East. What usually caught my interest or attention were lights flying too low to the ground. FAA regulations require pilots to fly no lower than 1,500 feet altitude unless the plane is descending for a landing. These lights were much too low to the ground and too far from Stewart International Airport (eight miles to the northeast) to be on a landing approach, and they were headed in the wrong direction. So I oriented and focused my camera for the first time exposure and opened the shutter.

After taking the first picture, with pocket audio recorder recording the times for each picture, which I called out while reading my Indiglo wrist watch, I moved the camera for the second shot, anticipating the next viewing field through the viewfinder. No sooner than I opened the shutter, the craft turned towards me. Now all I saw were two headlights, whereas before I had seen a couple strobes and white lights. The pilot not only turned on his brighter "landing" lights, but he was headed in the wrong direction, away from Stewart International Airport. He was headed straight for me.

As the craft approached my position, flying no more than a few hundred feet above the ground, it began to move back and forth, side to side and it flared its headlights many times brighter than landing lights. The first three time exposures of it coming towards me show this 'S' motion clearly, which I call a snaking motion. The craft climbed gradually in altitude after it turned so that the 'S' form would show up clearly on the photographs, which it did. Had the craft leveled off before the 'S' motion began, its lights would have overprinted its previous position and motion, and the 'S' form of its path would not have been evident.

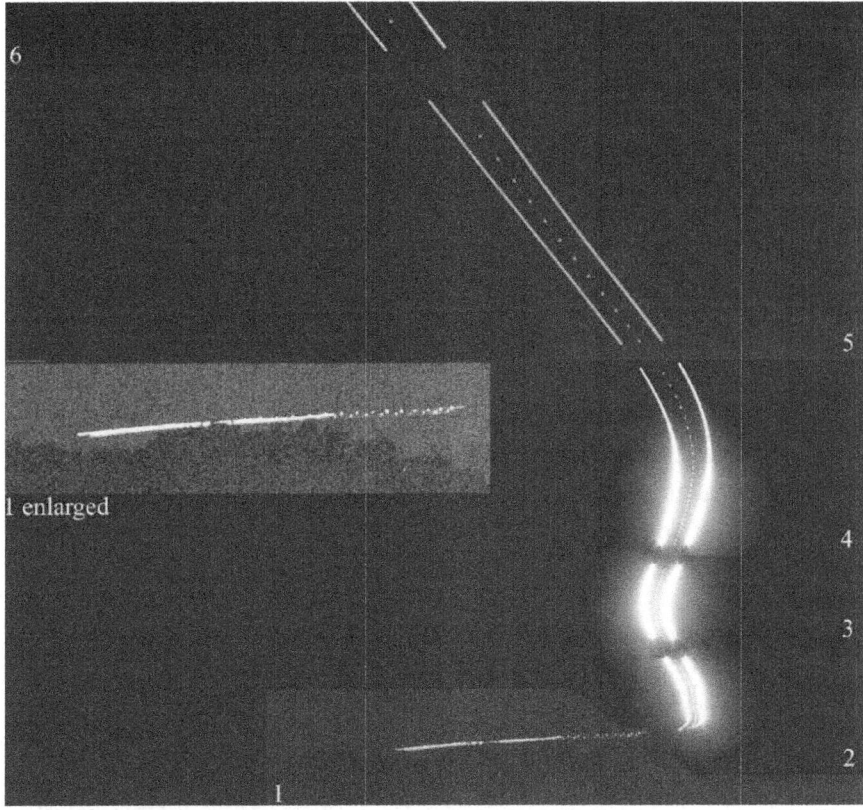

The lights became very bright, ionizing the air around them to form an orange-red aura or glow that extended out dozens of feet from the craft. How do I know they were plasma lights? Aircraft lights do not ionize the air, and the time exposures show atmospheric generation of brown nitrous oxide gas (orange-red when hot) very clearly.

About a quarter mile away I began to hear a sound. As the sound grew in volume, it sounded more and more like that of a jetliner. But as the sound grew louder, all its frequencies slowly dropped in pitch or frequency. Not only was the volume increase from so far away anomalous, but the distinctive descent in pitch – not by a little, but by a lot – was unexpected. I should not have heard any sound until an aircraft was almost upon me, and then the pitch should have increased slightly as the high frequency turbine wine of its engines became discernable. Once a conventional jet is over a listener on the ground, a sudden jump in volume or loudness – the roar of its engines – would drown out any Doppler pitch increase (even though that rise would be detectable on frequency spectrograms.).

As the jet-like sound grew louder, you can hear me comment on the audiotape, "Nice sound. You've got a lot of turbine sound." But my comment was sarcastic, suspecting that this engine sound was fake. It sounded like a jet taking off from an airport, but in reverse. Later in the late 1990s when I had computer software to digitize and analyze these sounds, I reversed the audio recording. And sure enough, it sounded like a Boeing 707 taking off (using a commercial audio file for comparison), not landing, and not flying at a constant speed over an observer.

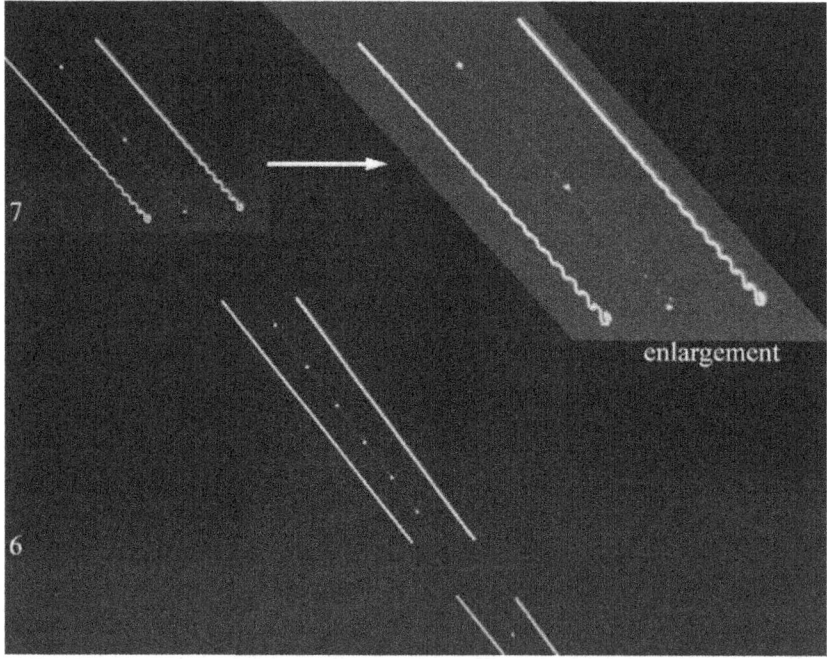

As the craft flew over me I looked up and saw its Manta Ray shape with narrow tail. I turned my camera around and cocked it up as much as possible on the tripod to get pictures of it overhead. My pocket audio recorder was still recording. I knew I could calculate the times for the quick photos using the sound of my camera shutter and film driver. But then something unexpected happened, and in retrospect shocking occurred.

The Manta Ray had climbed to at least 600 feet above me during its approach. It was clearly targeting me on the ground. The pilot knew I was standing there in total darkness ready to record what I saw – initially from one mile away! As the Manta Ray flew directly above me my camera jammed. Its electronics failed. The sound from the Manta Ray simultaneously stopped for a least a few seconds. Then audio recording resumed, and this gap in sound was clearly recorded, followed by a clarinet-like sound and a melody not unlike the X-files theme song! As soon as the new harmonic sound began, my camera worked again. That is, my electronic cable switch opened the shutter for another picture. But the ET-UAP tune was a surprise, perhaps indicating a little alien humor. Do you still have any doubt that the sounds produced are fakes?

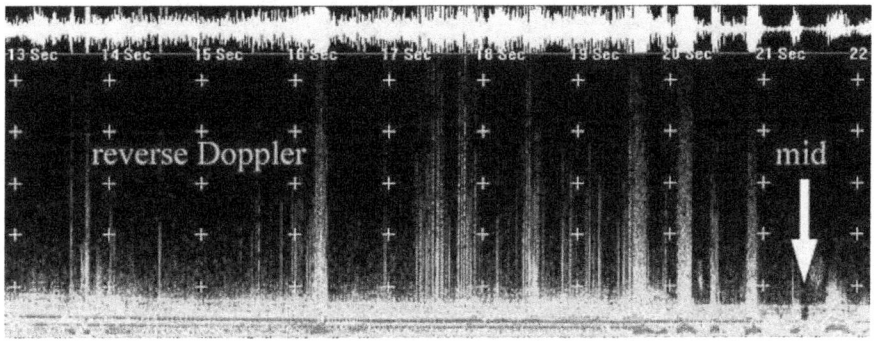

29 September 1995 sound spectrograms of Manta Ray on approach to the microphone, and on departure. Arrow (mid) shows relative position of microphone.

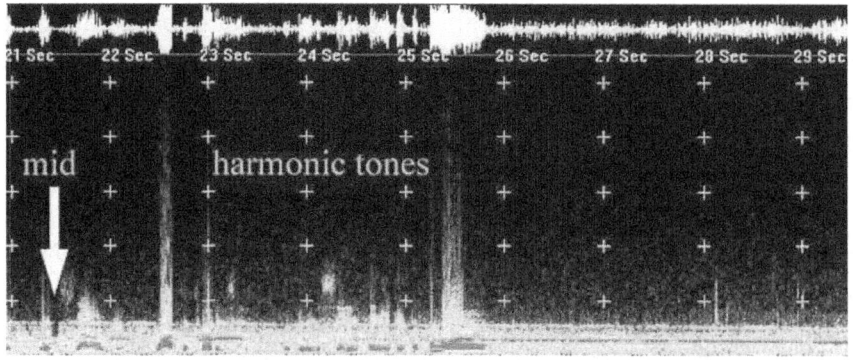
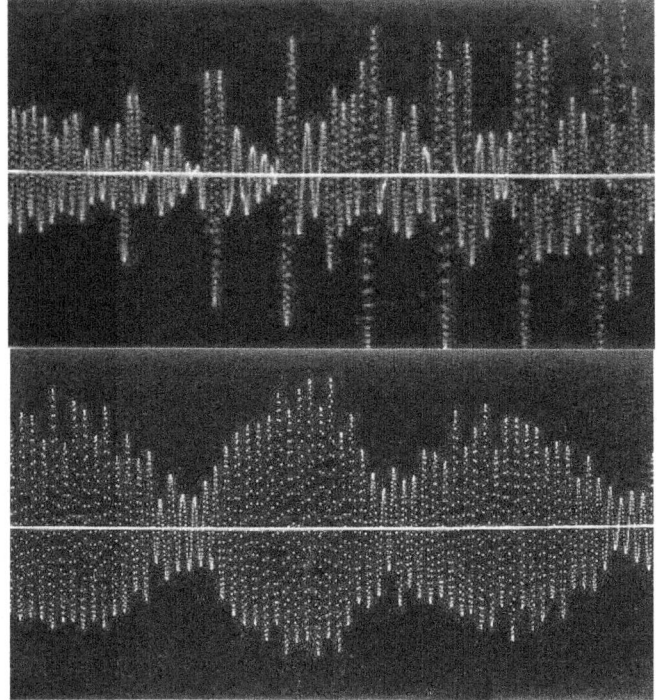

Jet-like sound before reaching me (top), followed by harmonic melody as it flew away (bottom). The spectrogram below represents the audio encounter.

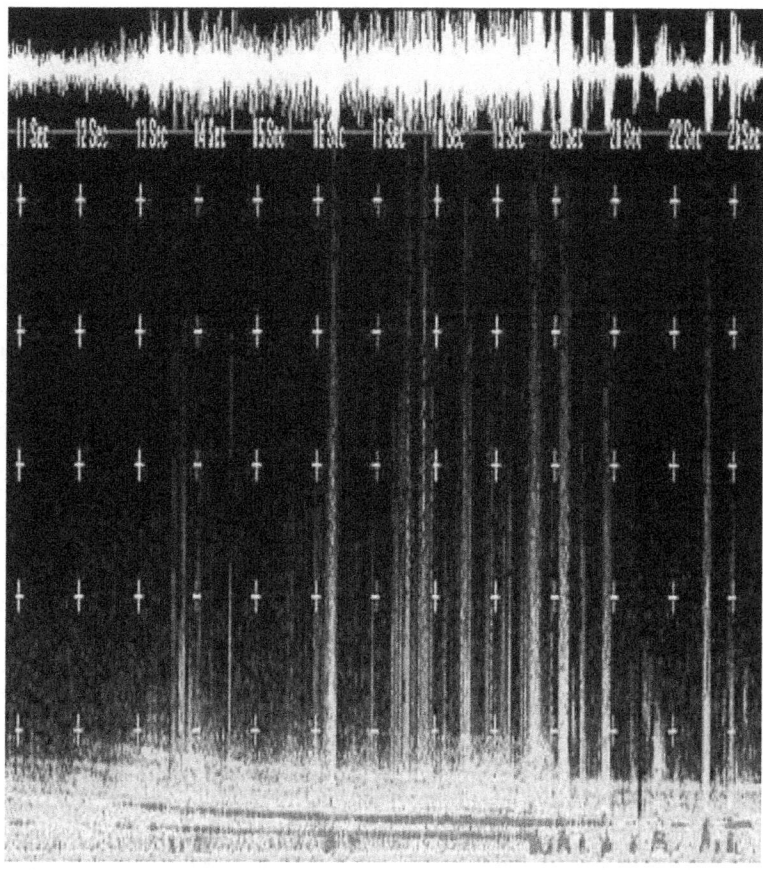

The above frequency spectrogram is compressed sideways to show more clearly how the frequencies dropped up to the vertical break directly over the microphone. The vertical spikes represent signatures of human voice as I spoke during the event.

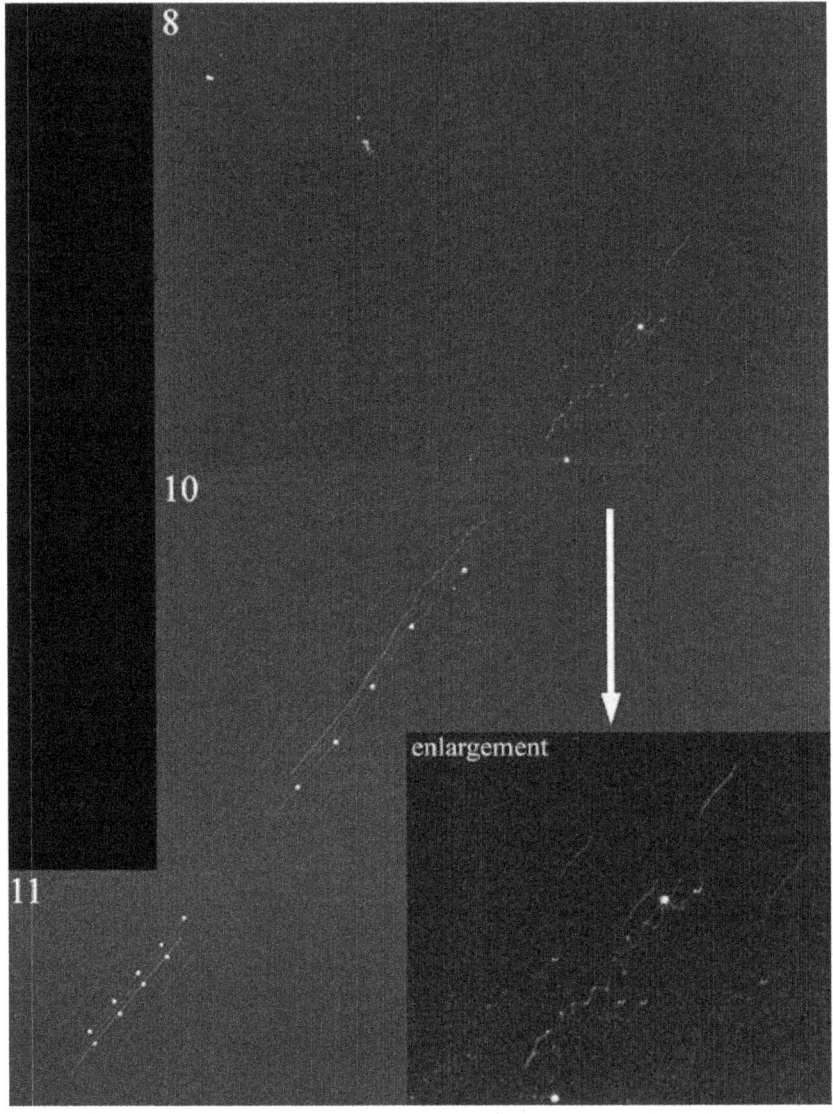

Above images: Time exposures of Manta Ray on 29 September 1995 descending to my West after arching over me. It then disappeared behind Thompson's Ridge, only to re-emerge through a gap to flying East back to where its detour had begun. Commercial and military pilots would not do that, especially flying only dozens of feet above the trees at night (shown in contrast-enhanced window below).

Unconventional Aerial Phenomena

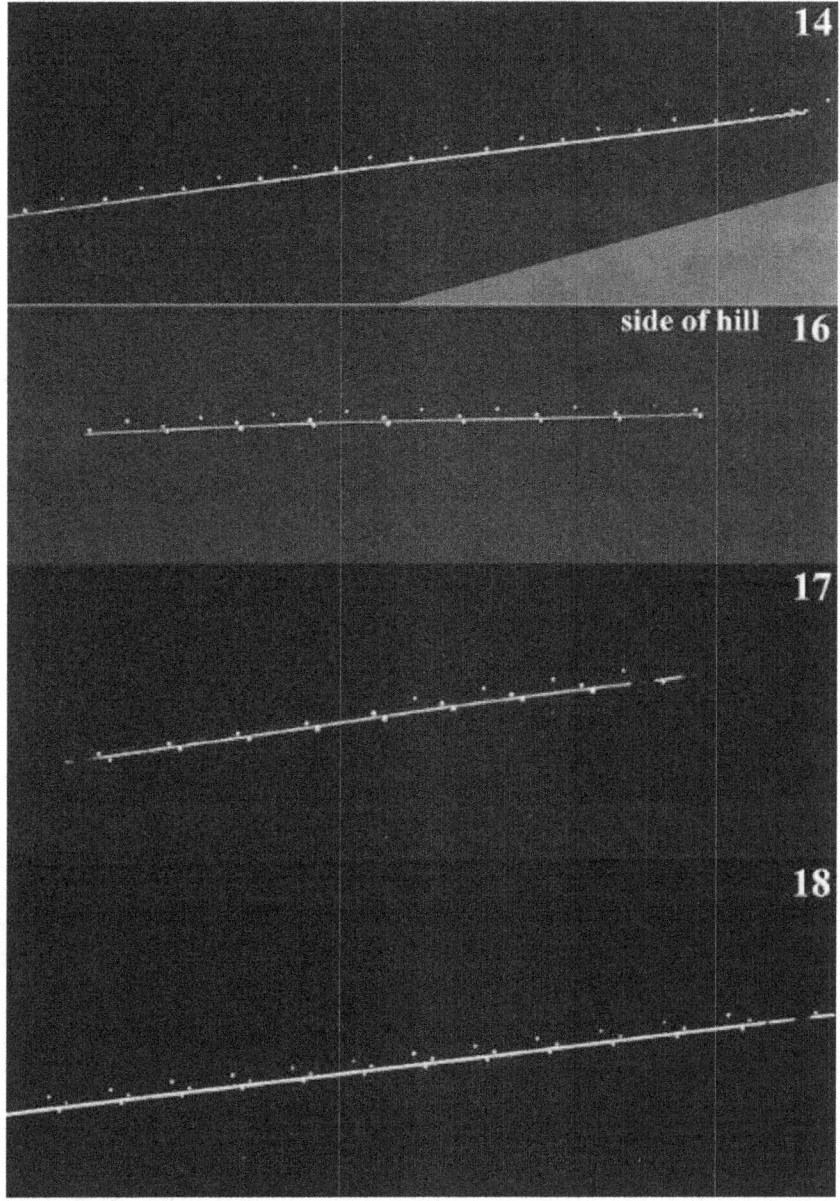

The time exposures 8, 10 and 11 above show the Manta Ray descending from its apogee over Cornet, as it headed for Thompson's Ridge to the West. Time exposure #8 (see enlargement) shows movement of the camera when it began to not respond to shutter closure, as I hit the camera, thinking that the shutter was stuck. Note how a second series of strobes appeared along the midline of the craft in time exposure #11. Time exposures 14, 16, 17, and 18 show the craft flying low over trees as it returned East. With computer enhancement on time exposure #14 you can see how close the Manta Ray was to the tree tops.

After descending to the West, the Manta Ray disappeared behind Thompson Ridge, which rises only about 200 feet above the valley floor at that location, then sharply turned to the South, briefly dropped down behind the ridge (Would a large jetliner fly that low to the ground?), reappeared or emerge through a gap in the ridge, and flew East low over the trees back in the direction it had come. When it got back to the point in the valley where it had initially turned towards me, it turned sharply to the South again, and continued its merry way down the valley. Go figure!

A daytime picture of the area it flew over to the southwest is shown below, along with a flight path over that area simulated for nighttime.

Below that is a contour map showing the looped path of the Manta Ray back to its original course down the valley.

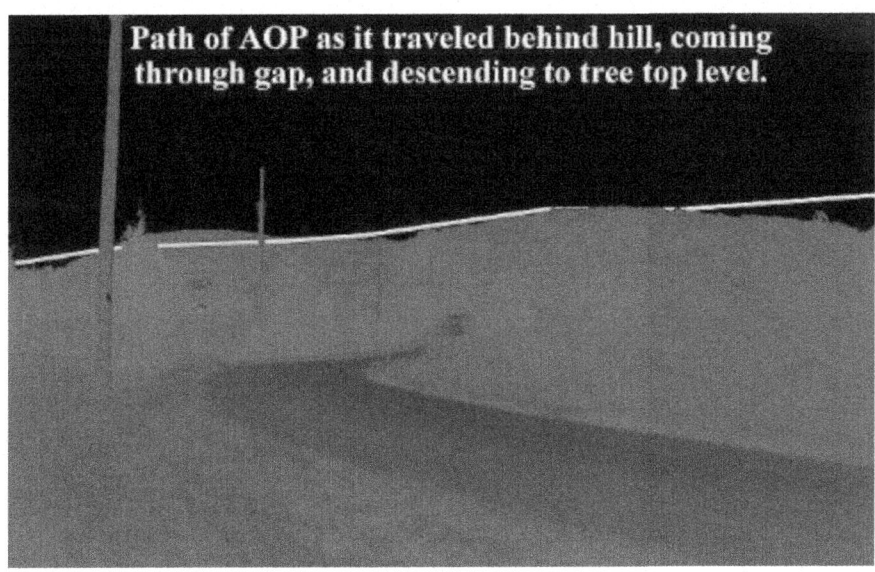

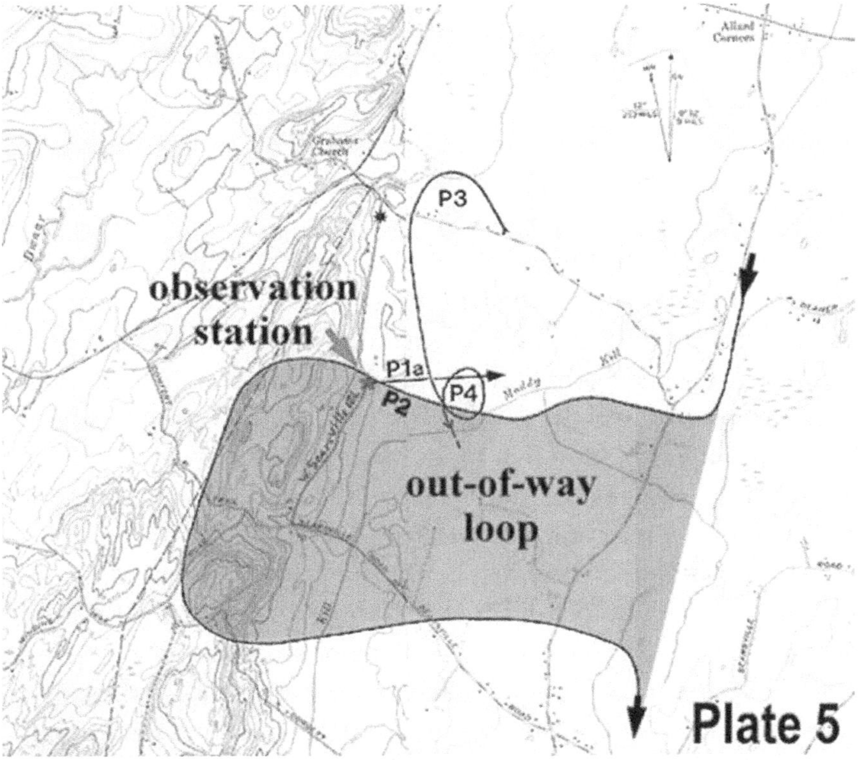
observation
station
P1a
P2
P3
P4
out-of-way
loop
Plate 5

Path of AOP as it traveled behind hill, coming
through gap, and descending to tree top level.

In summary, the behavior and flight characteristics of this flying object was anything but conventional.. The sounds it produced were clearly manufactured and played through a speaker-like system. Later analysis enhanced the synthetic nature of the sounds, confirming Ellen Crystall's statement to a group of sky watchers, "They make all kinds of sounds."

The fourth and fifth examples of Reversed Doppler described here occurred during the most intense period of encounters with the Manta Ray on 9 and 10 August 1996, in which the ship flew directly over Bruce Cornet and Tommy Sinisi several times between 11:12 pm on Friday night and Saturday morning. During this period the Manta Ray was seen five times, either taking off or landing in a forest just to the east of West Searsville Rd., or flying around the valley. Tommy Sinisi and I drove from Red Bank, NJ, to Pine bush, NY, on the evening of 9 August in my Chevy van. We found the gate open to the farm fields on the east side of the ridge that parallels West Searsville Rd. (That area has since been converted to estates with houses on top of the ridge and the side of the hill landscaped). It was an opportunity that I had wanted: To park and camp on the slope of Thompson ridge all night to observe UAP activity in the valley below. We didn't realize that we were expected, and that some of the most important data (evidence) would be collected that night. The pictures below, looking east, were taken from the slope of the ridge, and show the valley and forest where the Manta Ray took off behind the farm fields and new houses, built since 1997.

Four times we saw the Manta Ray as it few over us or near us as it took off from the valley below, or silently flew down into the valley from the west to land or disappear into the same forest from which a previous Manta Ray had emerged.

Path of Manta Ray lifting off from forest, then turning towards us.

Time TE = Time Exposure.	Event
11:12:33 - 11:15:08 pm (7 TE only)	Manta Ray comes over ridge, descends slowly and silently to forest in valley, drops out of sight into forest.
11:42:27 - 11:45:27 pm (15 TE and sound)	Manta Ray lifts off vertically to about 300 feet, turns on lights, then flies (climbs) over observers while producing a reverse-Doppler mechanical sound.
1:23 am (no TE or sound)	Manta Ray comes over ridge, descends slowly and silently to forest in valley, drops out of sight into forest.
2:42:10 - 2:44:14 am (sound only) TE = Time Exposures	Manta Ray lifts off vertically to about 300 feet, turns on lights, then flies (climbs) over observers while producing a reverse-Doppler mechanical sound.

The third time we saw the Manta Ray it was traveling east over Thompson's ridge. It descended slowly and silently into the valley below, and dropped out of sight into the same forest it had emerged from earlier. It was 1:23 am. The pictures below show just

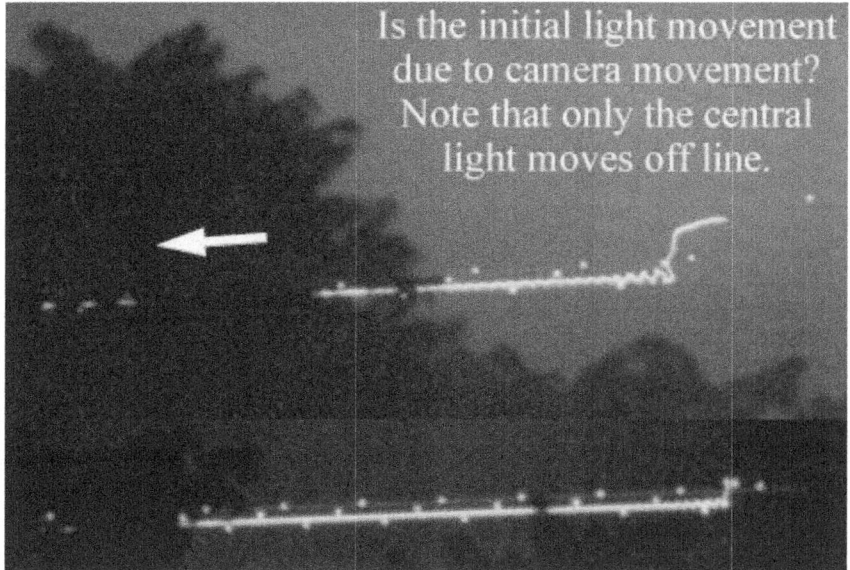

how low this craft was to the trees in the forest in the valley below as it descended and disappeared, traveling at a calculated speed of only 46 mph.

Twenty two time exposures were taken of two take-offs and one landing before Midnight (see pictures in Appendix I). The fourth and last time we saw the Manta Ray it took-off from the forest in the valley at 2:42 am on 10 August 1996, and disappeared beyond Thompson ridge to the west at 2:44 am. Only sound was recorded. We were sleeping lightly in back of the van when we heard its noise coming from the valley below. I was able to get a nearly complete audio recording of it from the back window of the van as it flew directly over us and over the low ridge we were camped on, as it traveled west.

The last time we heard the Manta Ray, it was behind the ridge after 3:00 pm, but it was recorded only on audio tape. We only heard its loud sound coming from behind the ridge, wondering why the craft was producing so much noise, which surely would have rattled windows and woken up Pine Bush residents on that side of Thompson ridge.

Most spectacular part of this event was our witnessing the Manta Ray lift off vertically from a dense forest to the south of the Indian Mound east of West Searsville Rd. It would have to have passed through trees, as Vinney Polise and another person witnessed nearby on 6 May 1995 (Polise, 2005: The Pine Bush Phenomenon, p. 74-75), which we all know conventional aircraft cannot do, which is why Dr. Ellen Crystall thought there is an alien base underground (Crystall, 1991).

I saw a red strobe rise silently above the trees to an altitude of about 300 feet, then change to white strobes as it began moving south. Then we heard a distant sound like that produced by a C-5 Galaxy military transport, which flies out of Stewart A.F.B. The lights continued south for about a thousand feet, then turned sharply west towards us. I caught all movements of this craft, except its initial lift-off with only a red strobe visible, on 15 time exposures. Once it turned towards us, it turned on a pair of headlights and began climbing towards us. Its sound resembled that of a C-5 taking off, but when we began hearing it, the craft had already taken off silently, and we were hearing its sound clearly from a quarter mile away! No or little engine sound should have been heard once the craft turned towards us, because if this had been a real C-5, most of its engine sound would have been directed behind the aircraft and away from us. Later sound analysis showed the sound to be synthetic and exhibiting Reversed Doppler, proving that this was not a military aircraft.

During that evening we recorded the audio for two complete runs of the Manta Ray at and over us. In both events the path it took was the same. Both times the sound it projected resembled that coming from a C-5 Galaxy transport taking off. In March of 2008 I had the opportunity to videotape and audiotape the sounds of a C-5 taking off from Stewart (A.F.B.) International Airport near Newburgh, NY. The sound was unmistakable, but it had none of the distinctive or anomalous characteristics of the sound coming from the Manta Ray on 9-10 August 1996. The authentic C-5 sound had white noise, normal Doppler, and a continuous spread of frequencies, unlike that of the Manta Ray. Clearly, the Manta Ray, because of its large size, was being camouflaged to mimic the C-5 by matching its nighttime navigation lights and sounds. It is also clear in hindsight that we may have been used for quality control. Because I knew that they could read my mind and track me (they later visited me in red Bank, NJ, flying over or near my house), they would know how well their camouflage worked. They would always give me data that upon close analysis would expose their identity and trickery, but how would I react to hearing the sound of a C-5 or jetliner pass overhead? Because I had Tommy Sinisi with me that evening, they had an inexperienced witness for control. Was this the reason I was given so many opportunities to view and photograph their hardware? More later in Chapter Seven on abductions.

The Sixth example of Reversed Doppler occurred on 4 May 2000 at 11:30 pm. The location was just south of the dogleg bend on West Searsville Rd. at the gate to a new (at the time) estate and houses on the slope and top of Thompson ridge. I doubt that the new property owner knew that his property was directly under the flight paths to UAP activity out of the valley (e.g. 9-10 August 1996 accounts above). Barbara Hartwell, a former CIA employee, UAP researcher, and experiencer, accompanied me on that night. In 1993 I was invited by Kenny Lloyd, Director of a Public Broadcast Service PBS) program in southern Connecticut, called "Intergalactic Television Network" (cable TV). Barbara Hartwell hosted that program when it was taped in January 1993. It aired in April-May 1993 as a Three Part Series on Pine Bush UAPs. I was later told by Hartwell that this TV program was a front for the CIA to gather intelligence on non-military or non-government people doing research in the Hudson Valley and Pine Bush phenomena. In 1995 Hartwell began her own research on abductions in the Pine Bush area, and she met me in the field on several occasions. We became friends, which is why we continued to work together and went out to look for craft at the hotspot on 4 May 2000. We were not disappointed.

We had just witnessed and recorded on video a large two-tiered ship flying to our east, turning south after it anomalously flared its paired headlights to the intensity of sun bursts (see description on page 32 under The Hudson Valley Mega Ships). We didn't have long to wait for the next event. At about 11:30 pm I spotted what looked like the navigation lights of a commercial jet approaching from the northwest at an altitude estimated at no higher than 500 feet. It was too low to be a commercial jetliner. Even though you can hear me say on the video, "It's probably just a jet," I sensed that there was something odd about it. And I even said that on the videotape.

First, if it was a jetliner, it was not on a normal flight path to Stewart. The craft should have followed the low ridge to our west, and turned east at the red beacon on that ridge. But these lights were crossing the ridge at an angle and headed straight for us!

Second, the jetliner was flying well below FAA regulated altitude that far from an airport, and what we had observed for aircraft heading to Stewart nine miles away.

Third, there was a cluster of small lights where a jetliner's nose should be, and that is not standard lighting on an aircraft.

The craft was almost upon us, a few hundred feet away, when we first detected a sound. It did not get loud until the craft was almost upon us – **an improvement in mimicry**. The sound was that of a jetliner. It had me fooled until it got directly overhead. Then as it flew directly overhead, two white lights at its outboard corners – where the wingtips of a jetliner should be – turned on pointing downwards, and sending long beams of light to the ground! Those beams then rotated forward until they pointed ahead. Headlights on a jetliner are not positioned out at the wingtips. They are positioned close to the fuselage so that they can illuminate the runway in front of the pilots. And there is no functional reason to engineer them to rotate forward **after** they are turned on, like headlights on sports cars that are hidden under the car hood. And they point up before rotating, not down.

> As those light beams turned on overhead, Cornet exclaimed:
> Cornet: "It's got search lights!"
> Hartwell: "Search lights?"
> Cornet: "Yah, pointing down."
> Hartwell: "They're looking for you, Bruce."
> Cornet: "That's the damned Manta Ray!"

When I looked up at its stealth black silhouette, I saw a familiar shape. There was no cigar-shaped fuselage sticking out front, or tail assembly in the back. There were four small lights forming the margins of what could have been a window at the blunt nose of a delta-shaped front. The video I took captures this feature in detail. These small nose lights also had short beams projecting from them. It wasn't until I had to zoom out and reacquire the craft in my video viewfinder that I looked up after turning around and saw the distinctive shape of the Manta Ray. Its shape and lighting, including a tail light, were the distinctive hallmarks of the Manta Ray. I told Hartwell that I had other video of the Manta Ray that could be substituted for this one, the comparison in shape and lighting was so similar. Later I checked my videos, and the similarity was exact.

Below is a comparison of the Manta Ray videotaped on 4 September 1992 and then seven years, eight months later on 4 May 2000.

When the new video images were compared with a conventional aircraft lights, it became apparent that mimicry or camouflage was the objective. We were being used over the years for quality control, so that these crafts could fly anywhere in the United States, and be mistaken for conventional aircraft. But distinctions still existed, and could be identified only with careful recording and analysis of data. The four small lights at the nose of this craft turned on only as it flew overhead, as if to say: "Fooled you!"

The video images below clearly show the four lights at the nose of the Manta Ray's chevron-shaped anterior, and how the two top lights turned on only directly overhead.

Below is a comparison between a commercial jetliner videotaped on 1 September 1999 and this Manta Ray to show how close the pattern of navigation lights and bright lights had become during the 1990s. Note that the top "beams" in the jetliner images are reflections off of the cigar-shaped fuselage, which the Manta Ray does not have. Therefore, the Manta Ray has to generate look-a-like reflections with a second pair of small lights. Without careful and detailed observation and comparison, it would be easy for an unsophisticated and sloppy skeptic to conclude that the Manta Ray was just a conventional aircraft and that Cornet and Hartwell were mistaken that they observed something that was unconventional. So were we being used as quality control observers in order for the owners and builders of this craft to perfect their camouflage? It was become more and more clear that this UAP activity over a populated area with nearby military bases and airports was trying to stay below human radar detection.

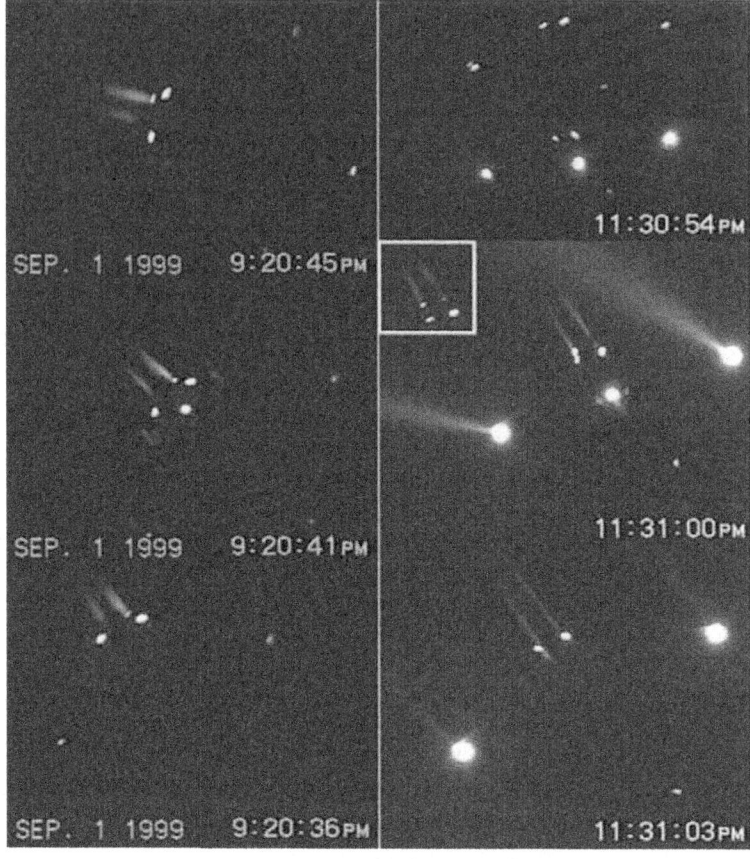

The sound the Manta Ray produced would fool any skeptic. When I analyzed it, all the distinctive characteristics of a synthetic alien sound were exposed: No white noise, multiple pure frequencies grouped together, and most importantly, Reversed Doppler! This sighting was the last one I had of the Manta Ray in the Wallkill River Valley. The next time I would see the Manta Ray would be east of the Hudson river northeast of Poughkeepsie (see Appendix I). Sightings shifted after the year 2001 to the eastern part of the UAP Corridor, and have continued over Long Island, NY.

The frequency spectrogram for the 4 May Manta Ray is given below. Note voice signatures, which appear as spikes or towers of grouped frequencies. They are not part of the Manta Ray sound:

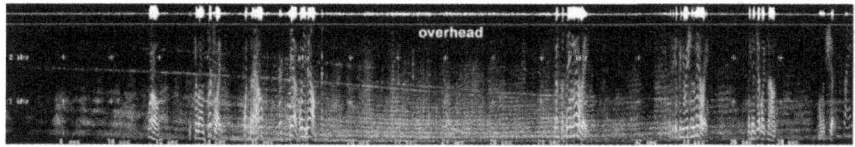

The first and most apparent characteristic in this spectrogram is the absence of white noise. The second red flag is the division of all the frequencies into 16-17 individual frequency bands that run parallel with one another, a signature of a sound synthesizer. The third indicator is that most of the frequencies fall below 2 kHz! Conventional aircraft engine frequencies usually range up to 4 kHz and even be as high as 8 kHz (and that doesn't include white noise, which can be as high as 10 kHz). For enlarged spectrograms, see pictures in Appendix I.

The Seventh example of Reversed Doppler: One of the most unusual recordings I made occurred on 22 September 1995. I went to Muddy Kill Lane on the eastern side of Thompson ridge, and positioned myself on the southern Muddy Kill part of that lane, about 1.5 miles northwest of Montgomery, NY. That locality provided an excellent view of the valley and air traffic going into and coming out of Stewart International Airport. However, it was an overcast night, and visibility was poor. I didn't think I would have any sightings. Then at around 5:43-5:44 pm, not long after sunset, I heard a distant sound of what resembled a C-5 galaxy taking off from Stewart Air Force Base (also Stewart International Airport) on the eastern side of the valley, 7.2 miles nearly due west of that airport. I turned on my camcorder and pointed it in the direction of the sound coming from the region of Stewart airport, hoping that I would see lights on the aircraft breaking through the low clouds as it got closer. The sound was loud and coming towards me, indicating that whatever it was, it was traveling west from the main runway at Stewart, which is oriented close to West-East. I never did see any lights. All I heard and recorded was this very loud sound of multiple engines throbbing in the moist night air, but from across the valley, which didn't make sense. The sound of a C-5 should not travel that far nor be that loud from that distance away. I have heard and recorded the sound of a C-5 taking off while I was next to the runway at Stewart, and it was not that loud. The low clouds and fog might aid sound propagation, but a C-5 would not be accelerating at full throttle for seven miles after lifting off the runway. And that didn't explain why all the dogs would start barking because of this noise.

I thought the C-5 should be closer, because the sound kept increasing in volume until it was so loud it vibrated the air around me. I heard the sound source pass over me, invisible in the overhead clouds, and I heard it fade into the distance to the west. Then I heard all the dogs in the area barking around me. The sound was so loud it scared the local dogs who were outside homes along Muddy Kill Lane. Soon after the sound abated, a pair of bright lights lifted off from the valley between Stewart and my location, and climbed silently into the clouds. I give a series of video frames below to record that anomaly. The steep climb is unlike that of any aircraft coming out of Stewart that didn't have rocket propulsion to assist its ascent. And such a take-off wouldn't have been silent.

When I got home I digitized and analyzed the sound on my computer. It was recorded in stereo. When the frequency spectrogram was forming on my computer screen, instead of seeing the indicators for a conventional turbofan engine sound, I saw the hallmarks of a synthetic sound from a craft. There was no white noise. The sound was broken up into a bundle of 24-25 discrete frequencies. But most important, the spectrogram displayed Reverse Doppler exquisitely and unmistakably. The frequencies had decreased to the point when the sound source reached me that the low bas vibrations reverberated and vibrated the air around me, scaring the local dogs, who barked incessantly.

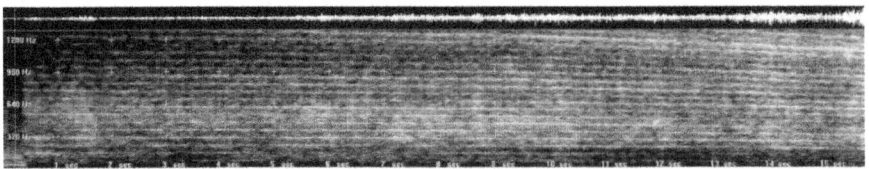

Above spectrogram record for 22 January 1997 displayed from left to right, even though sound propagated from right to left. All frequencies clearly decrease to the right, as the sound source got closer to the microphone.

Thus, even without seeing the craft or its lights, I was able to identify that the sound did not come from any conventional man-made aircraft. The paired lights climbing into the overhead clouds are shown below. The flashing and flaring was atypical of the lamps on conventional aircraft. What was this object if I was not an aircraft?

On 7 November 2007 Cornet was at Stewart International Airport (also Stewart Air Force Base) when a C-5 transport plane took off and circled around the airport before flying away. To the human ear it made a sound as similar to that made by the Manta Ray craft on 10 August 1996 and on 22 January 1997. The C-5 makes this type of sound only when taking off - possibly a result of all four engines at maximum throttle. As the C-5 came towards him, it displayed a normal Doppler sound: Frequencies rose until the plane was overhead, then the frequencies dropped as the C-5 moved away. He recorded the sound the C-5 made on his camcorder. Because this sound is only produced when a C-5 accelerates during take-off, for the Manta Ray to be confused with a C-5 taking off would mean that such an enormous aircraft had taken off from a nearby farm field nine miles from the airport - an impossibility. Therefore, was the Manta Ray trying to mimic a C-5? Its navigation light also mimicked those of a C-5.

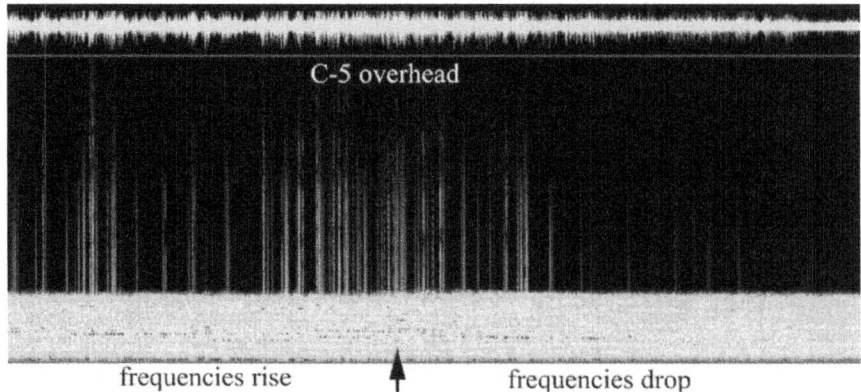

Introspection

In summary, the UAP phenomenon was different in the Wallkill River valley, which lay southwest of the Hudson River, than it was to the northeast of that river. Both areas had commercial and private air traffic, but only between Pine Bush and Stewart International airport (located adjacent to the Hudson River at Newburgh, NY) did the Black Triangles and Manta Rays attempt to mimic DC-9s and C-5s both with navigation lights and sounds. Part of the reason may be the high level of commercial and military air traffic over the UAP hotspot and ground magnetic anomalies between these two locations (I.e. Pine Bush and Stewart). I found through documenting the time, compass location or direction, and relative altitude of every light in the night sky from one or more observation localities (between July 1992 and August 1996) that UAP activity waxed and waned, seemingly in coordination with conventional air traffic volume or activity. As the number of airplanes increased, many times so too did the UAP activity. When few aircraft lights were seen (many documented with time exposures and recognized after analysis as not having or generating alien technology), very few sightings or time exposures of UAP lights were made. It became clear that these craft were using the "smoke screen" of air traffic to hide their nighttime activities. This relationship or correlation would surely enforce the skeptic's viewpoint that these UAP sightings were just misidentified conventional aircraft.

My method of distinguishing between IFOs and UAPs was the use of anomalous light traces on my time exposures. Frequently, UAP light traces had anomalous oscillations or movements to them, whereas conventional aircraft lights and strobes never had such oscillations, loops, right angle turns, flaring lights, tree-top flight paths, or changes in light number, position, and color. Or dive into the ground, or pass through trees without interacting with the matter in our world. Most man-made aerial vehicles cannot do these maneuvers or change their lights. If I couldn't identify some distinctive light anomaly or craft movement on my photographs, I counted that light trace as normal or conventional. Even though I probably underestimated the number of anomalous aerial vehicles on any active night, the highest volume of overall UAP activity for any night analyzed was 31%. Most active nights saw no higher than 10% UAP activity. Ironically, that is the value commonly given for valid UAP reports by UFOlogists.

During one night in 1992 at the dogleg bend on West Searsville Rd., I witnessed a commercial jetliner turn at the red beacon on the low ridge just south of my location (an established flight path across the valley to Stewart International airport). No sooner had the aircraft traveled east of the beacon and ridge, than a set of lights came through the gap in the ridge just north of the beacon and tailed the commercial plane just behind it but a few hundred feet lower in altitude. The jetliner was below 1,000 feet, because it was on approach to the main runway at Stewart. The UAP seemed to be flying in its radar shadow, or just low enough in altitude to be undetectable by the airport's radar or regional radar.

I was later told by Stewart radar operations that their radar system (radar tower located on top of a hill at the airport) did not keep records of information from the sector of the valley around Pine Bush where UAP activity is highest. Is this how the government covers up UAP hotspots so that there is no official record of all the activity skywatchers observed around the Pine Bush hotspot.

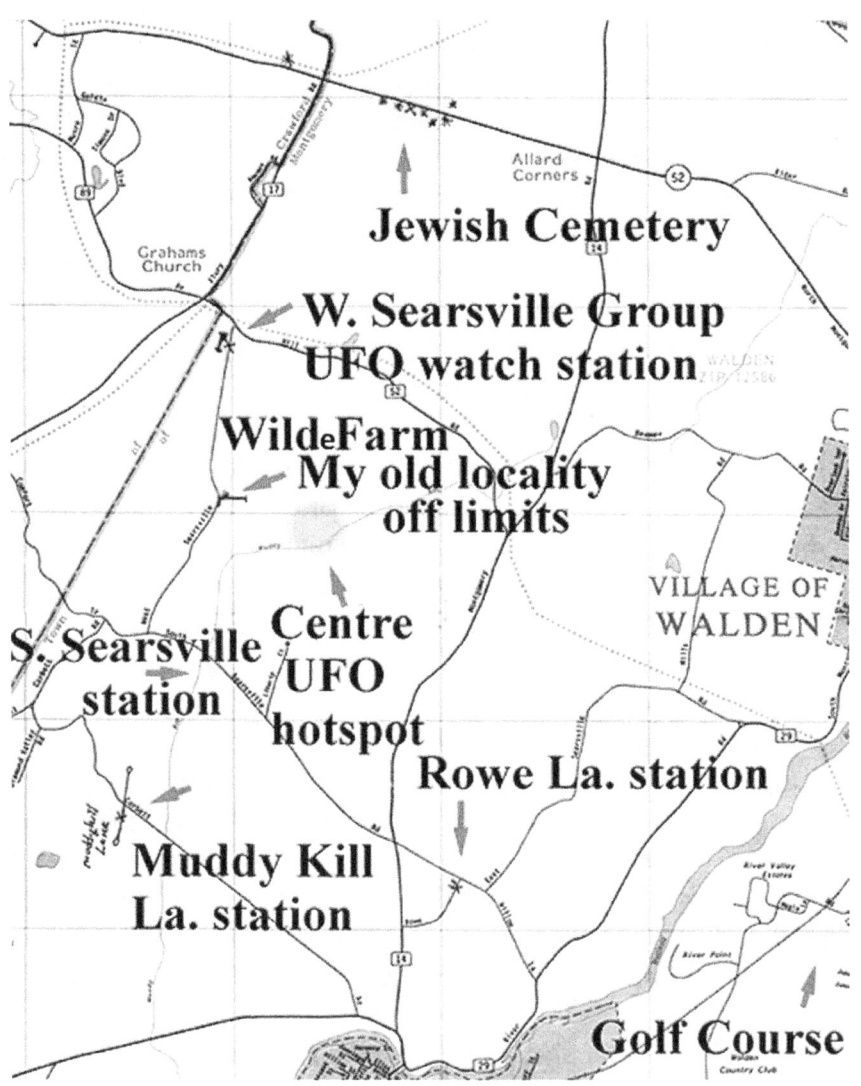

Chapter 6 Signs

The Contactee phenomenon has been going on for thousands of years (History Is Wrong: 2009; Twilight of the Gods: 2010, by Erich von Däniken). Descriptions and paintings with images of saucer-shaped or tetrahedral-shaped craft go back at least two thousand years, and if some of the descriptions in the Old Testament (Torah) represent contact and anecdotal accounts of sightings, the history of alien visitation is as old as the Bible. There are many books connecting UAPs with the Bible and religion: e.g.

- Rev. Barry H. Downing, Ph.D., 1968: The Bible and Flying Saucers;
- Steve, M.D., 1999: Extraterrestrial Contact, The Evidence and Implications;
- Curt Sutherly, 2001: UFO Mysteries, A Reporter Seeks the Truth;
- John E. Chitty, 2002: The Broken Bible, Picking Up the Extraterrestrial Pieces;
- Patrick Cooke, 2005: The Greatest Deception, The Bible UFO Connection;
- Sherry Shriner, 2005: Bible Codes Revealed;
- Steve Canada, 2006: Bible encoded Crop Circle Gods;
- Erich von Däniken, 2009: History is Wrong;
- Fred R. David, Ph.D., 2010: UFO Christianity Connection;
- Richard M. Dolan and Bryce Zabel, 2010: A.D. After Disclosure;
- Erich von Däniken, 2010: Twilight of the Gods;
- Gerardo Santos, 2012: UFO: Angels and the Mayan Calendar;
- A.R. Roberts, 2012: From Adam to Omega, An Anatomy of UFO Phenomena;
- Alan Dale Daniel, 2013: Tracking Ancient Legends, How the biblical Flood, Sky Gods, and UFOs Fit Into Prehistory;
- Michael J.S. Carter M.Div., 2013: Alien Scriptures: Extraterrestrials in the Holy Bible;

- Philip Chidi Njemanze MD, 2015: Igbo Mediators of Yahweh Culture of Life;
- Valerie Allen, 2015: UFO's and the Paranormal Are Normal;
- Rev. Barry H. Downing, Ph.D., 2015: UFO Revelation.

The work by Imbrogno and Horrigan (Celtic Mysteries: 2000) on the stone chambers, their association with strong magnetic anomalies, and the documentation of paranormal activity at many of these stone chamber sites in Dutchess, Westchester, and Putnam counties, (lower) New York State, USA, goes back at least 4,000 years (archaeological evidence) indicate controversial history of humans in North America long before Columbus or the Vikings.

Witnesses have described hooded beings (and even photographed them at night using flash) and have described witnessing ships "materializing at some of these magnetic stone chambers, implying the existence of interdimensional portals (Imbrogno and Horrigan, 1997 and 2000). Other magnetic anomalies in the Wallkill River Valley, at the center of the Pine Bush UAP hotspot, just 2.5 miles west of Walden, NY, for example, are also associated with interdimensional portals where craft have been seen, and even photographed diving through trees and into the ground **without** any accompanying explosion or wreckage (see pictures below and in Appendix I.

Above: 24 October 1994, 9:28 pm; photographed from West Searsville Rd. at Hill Ave., looking South (see more pictures in Appendix I). The arrow on the right shows location of the last light just above the ground.

Above: 12 November 1993, 9:04 pm; photographed from West Searsville Rd. at dogleg bend, showing triangular craft diving behind trees in foreground, without exploding (see more pictures in Appendix I).

On 5 October 1993 Brock and I were standing by the dogleg bend in West Searsville Rd. waiting for something to happen. I had infrared film in my Minolta XG7 SLR so that I could capture evidence for the temperature, and therefore plasma nature of their lights. I had taken one time exposure of a single light to the east over the tree row. The photo shows the light performing a twisting tight and loose spirals that resemble DNA (see pictures in Appendix I). Below is a negative image of the spiral. Was the purpose of this spiral to convey to us that those involved are EBEs, or Extraterrestrial Biological Entities with DNA similar to that of humans? Or were they just trying to show us the flight characteristics of their craft? I don't have an answer.

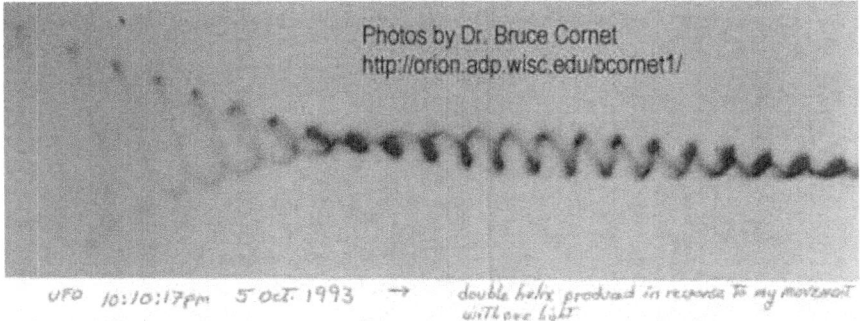

Just before 9:00 pm Brock alerted me to a pair of lights approaching from due west. The lights were too low to the ground and ridge to be a conventional aircraft, so I turned my camera around on top of the tripod and opened the shutter. I wanted to capture the light overhead, and they were headed directly towards me. But the pilot spotted us and apparently didn't want to give me any close-up shots. He may have read my mind and intentions. In hindsight, that is what seems obvious.

All of a sudden the lights stopped dead in place, and the craft hovered directly over the low ridge (Thompson's ridge) to the west. Then the lights (craft) turned right "on a dime" towards the south and began to move south along the top of the ridge. When those lights got to the red beacon on top of the ridge (used by commercial aircraft to orient in their turn towards Stewart airport to the east), they turned left to the east as if heading for the airport. But conventional aircraft cannot stop and hover, or turn "on a dime." The lights couldn't have been more than 300 feet above the valley floor (100 feet above that ridge). They moved east across the long field that parallels the east side of West Searsville Rd. When they got to the opposite side of the field and over the edge of a forest, they turned north (left) and started coming back towards me.

When they got to opposite from where I was standing, across the field and opposite from where the craft had stopped and turned over the ridge, the craft began to slow down. I opened my shutter at 8:57 pm.

By this point in its path, flashing strobes were apparent on either side of the back, and the front headlights were no longer visible. With my shutter open, the craft slowed to a stop over a field just beyond a tree row. It appeared to be a Black Triangle, but I could not be sure because it did not fly over me. The pilot had deliberately avoided Brock and me on the ground.

I kept the shutter open hoping to get a picture of it landing or disappearing into the ground just beyond that tree row. My camera had a motor drive attached, and all I had to do is release my finger from the electronic cable button and the shutter would close. Press the button again and the shutter would open. The film would be automatically advanced right after I closed the shutter. As fast as I could push, release, push, and release the button, I could take as many pictures as the roll of film would allow. The speed was less than a tenth of a second per shot advance, or to the next time exposure.

As I watched this craft hovering over the distant field, I had the thought that the time exposure being taken was long enough, and the flashing strobes would "burn a hole in the negative," because they were firing in the same place on the negative. I wasn't smart enough at the time to think that this was not my thought, but the alien's projected thought to make me close the shutter. So I responded out of fear, and decided I could refresh the film by advancing it to the next frame quickly enough before the craft could move. But the pilot was waiting! He was waiting for my move to close the shutter before he would move. To my surprise, as I released the shutter button and the film advanced, the craft dove directly into the ground and disappeared before the next film frame was ready to be exposed. It moved so fast that it is unlikely the craft just stopped on top of the ground and became invisible. It had to have gone through a portal or dematerialized into the ground. It was a classic Mexican Standoff, and the UAP pilot won! Although I didn't capture a picture of its lights intersecting the ground, that is what I visually saw happening between the tree trunks.

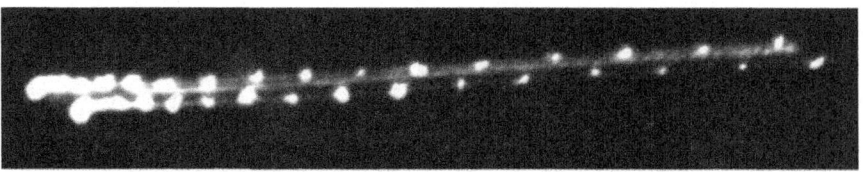

B&W infrared picture above. Photo of negative below.

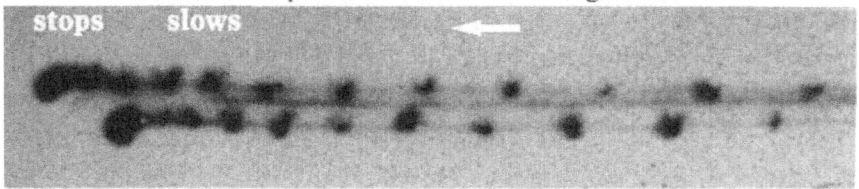

5 October 1993, 8:57 pm, West Searsville Rd. at bend. HSIR film.

In the picture above, not how the far left light is larger than the ones behind it, because the strobe was firing in the same place as the pilot hovered and waited for me to close my shutter. This photograph demonstrates that the lights stopped and remained in place, something only a helicopter or jumpjet could do. And it did it silently. Note also how the strobes along its path are irregular and not circular, indicating that the strobe light is not a bulb with a contained diameter. The strobe bursts are irregular, and therefore support the evidence on the IR film of hot plasma discharge.

Invisibility

Crystall (1991) rejects the idea of interdimensional space travel to explain the sudden disappearance of craft, which 28.3% of witnesses report in the Hynek et al. (1998) study. 62.2% of those who witnessed the craft disappeared, said it just vanished in thin air, while 37.8% said they saw it shrink until it disappeared.

Crystal says, "So many people seeing UFOs have reported that the craft "just vanished" or suddenly disappeared that the assumption developed among UFOlogists and the public at large that UFOs can dematerialize and must be "interdimensional," or "ultraphysical" or "Mexico City

." A host of terms has been created over the years to describe the capability of UFOs to move in and out of our familiar three-dimensional space from other dimensions or space-times or, to use one of the most exotic terms, parallel universes." (Crystall, 1991; p. 53-54).

By 1995 Michio Kaku had published his ground-breaking book in theoretical physics called, "Hyperspace, A Scientific Odyssey Through Parallel Universes, Time Warps, and the 10th Dimension." Stephen Hawking (1988) had published his book, "A Brief History of Time," on the possibilities for multiple dimensions and interdimensional travel. So the public, and especially UFOlogists and New Agers, were being educated to concepts and terminology that had previously existed mostly in closed scientific circles. The subject of interdimensional was new to UFOlogists, and as with any paradigm shift, there was resistance to change.

Crystall continues: I never accepted that theory. I still don't. If a person were looking at a fully lit UFO when it suddenly turned off all its lights, it would appear to vanish into thin air. But it would not have dematerialized: it would not be a psychic event. Its disappearance would be the operation of advanced technology – a kind of extraterrestrial "stealth" aircraft." Crystall, 1991: p. 54).

Crystall adds: "On some occasions when ships turned off their lights within fifty feet or so of us, we couldn't see the metal. It was disconcerting to know that a UFO was there right in front of us yet invisible. We even had some sightings at dusk, when the sky was still light, in which a UFO stopped in the sky above us with one large light on – and the bright light was all we could see!" (Crystall, 1991: p. 55).

On page 116 I talk about seeing a UFO with Fred Brock on 2 July 1992, where all we could see is a bright white light followed by a small red light. The day before I saw the same craft at the Harriman exit on the New York Freeway (south of Newburgh, NY). When it passed directly in front of a set of toll plaza lights supported at the top of a tall pole, I could see no structure connecting the lights. So I know for a fact that these visitors have a very sophisticated cloaking capability. Humans have been told this and shown this technology during the Contactee period between 1920 and 1960. Daniel Fry was shown this technology in July 1949 at the White Sands Proving Grounds in New Mexico by "Alan" (Ahlahn) not long after the Roswell crash and cover-up two years earlier (Fry's account is disputed by Ray Sanford, but occurs so early in the history of the phenomenon that Fry could not have known about this information unless he had been told by someone who already knew or suspected that this technology existed prior to Roswell). His account follows:

Fry took the seat nearest the door to the remotely-controlled sampling craft. Fry heard the voice of Alan directly inside his brain. Alan said, "I will now turn off the compartment light and activate the viewing beam." Fry said that the room became completely dark, and a beam of deep violet light came out of the projector above and behind him, and illuminated exactly the dimensions of the door in front of him. Under its influence the door became totally transparent. Fry said, "It was as though I were looking through the finest type of plate glass or Lucite window." (Good, 1998: p. 63).

Alan continued: "As you can see, the door has become transparent. This startles you because you are accustomed to thinking of metals as being completely opaque. However, ordinary glass is just as dense as many metals and harder than most, and yet transmits light quite readily. Most matter is opaque to light because the photons of light are captured and absorbed in the electron orbits of the atoms through which they pass. This capture will occur whenever the frequency of the photon matches one of the frequencies of the atom. The energy thus stored is soon re-mitted, but usually in the infrared portion of the spectrum, which is below the range of visibility, and so cannot be seen as light. There are several ways in which matter can be made transparent, or at least translucent.

"One method is to create a field matrix between the atoms which will tend to prevent the photon from being absorbed. Such a matrix develops in many substances during crystallization. Another is to raise the frequency of the photon above the highest absorption frequency of the atoms. The beam of energy which is now acting upon the metal of the door is what you would call a 'frequency multiplier.' The beam penetrates the metal and acts upon any light that reaches it in such a way that the resulting frequency is raised to that between the ranges which you describe as the 'x-ray' and the 'Cosmic Ray' spectrums. At these frequencies, the waves pass through the metal quite readily. Then, when these leave the metal on the inner side of the door, they again interact with the viewing beam, producing what you would describe as 'beat frequencies' which are identical with the original frequencies of the light. As a rough analog, the system could be compared to the carrier wave of one of your radio broadcasting stations, except that the modulation is applied 'upstream' as it were, instead of at the source of the carrier." (Good, 1998: p. 63-64).

The question that Crystall (1991) raises is whether disappearing acts of UAPs are a result of interdimensional movement, or a result of invisibility due to advanced technology. It is probably that when a craft's lights blink out and it appears to disappear that invisibility is the modus operandi. But when a craft with its lights still on passes through a dense forest and disappears into the ground, as the Manta Ray did before four witnesses on 24 October 1994 and two witnesses on 6 May 1995 (see page 65 and 69), was it "dematerializing" or changing its atomic vibrations so it could pass between atoms of our world, or did it enter a parallel dimension? The craft (probably a Black Triangle) that waited for me to advance my camera film so that it could dive out of sight on 5 October 1993 (see page 170), did it dive into the ground and go through a portal or Stairgate, or did it stop just above the ground as it turned out its lights? Because so many craft have been seen emerging or lighting off from fields and forests immediately around the "Indian Mound," it is hard for me to believe that they were all just hovering above the fields of wheat with their cloaking devices operating. The sky glyph of a "vortex" photographed on 28 April 1993 over the "Indian Mound" (described below) is certainly suggestive of a portal or Stargate in that location. The unstable magnetic field there is also indicative of something unnatural beneath the ground. The association between UAPs plus paranormal activity, and magnetic anomalies in southeastern New York, many associated with stone chambers, has also been recognized by Imbrogno and Horrigan (2000) as locations where these craft and phenomena originate.

Photographs taken on 28 April 1993 during five sightings of UAPs include images of a strange building-like structure within the forest across from the dogleg bend in West Searsville Rd. This structure is near the "Indian Mound," appears bluish gray in the time exposures, and appears to be deliberately lit by lights. It resembles an aircraft hangar. But this structure does not exist in that forest, at least not in our three dimensions. Could the photos have captured a parallel dimension "showing through" under a certain type of illumination? Obviously more scientific observation and testing needs to be dome there.

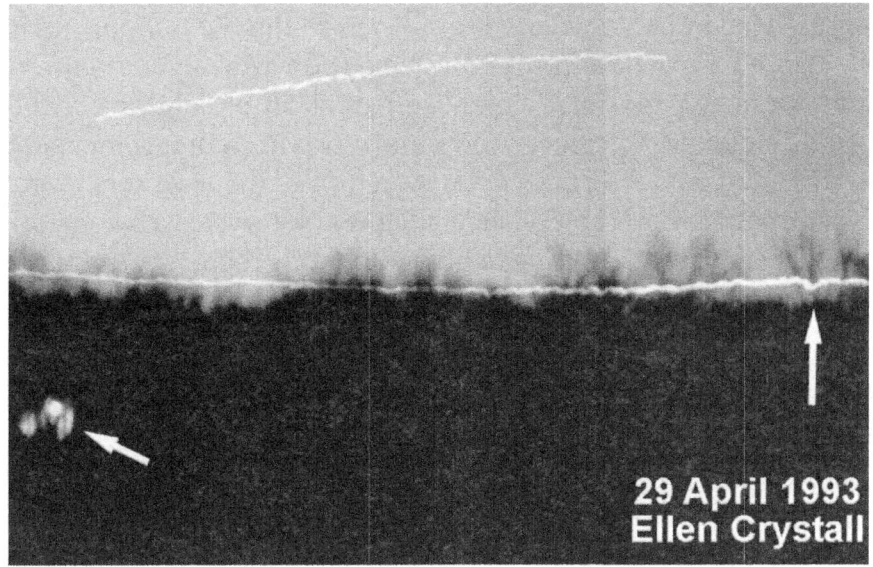

The "vortex" picture below taken on 28 April 1993 between 9:01:48 pm and 9:03:50 pm (2 minutes and 2 seconds) is comprised of two time exposures. The white light appeared suddenly below the distant tree line, and rose at about a 45 degree angle above a farm field on the Owen Dairy Farm. The craft generated a vortex-like image in the middle of the first time exposure, which collapsed and disappeared. Then at the beginning of the second time exposure, the main light flared to the extent of producing an orange-brown halo of nitrous oxide plasma gas at it rose suddenly. Then the pilot dimmed the main light and turned on red and white strobe lights, and flew away due South. The vortex image represents a holographic display, rather than a portal into another dimension, although the History Channel gave a presentation on 26 March 2008 called, UFO Vortexes, in which the UFO Hunters speculated that this might be an example of an interdimensional portal or wormhole opening up.

If you examine an enlargement of that "vortex," you will see that it is comprised of a central white light surrounded by 10 rings, each one having a colored sphere somewhere in it, which could represent a planet. In other words, the "vortex" could be a stylized representation of our solar system. Then to the right of our sun is what appears to be a blue orb, which may be a symbol for a neutron star. There is a four-digit hand-like extension of light below the "hand," possibly indicating energy emissions from that companion star which would affect our solar system. Were our visitors giving us astronomical information indicating that we exist in a binary star system? Or were they trying to holographically represent the Sirius binary star system from which they came?

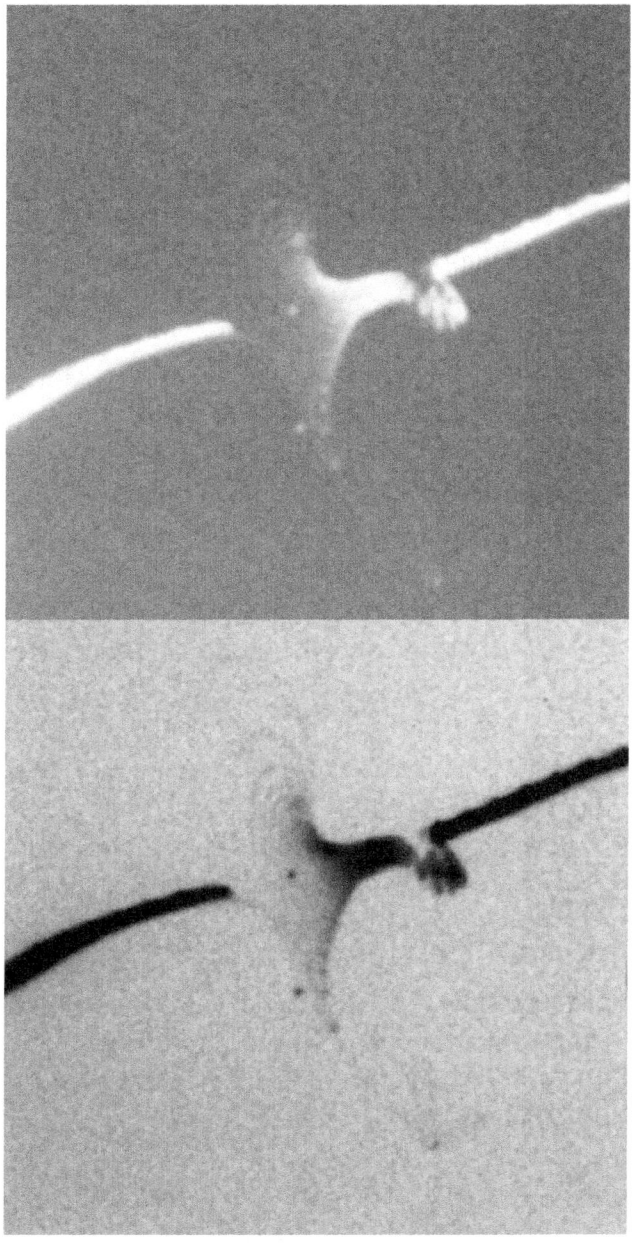

The "vortex" picture above represents a sky glyph or image "painted" on my photographic canvass by our visitors. Even though Crystall and I did not recognize the significance of this picture at the time, as more studies, research, and revelations were made, the pieces to the puzzle grew in number. The research done by Hynek, Imbrogno, Crystall, Horrigan, and myself provide numerous clues to what is going on and why. Without these different perspectives and data, all of my time exposure would just be pictures of lights in the night sky. Without providing a photographic canvass for symbols and sky glyphs to be "painted," some of the most important evidence would be missing.

The most common sky glyph or symbol produced on my photographic canvass was the 'S' sign or symbol. Seven crude to perfected 'S' signs were captured on film between September 1992 and October 1994, a period of 25 months. Five of those are significant in their distinctiveness, with the ones on 8 April 1993, 2 June 1993, and 16 October 1994 being unmistakable. The 8 April sky glyph performed by the Manta Ray is the most perfected, and resembles the shape of the Superman logo 'S'. That performance is described in detail on pages 50-54.

There are many words in the English language that 'S' could refer to, such as Sign, Signal, Symbol, Super, Sun, Savior, Self, Sapiens, Serpent, Satan, Secret, and the list goes on. Because the crafts typically departed to the South, one might suspect that 'S' could refer to South. The 29 September 1995 performance of the Manta Ray began with a Sinuous Snake-like motion of the craft as it came towards me in a shallow climb. Thus, 'S' could stand for Sinuous or Snake. That reference might conjure up images of Satan to those who think the devil rules the air of our planet and who think abductions are the work of the devil.

The Contactee Daniel Fry tells us in his "White Sands Incident" that began in July 1949 that's there are common symbols between intelligent life forms that evolved on different water planets. The extraterrestrial "Alan" (Ahlahn) said to him that the symbols of the tree and the serpent are not unique to Earth. They are natural or common ones relating to the origin of life in the waters of a planet.

The symbol of the serpent is not a reference to the snake, but to the motion of water, the waves of the sea. If you can accept his information, the sinuous undulations of the Manta Ray are therefore not a reference to a serpent, but to the origins of plant and animal life in the sea. The tree is described by Alan as a common symbol of life. Plants arose in the sea back in the Ordovician Period of Earth History, 400 million years ago, then spread onto land back in the Silurian Period – another 'S' word, and then developed into the first trees back in the Late Devonian Period 350 million years ago: example, *Archaeopteris*. The tree symbolizes the overall evolution of life on a water planet from the sea, to the atmosphere, and finally up into space (Good, 1998: p. 69). Interestingly, my education in college took me into the science of paleobotany (the study of ancient or fossil plants) at the University of Connecticut under Professor Henry N. Andrews (1970-1972) and later into geology where I learned about magnetic pole reversals and studied them in Triassic-Jurassic cores at the Lamont-Doherty Earth Observatory of Columbia University (1988-1993).

One could take this symbolism even further with the mention of the Late Devonian *Archaeopteris*, a medium-size tree with fern-like fronds. There is a similar name in vertebrate paleontology: *Archaeopterix*, which is the name for the oldest known bird from the Late Jurassic Period 150 million years ago. Archaeopterix retains many features of its reptilian ancestry, such as a long tail and teeth, combined with wings. Birds retail reptilian scales on their feet. Its flight through the air might also be compared to a sinuous motion – that of a winged serpent. In the Postclassic period (900-1519 AD) the Aztec god Quetzalcoatl was called the "Feathered Serpent," implying "winged serpent," because it could fly (even if only by means of technology). It has other names in Mayan (Kukulkan and Gukumatz) that also roughly translate to "feathered serpent."

In the Old Testament (Torah) the serpent in the Garden of Eden is linked to consciousness, awareness, and knowledge (*sapiens*), and as in "as cunning as a serpent." Its association with the tree represents life evolving into consciousness or intelligence: Hence the birth of NAR. Nar is the root word for narcosis, narcotic, and narcolepsy, meaning "of the mind." Unless all intelligent life forms are evil as interpretations of the Garden of Eden serpent imply, the tree and serpent have been grossly misinterpreted. Consciousness leads to questions, which lead to gnosis or knowledge, and a natural progression and evolution of life from the Sea to the Sky, and from ignorance to knowledge and awareness. Some would say that this was undoubtedly a part of God's plan.

On 2 June 1993 at 11:11 pm Crystall and Cornet witnessed a single white light blink on and rise slowly from just on top of the trees over the "Indian Mound" at the center of the UAP hotspot (2.5 miles west of Walden, NY). This event is described on page 9 (see pictures in Appendix I). The shape of the light path captured on four time exposures is that of a lazy 'S'. At the top of the first time exposure a small red orb emerged from the white light and moved quickly downward towards the ground where it looped to form a crude, vertical figure eight design just above the tree tops. A vertical figure eight symbol represents **life**, whereas a figure eight symbol on its side represents **eternity**.

Here we have two symbols "painted" on my photographic canvass, one an 'S' that probably represents both the wave for water and/or the Serpent, meaning NAR or consciousness (implying intelligence), and the other a figure eight that represents life! How appropriate and non-threatening for an extraterrestrial visitor to make its presence known, and to give us symbols that tell us other intelligent life is present in the Universe. These symbols and lights also tell us that life form has reached our planet. ET is here! Skeptics probably won't be satisfied until these visitors land on the White House lawn, and will dismiss and reject this evidence because it does not meet their fear-based litmus test for "proof." See Appendix I for colored images.

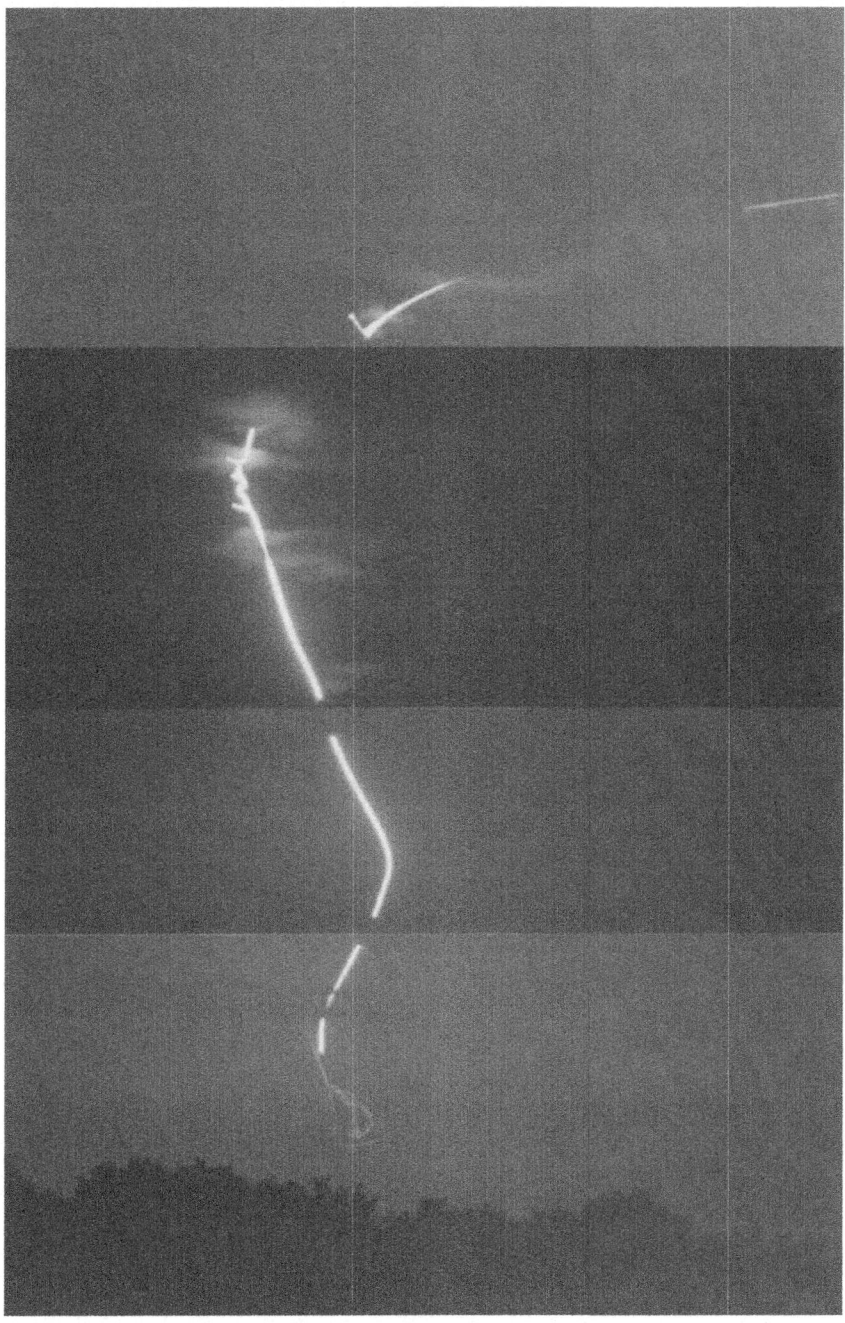

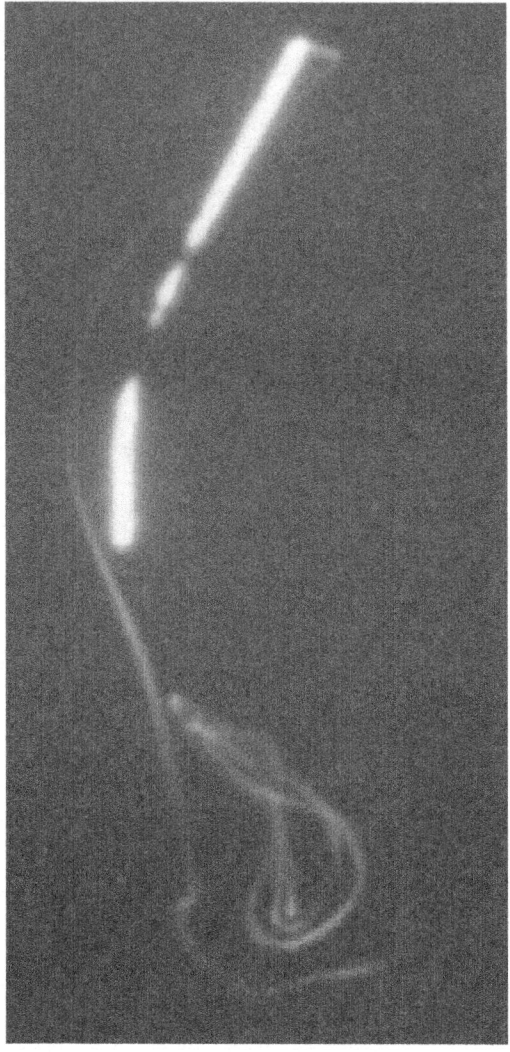

Witnessed and photographed by Cornet and Crystall on 2 June 1993 at 11:11 pm at the dogleg bend in West Searsville Rd., Montgomery, NY.

Bob Tarantino of Whitehall, PA, who became interested in the Pine Bush phenomenon in the mid-1990s, contacted me and we began sharing experiences. He is an electrical engineer. We both have had synchronistic encounters with UAP, and Bob has studied religions and symbols and archetypal meanings for decades. He recently had a profound insight into the meaning of that looped, vertical figure eight design just above the tree tops. With my focus on reversed Doppler sound signatures produced by the UAP in the valley, he recognized this sky glyph as a reversed treble clef symbol in music. His comparison below is not only very close to the treble clef reversed, but it also could symbolize the reverse Doppler sounds I had recorded in the field. So, was this a clue given to us by our non-human residents that they produced the above sky glyph image of a lazy S symbol rising into the overhead clouds?

 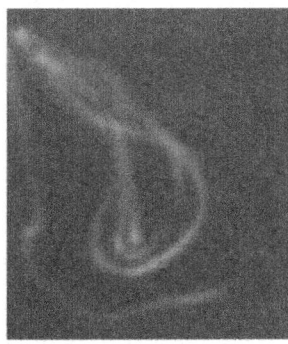 

treble clef   reversed

Contactee Information: Rules of Contact

On the subject of direct contact, which many people demand before they will accept the presence of ET, Fry asked "Alan," an alleged extraterrestrial (good, 1998: p. 72): "'Why don't you just set a small landing craft down on the White House lawn some morning, ask for worldwide communications facilities, and give your information and advice to the whole world at once?' asked Fry. Such a simple solution is only wishful thinking on your part, replied Alan. 'If you think a little, you will see that there are many reasons....why such a course would not be successful.'"

"...If we were to appear as members of a superior race, coming from 'above' to lead the people of your world, our arrival would

seriously disrupt the ego balance of your society. Tens of millions of your people, in their desperate need to avoid being demoted to second place in the universe, would go to any lengths to disprove, or simply deny, our existence. If we took steps to force the acceptance of our reality upon their consciousness, about 30 percent of the people would insist upon considering us as Gods, and would attempt to place upon us all responsibility for their own welfare. This is a responsibility we would not be permitted to assume, even if we were able to discharge it….Most of the remaining 70 percent would adopt the belief that we were planning to enslave their world, and many would begin to seek means to destroy us. If any great and lasting good is to come from our efforts, they must be led by your own people, or at least by those who are accepted as such..." (Good, 1998: p. 72).

The Brooking Report Effect on Government Policies

In December 1960 the Brookings Institute completed a study commissioned by the U.S. government concerning the implications of human travel in space and the consequences of meeting an extraterrestrial civilization. This institute is a high level think tank founded in 1916 and located in Washington, D.C. The report coming out of this study made a fear-based conclusion that there would be social disintegration if humanity came in contact with an intelligent extraterrestrial life form.

"The implications of a discovery of extraterrestrial life: Recent publicity given to efforts to detect extraterrestrial messages via radio telescope has popularized – and legitimized – speculations about the impact of such a discovery on human values. It is conceivable that there is semi –intelligent life in some part of our solar system or highly intelligent life which is not technologically oriented, and many cosmologists and astronomers think it is very likely that there is intelligent life in many other solar systems. While face-to-face meetings with it will not occur within the next twenty years (unless its technology is more advanced than ours, qualifying it to visit earth), artifacts left at some point in time y these life forms might possibly be discovered through our space activities on the Moon, Mars, or Venus. If there is any contact to be made during the next twenty years it would most likely be by radio — which would indicated that these beings had at least equaled our own technological level.

"An individual's reactions to such a radio contact would in part depend on his cultural, religious, and social background, as well as on the actions of those he considered authorities and leaders, and their behavior, in turn, would in part depend on their cultural, social, and religious environment. The discovery would certainly be front-page news everywhere; the degree of political or social repercussion would probably depend on leadership/s interpretation of (1) its own role, (2) threat to that role, and (3) national and personal opportunities to take advantage of the disruption or reinforcement of the attitudes and values of others. Since leadership itself might have great need to gauge the direction and intensity of public attitudes, to strengthen its own morale and for decision making purposes, it would be most advantageous to have more to go on than personal opinions about the opinions of the public and other leadership groups.

"The knowledge that life existed in other parts of the universe might lead to a greater unity of men on earth, based on the "oneness" of man or on the age-old assumption that any stranger is threatening. Much would depend on what, if anything, was communicated between man and the other beings: since after the discovery there will be years of silence (because even the closest stars are several light years away, an exchange of radio communication would take twice the number of light years separating our sun from theirs), the fact that such beings existed might become simply one of the facts of life but probably not one calling for action. Whether earthmen would be inspired to all-out space effort by such a discovery is a moot question. Anthropological files contain many examples of societies, sure of their place in the universe, which have disintegrated when they had to associate with previously unfamiliar societies espousing different ideas and different life ways; others that survived such an experience usually did so by paying the price of changes in values and attitudes and behavior.

"Since intelligent life might be discovered at any time via the radio telescope research presently under way, and since the consequences of such a discovery are presently unpredictable because of our limited knowledge of behavior under even an approximation of such dramatic circumstances, two research areas can be recommended:

"...Continuing studies to determine emotional and intellectual understanding and attitudes – and successive alterations of them if any – regarding the possibility and consequences of discovering intelligent extraterrestrial life unfamiliar events or social pressures. Such studies might help to provide programs for meetings and adjusting to the implications of such a discovery. Questions one might wish to answer include: How might such information, under what circumstances, be presented to or withheld from the public for what ends? What might be the role of the discovering scientists and other decision makers regarding release of the facts of discovery?" (Brookings Report, 1960: p. 182-184).

The Brookings Report had a significant impact on NASA policies of disclosure, especially towards the discovery of anything on the Moon or Mars that could be interpreted or misinterpreted as alien artifacts. Richard Hoagland's books and video productions on the "Face" on Mars and on the "Cydonia Complex" of alleged ruins and monuments of an ancient Martian civilization are a case in point (e.g. The Monuments of Mars: A City on the Edge of Forever, by Richard Hoagland, 1987; Dark Mission, The Secret History of NASA, by Richard Hoagland and Mike Bara, 2007/2009). Plausible deniability has been largely the official response of NASA, and the U.S. military and government have gone to great lengths and effort to minimize and defuse any potential "threat" caused by UAP sightings and crashed alien craft, exemplified by the Roswell incident (e.g. Crash at Corona: The U.S. Military Retrieval and Cover-up of a UAP, by Don Berliner and Stanton T. Friedman, 1992; The Day After Roswell, by Col. Philip J. Corso (Ret.), 1999; The Roswell UAP Crash, What They Don't Want You to Know, by Kal K. Korff, 2000; Operation Roswell, by Kevin D. Randle, 2002; The Roswell legacy, The Untold Story of the First Military Officer at the 1947 Crash Site, by Jesse Marcel, Jr., and Linda Marcel, 2008). The Brookings Report goes on to say that professions most affected by disclosure of alien existence or worse, an alien presence on Earth, would be scientists and theologians, who represent opposite extremes in explaining our existence and how life began. On the one hand, the history of Ufology, Contactees, and Abductees would suggest that scientists and theologians are to a large extent in denial and have set their goal posts for validation or confirmation of an alien presence too high. On the other hand, if extraterrestrials are present, they have gone to great efforts and extent to remain hidden, except to a select few humans at any one time or place. So, which is it? In order to understand the dynamics and problems involved, I am going to refer you to the explanations of others who have considered the subject and its ramifications more than most scientists and philosophers.

A Possible Extraterrestrial Strategy for Earth

James Deardorff summarizes a cogent argument why extraterrestrial visitors remain just out of reach, and beyond sciences ability to validate. He reviews arguments "which hold that our Galaxy is nearly saturated with extraterrestrial life forms, that our existence requires in hind sight that they were and are benevolent toward us, and that our lack of detection of them or communication from them implies that an embargo is established against us to prevent any premature knowledge of the." (Deardorff, 1986, A Possible Extraterrestrial Strategy for Earth: Summary). If the history of extraterrestrial contact and sightings is even 10% accurate, his hypothesis has merit. The manner in which humans have reacted to this history and evidence, through skepticism, denial, and censorship gives credence to an inconsistency between the evidence and societal and national response, such that "any sudden lifting of the embargo in a manner obvious to the public would cause societal chaos and possibly touch off a nuclear exchange, while any communications received via radio telescope would likely be either quickly confiscated by government agencies and not revealed to the public, or heavily censored. The inconsistency is that the advanced civilization should be expected to have planned some other strategy, if it is actually benevolent, experienced and intelligent.

"It follows that any embargo not involving alien forces must be a leaky one designed to allow a gradual disclosure of the alien message and its gradual acceptance on the part of the general public over a very long time-scale." (Deardorff, 1986, A Possible Extraterrestrial Strategy for Earth: Summary).

Deardorff elaborates on how government agencies might react to enforce an embargo: "As pointed out by Bracewell (1975), were the extraterrestrials to communicate with us via radio waves from space, or via a probe sent to broadcast while circling the Earth, strenuous attempts would no doubt be made by the government agency concerned with national security of the country detecting the communications to keep them top secret. It would be quite naïve to reason otherwise. The secrecy would be in hopes of obtaining some military or economic advantage over other nations from the decoded information, especially over other nations deemed unfriendly (This would be especially true for alien craft crash retrieval). Even if the detection of the incoming communications by a non-government research group were announced over the news media, the government could easily disclaim it as an erroneous report or hoax the next day (as the U.S. government did with the Roswell crash debris).

He goes on to say that our failure (SETI) to observe some of their radio communications must mean that they have some other and better method of communication than electromagnetic waves in the "water hole" frequencies. (Deardorff, 1986, A Possible Extraterrestrial Strategy for Earth: p. 17). "The extraterrestrial communication could be emplaced in a manner easily accessible to the general public but in a form not acceptable or believable to scientists. Government agencies, upon advice from scientists, would then take no actions, and the embargo would more or less remain intact. Awareness of what was taking place would then proceed very gradually – no faster than humankind in general was inherently prepared to accept the extraterrestrial messages." (Deardorff, 1986: p. 98).

In a very real sense this is what has been happening in the Hudson Valley and Pine Bush areas. Even with six and journal articles books published since 1986 o the phenomenon in the UAP Corridor, scientists have remained distant and even hostile to the evidence (see Imbrogno and Horrigan, 2000, on their reaction to the stone chambers). The media has not been much better, taking an interest in the extraordinary claims, but balancing those claims with well-rehearsed skepticism in the 1980s and 1990s. Even when the Sightings and Encounters field crews came to Pine Bush to do stories and shoots about Crystall and me, and when faced with an extraordinary series of performances by ships for their cameras (e.g. on 29 April 1993), they failed to show that footage on their TV programs. Instead, they used a single captured frame of the Manta Ray, and with special effects, showed it rising up from behind some trees. Thus, even the Media acted to censor and reduce the information content of communications from our visitors.

By the 21st Century, the use of skeptics has fallen off considerably with the third wave of UAP sightings. The first wave in the early to mid-1980s was hardly mentioned in the news, and when it was, equal air time was given to alternative but false explanations. During the second wave in the early to mid-1990s, there was considerably more media attention as TV shows such as Sightings, Encounters, and X-Files became popular. There was always the skeptic's view or opinion presented, but the information on UAP sightings and even alien abductions was getting out to the public.

The third wave, underway in the first decade of the 21st Century, has received even more media attention and less government censorship as TV shows on the History Channel and Discovery Channel have served as documentaries, educating and informing the public of what transpired in the first two waves and earlier in the 20th Century. The History Channel, UFO Hunters, has been very popular, with little of the type of "canned" skepticism that shows in the 1990s had. I was featured in a show called UFO Vortexes that was first aired on 26 March 2008, while Phil Imbrogno and Marianne Horrigan talked about the stone chambers in the second half of the program. My time exposure of a UFO producing a vortex-shaped sky glyph over the "Indian Mound" was the centerfold on UFO Vortexes (see description and pictures above on pages 176-177, and in Appendix I: 29 April 1993). Ironically, that sky glyph was captured on film the night before a Black Triangle and the Manta Ray made spectacular runs at the Sightings cameras on that night, fifteen years before the History Channel crew visited the same area.

Even though the History Channel hired a below ground imaging company, which used a GPR (Ground Penetrating Radar) unit to image what is below the Beth Hillel Jewish Cemetery, and revealed for the first time an edge of an entombed Triangular Probe, they presented little or no information on that discovery on the TV program. Was it still too potentially explosive a revelation? Did it leave no room for plausible deniability? As with a full term pregnancy, if you don't like what comes out, you can't put it back! Thus, was it not the right time for such a revelation? Will this book help pave the way for even more spectacular documentaries during the fourth wave after 2016? See the Chapter below on Evidence

For An Underground Magnetic Focusing System, An Entombed Triangular Alien Robotic Probe.

Inflation Theory and ET Visitation

Deardorff, Haisch, Maccabee, and Puthoff (2005) update the implications for new theories in theoretical science: "It has recently been argued that anthropic reasoning applied to inflation theory reinforces the prediction that we should find ourselves part of a large, galaxy-sized civilization, thus strengthening Fermi's paradox concerning "Where are they?" Furthermore, superstring and M-brane theory allow for the possibility of parallel universes, some of which in principle could be habitable. In addition, discussion of such exotic transport concepts as "transversable wormholes" now appears in the rigorous physics literature." (Deardorff et al., 2000: Abstract).

A statement such as this coming from seasoned and well-respected scientists in their communities during the third wave of recent extraterrestrial contact is a major shift in thinking and attitude. It signals a relaxation in censorship by scientists and government that may be in large part responsible for television shows such as UFO Hunters, that explore past UAP sightings with scientific vigor, something that could not have happened two decades earlier.

Deardorff et al. (2005) state: "The extraterrestrial hypothesis (ETH), that intelligent life from 'elsewhere' in the universe could be visiting Earth, has become less implausible through suggestions that the velocity-of-light constraint – 'they can't get here from there' – is not as restricting as had been assumed previously. This restriction has its origin in the special theory of relativity, which we do not question. However, within the context of general relativity (GR) there are three approaches which may permit legitimately bypassing this limit, given sufficient advanced (perhaps by millions of years!) knowledge of physics and technology.

"One approach popularized by Thorne and Sagan concerns the possibility of wormholes, or cosmic subways, a form of shortcut through the space-time metric [Morris and Thorne, 1988]. Using the standard GR as a basis, certain mathematical requirements for transversable wormholes have been derived and published in the scientific literature and it appears that there is the possibility of engineering a wormhole metric, at least in principle [Visser, 1996]." (Deardorff et al., 2005: p. 44).

These authors go on to call for open scientific research on UAP sightings, with special attention to high quality UAP reports and data that yield indications of ET intelligence and strategy. "Given the highly advanced ET science and technology to be expected in considerably older civilizations, coupled with the many observational reports since WWII of highly advanced technology seemingly operating at will within Earth's skies, it is only logical to search for evidence of ET visitations in at least a fraction of the ongoing, unexplainable reports popularly referred to as 'UFO sightings.' Reluctance to do so could result in our failure to realize that observations of 'genuine' ET visitations have been occurring." (Deardorff et al., 2005: p. 44-45). In addition, "scientists should not feel reluctant to study these inasmuch as the [Condon] Report's executive summary stated that 'any scientist with adequate training and credentials who does come up with a clearly defined, specific proposal for study [of UFO reports] should be supported.'" (Deardorff et al., 2005: p. 46).

They attempt to explain why, from the viewpoint of the scientific community and society as a whole, the relative rarity in time and space of convincing sightings and because of the limited number of credible witnesses, the reality of an ET presence was not totally obvious or convincing. It took decades of recording and tabulating sighting reports by three different groups in three separate databases (NIDS, MUFON, and Larry Hatch records) and their compilation by Dr. Colm Kelleher of NIDS to give scientists a realistic picture of just how often sightings of Black Triangles have been made over the United States during a 13 year period from 1990 to 2003.

This may be the case for most sightings reports, but it is certainly not the case for the Hudson Valley sightings detailed by thousands of witnesses in a relatively confined region of three counties in New York State over a period of more than three years! And that explanation is also not relevant to the hundreds of recorded and documented close encounters (see Appendix I: 131 sightings) I have had over a period of eleven years, with dozens of sightings occurring in the same location at the dogleg bend in West Searsville Rd. These sightings include three huge holographic images projected over rural areas between Walden, Montgomery, and Pine Bush, something that humans today would find difficult to impossible to duplicate:

Vortex or Binary Solar System (sun with a neutron or dark companion star) over a hay field and Indian Mound on 29 April 1993; subject of the History Channel program on UFO Vortexes, which aired on 26 March 2008.

Unconventional Aerial Phenomena

Golden "Egyptian" Pyramid over Indian Mound with splay of golden light radiating up through it into the sky (Pyro-mid or Fire in the Middle) on 31 May 1995.

Unconventional Aerial Phenomena

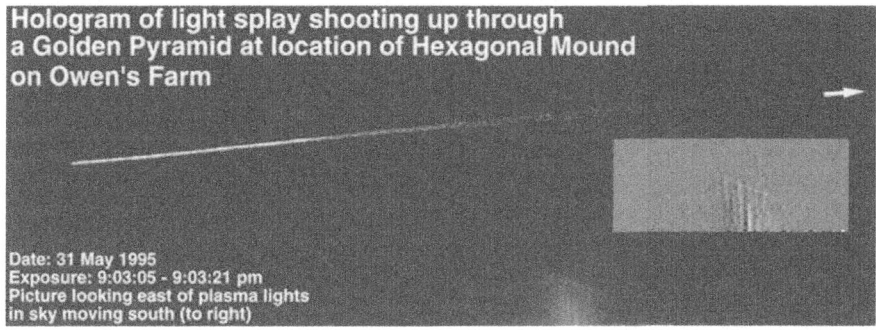

Green round "Greek" Temple on top of corn in field below it on 17 July 1996; green beams of light on left coming from the Manta Ray, which had flown over us just before this picture was taken.

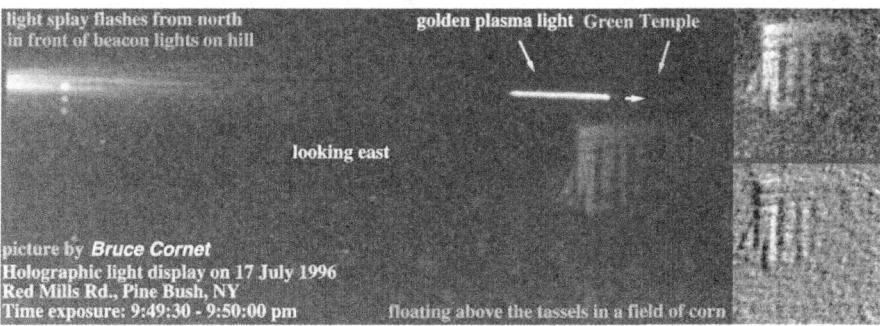

If these visitors had not cooperated with the rigorous demands of the scientific method by providing repeated appearances, performances, and sky glyphs at the same locations, and allowing tests to be performed to rule out competing interpretations and false concepts, as a scientist I would not have continued my investigations and experiments for as long as I did. I regard the Hudson Valley-Pine Bush evidence as being one of the largest scientific leaks of ET information that didn't just sneak under the government's radar and censorship, but lit up the sky and made headlines in plain sight! And this leak involved at least three scientists: Dr. J. Alan Hynek (astronomer), Philip J. Imbrogno (astronomer) and myself (geologist and paleontologist). Also involved indirectly in this research through study of the data were Dr. Bruce Maccabee (physicist), Dr. Colm Kelleher (molecular biologist), and Scot L. Stride (JPL engineer). To date more than nine books (including this one) have been written on the subject, plus numerous journal articles, qualifying it for consideration by the scientific community and society in general as evidence for 'genuine' ET visitation.

- Night **Siege**, by Hynek, Imbrogno, and Pratt (1986/1998;
- **Silent Invasion**, by Crystall, 1991;
- **Contact of the 5th Kind, The Silent Invasion Has Begun**, by Imbrogno and Horrigan, 1997;
- **Celtic Mysteries**, by Imbrogno and Horrigan, 2000;
- **The Pine Bush Phenomenon**, by Vincent Polise, 2005;
- **UFOs Over New York, A true History of Extraterrestrial Encounters in the Empire State,** by Preston Dennett, 2008;
- **Interdimensional Universe**, by Imbrogno, 2008;
- **In The Night Sky**, by Zimmermann, 2013.

DNA Symbols

The symbol 'S', which may be a sign for "water" via the meaning of Serpent (wave), was performed seven times beginning in September 1992, but some of the sky glyphs were crude and imperfect, while others were unmistakable. However the symbol for the DNA double helix was first created as a sky glyph on 5 October 1992, and again on 7 October 1992 only 15 days later. The first unmistakable 'S' sign was not performed until April 1993, and the first figure 8 symbol for life not until June 1993. Communication began with the Serpent symbol for "water," then "DNA, and finally "life." What did the symbol for DNA look like, and how was it produced?

On 5 October a single white light was spotted to the east of the dogleg bend in west Searsville Rd. (see below, and Appendix I,I had infrared film in my SLR camera. I took one time exposure at 10:10 pm. The picture shows a small hot light performing a linear series of alternating tight, then loose spirals. The first impression I had when seeing this image was that it represented the DNA helix, but it was too simplistic to convince skeptics. It was just a spirally light, that by itself implied non-human technology and non-human intelligent control, but that was apparently not the desired intent.

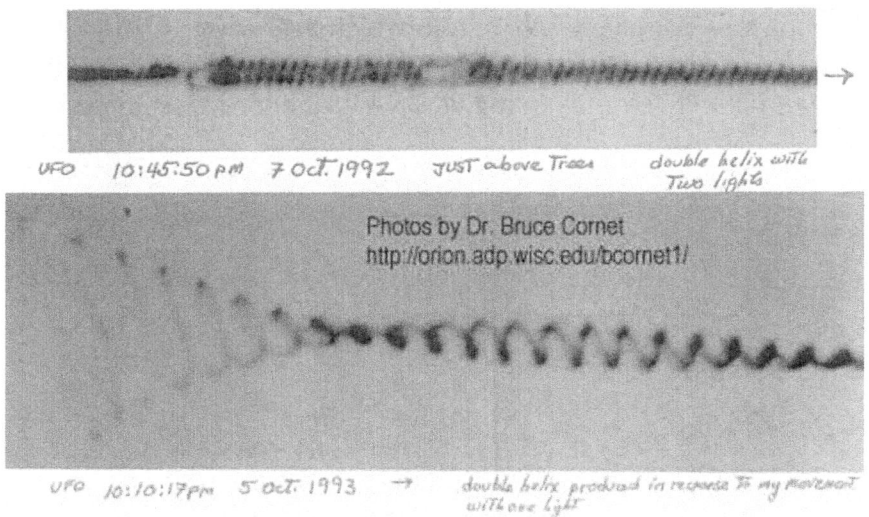

Then on 7 October 1992 with Kodak 400 ISO color film in my camera I photographed at pm a double helix pattern produced by a pair of lights moving south at about tree top level between the "Indian Mound" and my position on the east side of the dogleg bend in West Searsille Rd. The spinning of the two lights about a common center produced an unmistakable double strand with numerous interconnecting "bonds" looking at the pattern from the side. The sky glyph on my photographic canvass (time exposure) even has "nodes" embedded in it at intervals, produced by the flashing of its headlights. The second sky glyph is unmistakable and should convince skeptics. It should also dispel any notion that this sky glyph was produced by a "Stormville" stunt pilot, at night, silently, only 200-300 feet above the ground!

In 1992 there were various ideas and hypotheses being expressed and shared by the researchers and sky watchers who gathered along West Searsville Rd. at night. I was so busy collecting data that I literally could not see the bigger picture unfolding. It was the proverbial forest for the trees comparison, and it would take years and even a decade for researchers to put together all the data, and to make sense out of it as it may have been intended. Therefore, it is now not surprising that a little more than a year later another attempt at a sky glyph was made, which was much closer to the camera and which showed (symbolically) the hydrogen bond links between double strands that give DNA its ladder-like appearance.

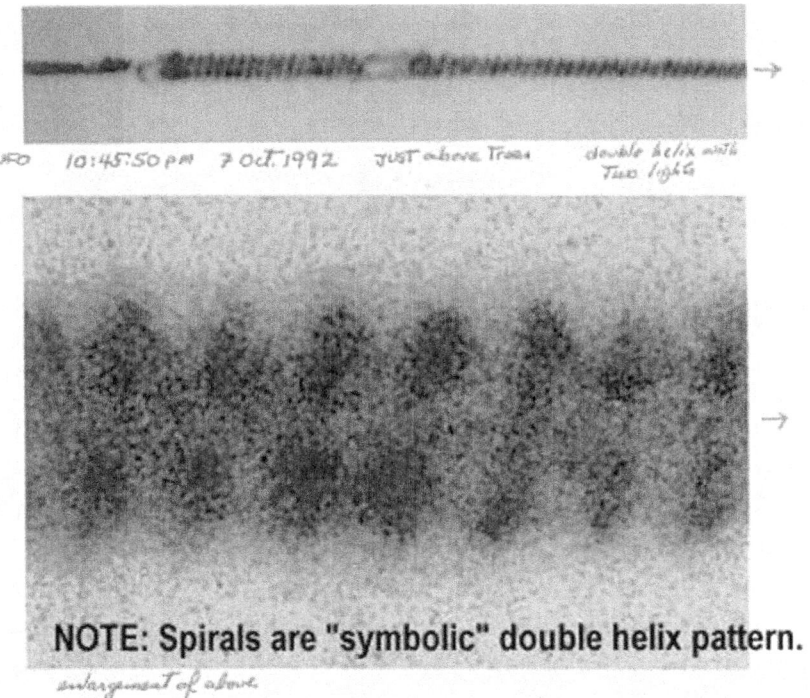

http://www.sunstar-solutions.com/AOP/SOW/illusion.htm

On 12 November 1993 I was alone at my primary observation station at the dogleg bend on West Seasrsville Rd., waiting for something to happen. At 9:04 pm I spotted a pair of headlights coming towards me from the west. The lights were low and barely cleared the tops of the trees on the low ridge to my west. I turned my camera and pointed it at the lights, beginning a series of 10 time exposures taken over four minutes and 53 seconds. The craft was a Black Triangle. It flew low over me traveling east, diving below tree level to my east. At most it traveled 1.5 miles (possibly as little as 0.5 miles) from the top of the ridge to Muddy Kill (stream) in the valley below, giving an average maximum speed of 20 miles per hour!

I captured an incredible time exposure while it flew directly over me, something I had not been able to do before. When the Triangle reached the forest on the other side of the field, I spotted another set of lights flying south at the same altitude. My time exposure captured both lights in the same image, along with a clear outline of the trees below. But the lights traveling south did something unexpected. They went out when the craft passed in front of the Triangle, then came back on immediately after passing beyond the path of the Triangle. The message was clear: Pay attention to the cross connections.

When I had the images printed, I was surprised to see something I had never seen a craft do before (or since). As it flew directly over me, ladder-like stripes of light crossed between the two headlight traces. The color of all the lights was golden, not white. The stripes were wide, and therefore did not bring to mind the stylized graphic representation of nucleic acid bases joined across two DNA strands. But now in hindsight, that appears to be exactly what the image on my film was intended to represent. The images below are shown as negatives to highlight the ladder-like details.

Unconventional Aerial Phenomena

12 November 1993, West Searsville Rd.

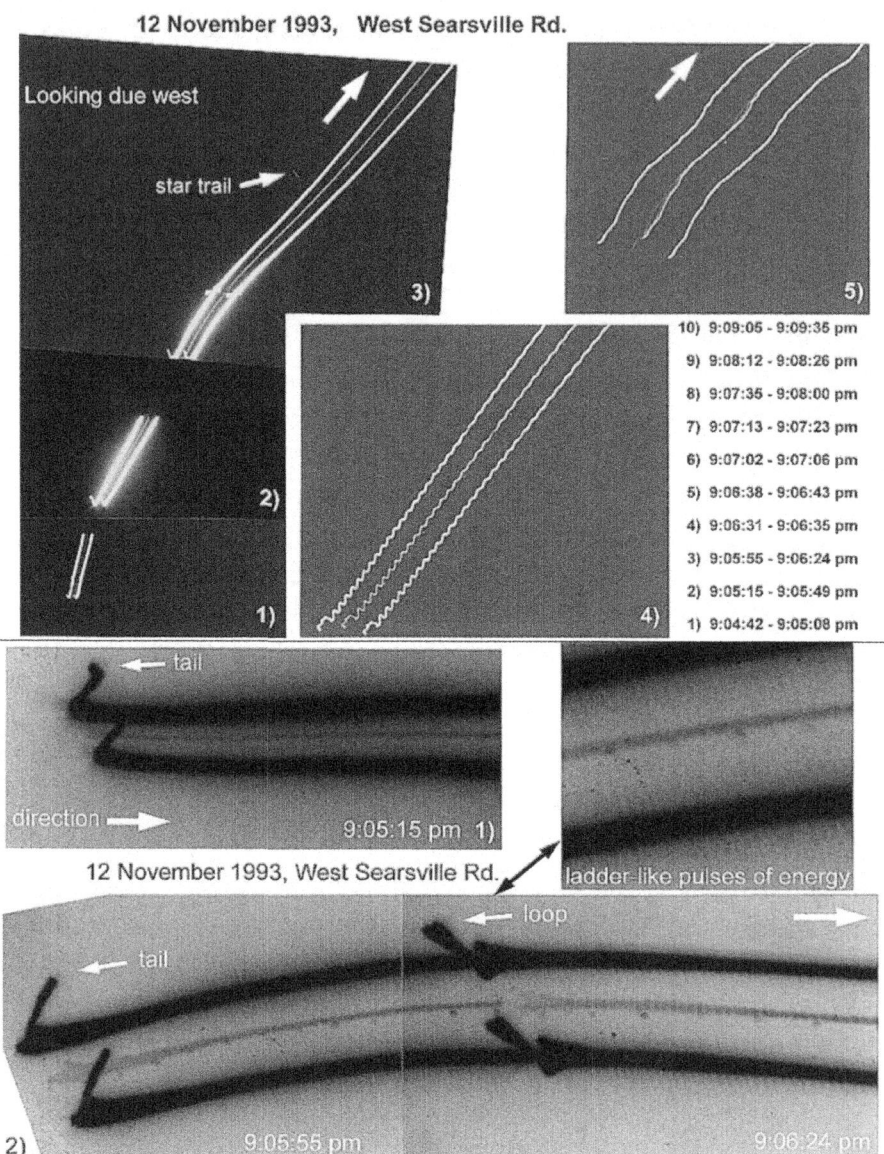

Flew over us, then banked hard and dove towards the ground; disappeared below tree line 1/2 mile to the east.

Unconventional Aerial Phenomena

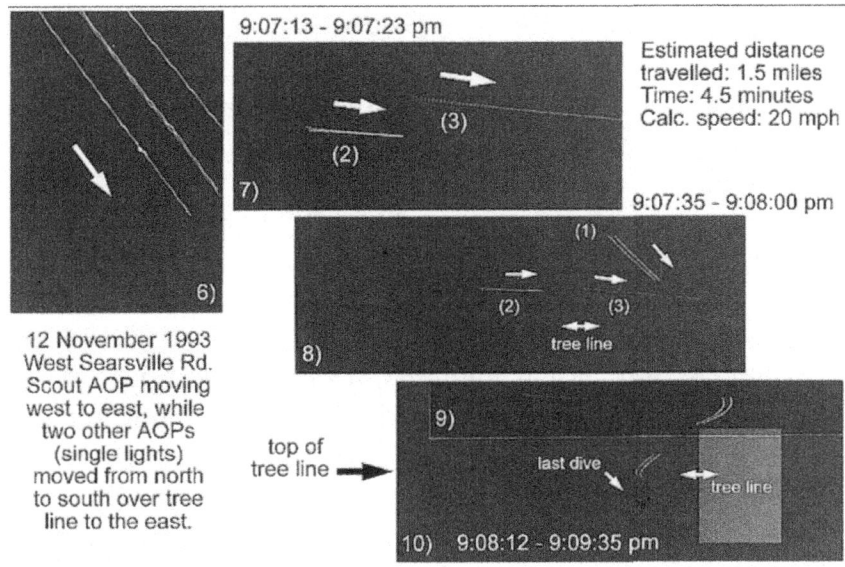

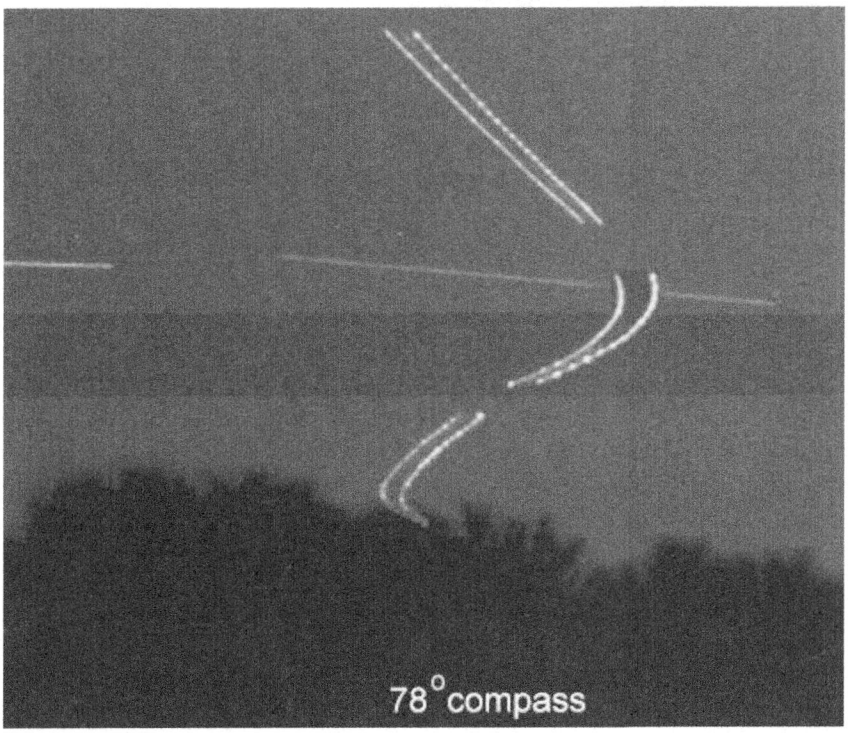

The (positive) image above clearly shows the Triangle diving behind the trees in the foreground, not disappearing into the distance as a function of perspective: the craft had not traveled far enough away to appear to fall below the horizon.

A skeptic could argue that a spiraling pair of lights when viewed or photographed from the side only resembles a DNA molecule, and the "chromosomal strands" are but an illusion of overlaying spirals viewed from the side. The ladder-like connections in such a sky glyph are not visible, and therefore the sky glyph is not convincing enough to prove the identity of a DNA symbol. But what that skeptic would miss is the fact that we have no aircraft that can duplicate such a tight spiral, and only a suicidal pilot would perform such a figure in the night sky less than 500 feet above the ground and forest.

The pattern created by the Triangle lights on my film, however, is another story. The resemblance to the structure of a DNA molecule is much more convincing. It is that unique structure of life that was being conveyed. The message was not: Humans possess DNA, but rather that DNA is shared or common between humans and the visitors. This is nothing new. The stories of Contactees described in Alien Bases (Good, 1998) relate experiments between humans and visitors in ancient times to create hybrids and chimeras, as exemplified in Sumerian cylinder seals and bas reliefs (Twilight of the Gods, 2010: Chapter Two). The stories of Abductees who contain alien DNA in their genomes can be read in Meet the Hybrids (Mendonca and Lamb, 2015). Such hybrids would not be possible unless the DNA of the visitors is more similar to ours than that of the Chimpanzee (see The Language of God, by Francis Collins, 2002).

Some Contactees were even told that they are us from previous failed civilizations on Earth that had reached the stars. A rare case of hair samples collected from human-like visitors has demonstrated an ancient genetic link. But this is a subject requiring its own story and book (Genetic Analysis of a Hair Root from a Reportedly-Alien Blond Female: Mitochondrial and Nuclear DNA, by the Anomaly Physical Evidence Group, May 2000).

Holographic Projections - The Golden Pyramid

On two separate occasions in two different areas I captured what can best be described as holographic images projected above features or places on the ground. The first occurred on 31 May 1995. The location was the "Indian Mound" at the center of the UAP hotspot. The observation location was the dogleg bend on West Searsville Rd. As with most sightings something caught my attention. This time it was a pair of lights below FAA minimum flight altitude of 1,500 feet, which were traveling south. The lights were golden in color, which is unusual for lights on conventional aircraft. At 9:03 pm I opened my camera shutter and began a 16 second-long time exposure. But just as the lights reached the compass direction of the "Indian Mound" (they were behind that mound and more distant), I saw a golden flash of light radiate upwards from the mound. I would not know what happened until I got my prints back from the photo shop.

In the photo you can see the UAP lights fade and sputter out as they reached the direction of the "Indian Mound." But there radiating up from the ground is a golden splay of light. The light increases in width upwards before fading out. Somehow, a brilliant light source within or below the mound generated that light, indicating some sort of device is buried there. However, that is not the most important aspect of this display. The light appears to illuminate a large otherwise invisible golden pyramid. The light splay passes up through and out the apex of that pyramid. And pyramid means: "Light in the Middle"! Pyro (light) mid (middle). For symbols this display couldn't have been more potent: A direct reference to Egypt was made, but why?

Even though this image of a pyramid might be an example of a hologram, later magnetic measurements revealed something extraordinary. The light created the image of the pyramid, but it appears that the pyramid exists there in magnetic form without the light. That was determined later using a fluxgate magnetometer over the "Indian Mound." And like the building-like structure illuminated in the adjacent forest (see page 176), which does not exist in our three dimensions, the light may have made visible in our world a structure that's exists only in a parallel world.

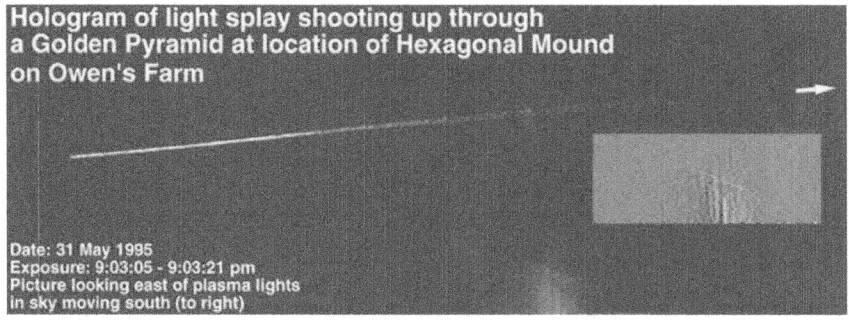

Hologram of light splay shooting up through a Golden Pyramid at location of Hexagonal Mound on Owen's Farm

Date: 31 May 1995
Exposure: 9:03:05 - 9:03:21 pm
Picture looking east of plasma lights in sky moving south (to right)

Circular "Indian mound" with hexagonal platform at center of Pine Bush UFO hotspot

Circle 2

ellipses define margins and center
hexagonal shape now imperfect due to erosion and slumping

Circle 1

The images above are duplicated in order to show how the golden pyramid was positioned over the Indian Mound. This connection is significant not only historically, leading to Native Americans chosing this location as sacred, building a huge sweat lodge over the second permanent crop circle (Circle 2).

On 28 October 1995 a group of six people got permission from the dairy farm owner to spend the evening on the "Indian Mound" sky watching. We were not disappointed, and were given a spectacular performance at night of a Black Triangle lifting off from an adjacent field (one of the fields from which numerous previous liftoffs and disappearances occurred). That ascent was marked by a flaring of the craft's paired plasma lights until it cleared a small cluster of trees at the center of a permanent "crop" circle, defined by trees around portions of its perimeter. The craft rose silently and arched over us, then descended and disappeared into a forest on the other side of the mound, circumscribing the mount's importance. Its arch was oriented northwest to southeast (see color pictures in Appendix I).

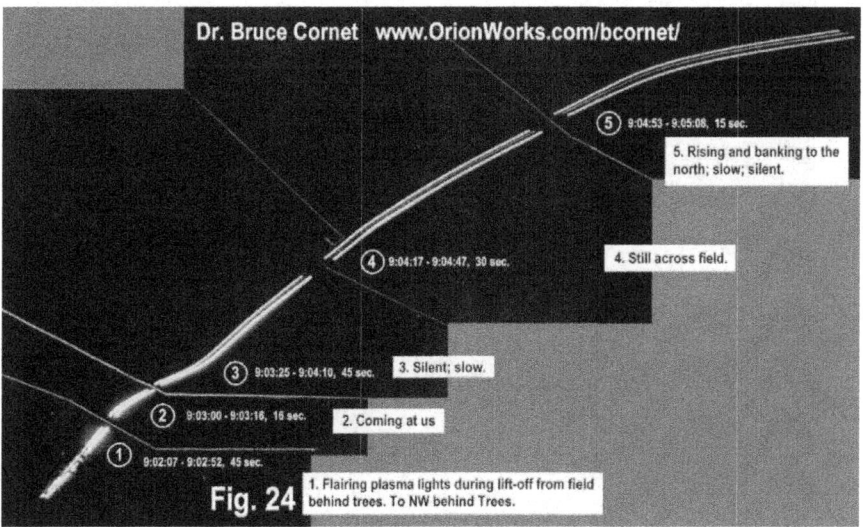

Fig. 24

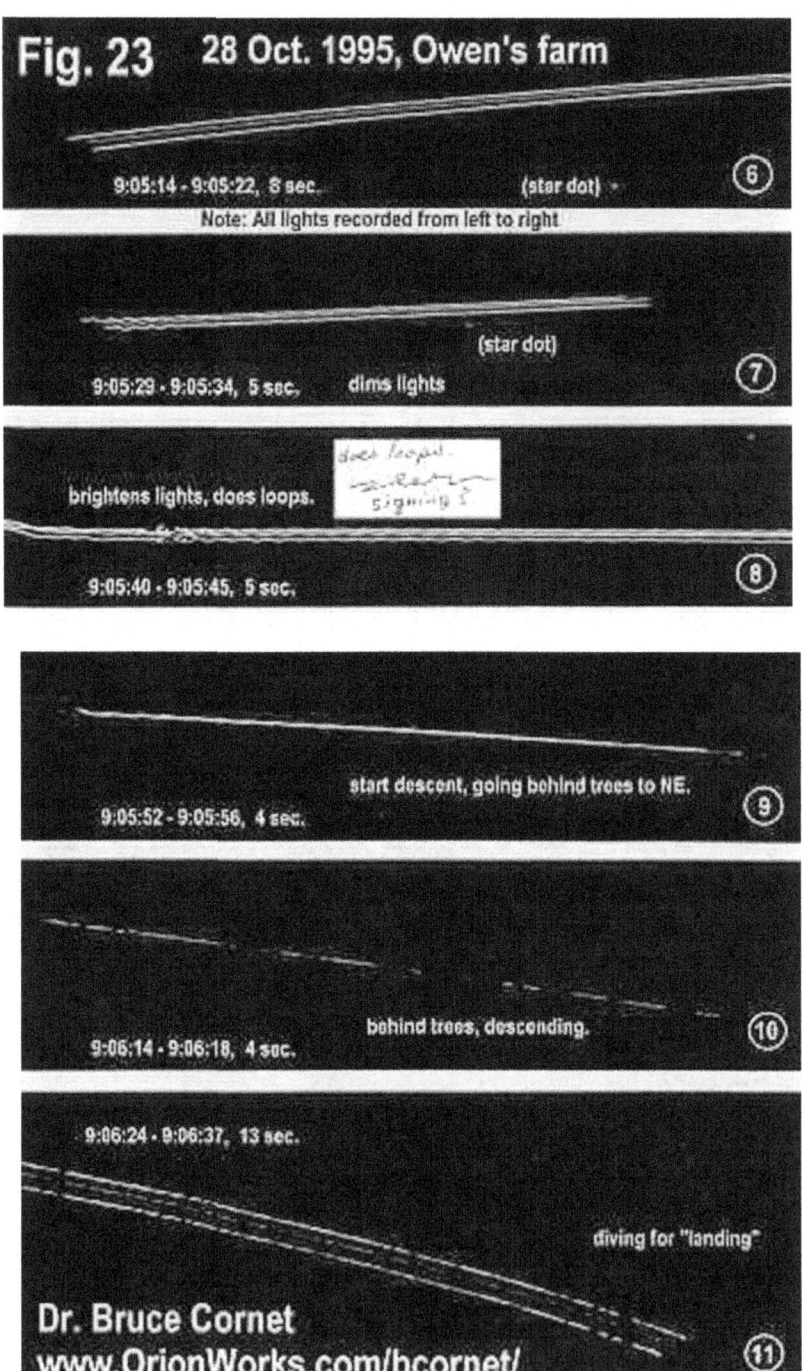

Before it got dark that evening, we measured the magnetism of the mound with a hand-held fluxgate magnetometer. We discovered that the mound is magnetic, but in a very unusual way. First we had to calibrate the magnetometer away from the mound. We tuned it so that the alarm would go off if the sensor encountered a level of magnetism greater than background. Then we approached the mound. We discovered that as we got to the top of the mound, climbing up its flank or slope, the alarm went off. But the alarm did not stay on as we walked to the center of the mound. Continuing across the mound we found another point at the outer edge where the alarm again went off. In other words, there was a magnetic wall that went around the six-sided mound at its top. Then I had an idea: The light had illuminated what appeared to be a pointed pyramid-like structure, the base of which was the size of the raised portion of the mound. So I held the magnetometer sensor three feet above the ground and made another transect across the mound. To my surprise the alarm did not go off at the mound perimeter as it had when the sensor was near the ground. The alarm went off more towards the center of the mound. With each transect, we held the sensor higher and higher, and as its elevation above the mound increased, the alarm went off closer and closer to the center of the mound until we could hold the sensor no higher with outstretched arms (the sensor was at the end of a pole). We were able to map a magnetic wall up to ten feet above the mound that was angled like the slope of a pyramid.

We had stumbled upon a pyramid-like structure defined by a magnetic field. The shape may not be four-sided, but six-sided. Regardless, there is no natural way magnetism generated by rocks in the Earth can produce a structured field like this one above the ground. It is as if we were detecting the outline of a feature in a parallel world. We may have stumbled upon the importance of the mound that marks something invisible in our world and dimensions, but which is not invisible in a parallel world. In other words, the pyramid-like feature in a parallel dimension may define the location of something artificial buried below the mound in our dimension. More will be discussed on the significance of this discovery in Chapter Six.

In Susan B. Martinez's book, The Lost History of the Little People, Their Spiritually Advanced Civilizations around the World (2013), she describes in Chapter Seven on Mounds of Mounds the ground layout of a Native Sweat Lodge in keyhole configuration. When her drawings are compared to the circle and mound in the above aerial photograph, a distinct similarity can be noticed. It would appear that this location on the former Owens Dairy Farm may represent a previously undiscovered Native American site of worship. Note that the golden pyramid occurred over the mound, in the position of the bonfire with teepee-like arrangement of logs, making it look pyramid-like. The sweat lodge would have been located over the ground circle defined by (white) circles directly to the west of the mound. A sacred path would have joined the mound (Unchi) and circle. And like Martinez's drawing, the features on the ground are oriented in an east-west direction. Is it possible that local Native Americans recognized this site as having spiritual significance due to all the lights that emerged from and disappeared into the ground there and around this site? The Black Triangle that lifted off the ground as we watched on 28 October 1995, did so from the middle of the sweat lodge circle just behind a cluster of trees marking its center, and it arched over us, disappearing to the east in the woods. Did our visitors put on similar performances for those who built that sacred sweat lodge and Unchi?

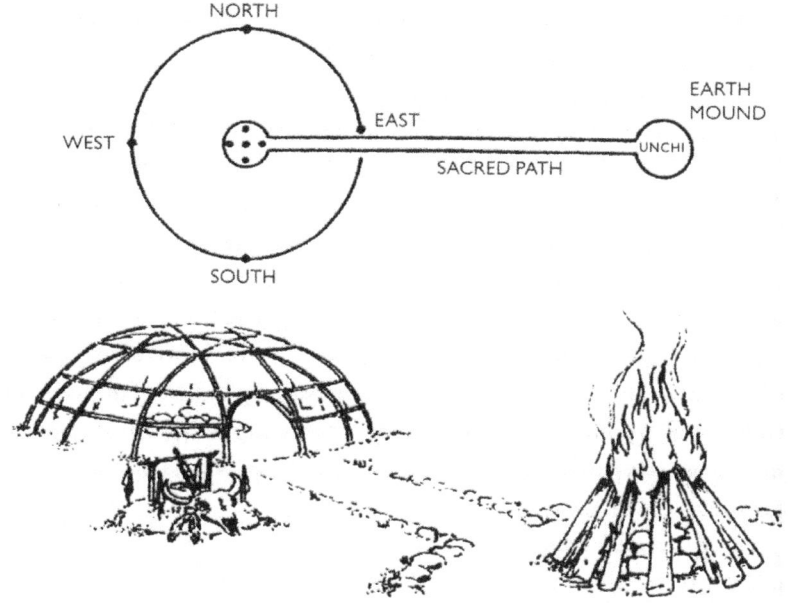

Above drawings from Martinez (2013: p. 252, Fig. 7:21) show a Native Sweat Lodge and above it the layout of its characteristic keyhole configuration.

The Green Temple

The second holographic projection occurred on 17 July 1996 off of Red Mills Rd., just northeast of Pine Bush, NY. It appeared over a field of corn behind a small church building. I had been waiting for other sky watchers to arrive. Barbara Hartwell arrived, and soon after Dawn and Keith. We were unloading our cameras when the Manta Ray flew directly over us at about 9:48 pm no more than 800 feet above us. I swore at it, because I was still setting up my camera and tripod, and couldn't get a picture of it. I ran to the edge of the parking lot adjacent to the lawn and field covered with mature corn. As I set up my equipment there, I watched in frustration as this huge black ship drifted silently towards the northeast in the direction of Newburgh on the other side of the valley. We could see the radio tower beacons flashing on the mountain tops (Ramapo Mts.) across the Hudson River, behind the Manta Ray.

As soon as I had my camera ready, I tried to locate the Manta Ray, but all I saw was a red flashing strobe. Then even the strobe disappeared. Disappointed at missing a great photo opportunity, I recounted what had just happened with the others. Then a set of lights appeared to the northeast, a bit further away and lower to the ground. I pointed my camera, placing the lights at the left of the camera viewing window, and opened the shutter at 9:49:30 pm. The lights moved southward. All of a sudden a long green-colored flash occurred above the UAP lights to the left, and seemingly coming from the location in the night sky where the Manta Ray had been last seen. I saw something briefly illuminated just on top of the corn in the field in front of me, but it was too quick to make out what it was. The object that appeared was located about halfway across the field, judging its distance to a barn and silo on the opposite side of the field. We didn't see much else that evening, and went home somewhat disappointed.

But when I got my pictures back from the photo shop, there in the photo of the UAP trace in the distance was this green object sitting on top of the corn stalks. To its left was a brilliant green splay of light pointing at the object, and coming from the direction of the Manta Ray, which cannot be seen in the photo. When I enlarged the object, it resembled a round marble temple that I had seen before. It also resembled the circular temples called Tholos or Dipteral in Greece. The round temple at Delphi, Greece, comes to mind, as does the Temple of Venus at Linderhof in southwest Bavaria. It also reminds me of the temple ruins on a hill along the Mediterranean coast, which my French class visited during the summer of 1963.

But the hologram most resembled the marble temple at the center of The Garden of Texas Liberty in Memorial Oaks Cemetery in northwest Houston, TX, where my late wife, Bonnie Lee Cornet, is buried. See **The Passion of Bonnie** at:

http://www.sunstar-solutions.com/AOP/Bonnie/bonnie.htm

Compare the images below of the green temple hologram and the temple at the center of the triangular Garden of Texas Liberty.

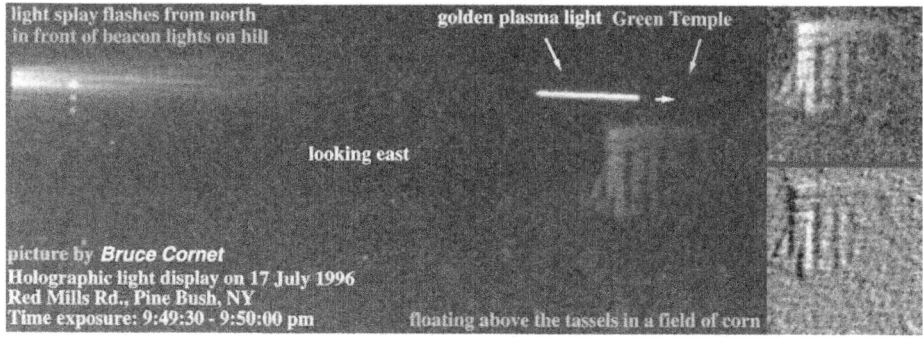

The hologram temple had about eight columns supporting a ring capstone. When I brought up a picture of the temple in the middle of a triangular memorial area in the Garden of Texas Liberty on my computer, it matched perfectly the temple in the hologram. Each had eight columns. So why did our visitors make a reference to that temple, and indirectly to my late wife's gravesite?

So why were our visitors making reference to that temple, and indirectly to my late wife grave? Or is this simply my wishful thinking? My exposure to the UAP subject and my first sightings began near her birthday on 18 June nearly a year and a half after she died in my arms. She finally succumbed to the ravages of cancer and heart disease on 12 January 1991. And there is that temple hologram picture with two more symbols: corn and ET. I couldn't help but make the connection: My last name is Cornet, which can be spelled CORN ET. Yet the symbolism doesn't end there.

My late wife Bonnie Lee Cornet was very psychic. She had demonstrated to me her telekinetic, precognitive, and telepathic capabilities. She seemed to know facts well beyond her high school education, and said she had an eidetic memory, commonly called a "photographic" memory. By leaving me clues and incontrovertible evidence that she knew when she was going to die, she demonstrated her psychic abilities. After she died, I continued to receive her thoughts. She gave me proof she knew where I would be on Friday, 18 January 1991 by telling a friend and bank teller in

Houston that I would be bringing her back to Texas for good (from New York), because I had gotten a job in Texas and would see her on a specific date and time. I confirmed later that she had spoken to her friend a week before she died (phone record). The job, it turned out, was to bury my wife at the Memorial Oaks Cemetery!

Bonnie is buried in The Garden of Texas Liberty, a triangular plot of land at the southeast end of Memorial Oaks Funeral Home & Cemetery in west Harris County, TX. It has a circular marble Temple with eight columns at the center of the triangle.

The "Garden" is located on the far southeast corner of the Memorial Oaks Cemetery land. Its architecture and layout indicates sacred geometry. In the center of the Temple is an obelisk with carved writing on all four sides. The inscriptions tell of the decisive battles between Texans and Mexico that led to Texas becoming an independent Nation in 1845. How Bonnie and I were able to purchase burial plots in this special commemorative area can only be described as "arranged." It was supposedly reserved for important native Texans, but I was able to purchase two plots next to one another on the west side of the Triangle inside the road that circumscribes and outlines the Triangle.

Now for some of my readers this story will be incredible or hard to believe. A grieving husband (widower) may be thought to imagine thoughts from a wife on the other side due to shock and trauma. But on two separate occasions Bonnie's spirit moved objects in front of me, once in front of another witness. The mystery of who Bonnie was/is deepens when her spirit appeared over her grave on Sunday, 20 January, before a video camera. My mother, Elizabeth, was a witness. Without that video, as a scientist I knew I would have had no evidence for such an incredible tale. You may ask, what does this event have to do with UFOs over New York? Answer: Everything, considering that it involves a giant white disc-shaped UFO embedded in a cloud over Bonnie's grave! Bonnie died in Orange County, New York, but our Visitors followed us all the way down to Houston, Texas.

With clear evidence of something extraordinarily anomalous on video, I had logical reason to share my story. I had photographic proof!

I took my bulky Sears VHS camcorder with me when my mother and I took one last trip to Bonnie's grave on the morning of 20

January 1991. I was flying back to New York that afternoon. I wanted to record where she was buried for her Harrington family, all of whom lived in the Northeast United States. I held the camcorder on my right shoulder as I drove through the cemetery to the "Garden of Texas Liberty," where Bonnie is buried in crypt #12. Next to it is crypt #11 reserved for me.

As I drove into the "Garden," crossing a bridge over a bayou, called Turkey Creek, I recorded the marble Temple in center view with a large cumulous cloud overhead. It was just after sunrise, and the Sun was hidden behind the trees to the right or east side of the "Garden." I did not notice anything unusual at first, because of the low level of lighting. However, later review of the video would reveal something spectacular.

I drove around the triangle and Temple, and parked on the west side before reaching a freshly covered grave, Bonnie's grave. My mother and I got out of the car and walked around, admiring the grave monuments lining the outside of the road and Triangle. I then began panning the area with my camcorder to capture as much of the "Garden" as I could from that vantage point. I zoomed in on the central Temple, thinking that the Sun behind the Temple was causing strange lighting effects on the shadow sides of the marble columns. Later review would show purple and green lighting illuminating the outsides of the columns facing me – an impossibility for the Sun to illuminate the shadow sides of those columns. The colors alternated from column to column: green, purple, green, purple, etc. The colors jumped in unison from column to column from left to right, giving the impression of a merry-go-round movement. Bonnie frequently told me that she loved merry-go-rounds. Was she giving me a clue to her spirit's identity?

I then panned my camcorder to the right over to Bonnie's grave. I saw a flash of green follow me from the Temple over to and above her grave. I began saying my good-bye or farewell to Bonnie with her grave and flowers in center view. As I said, "There is my love, love of my life, right up front where she would want to be..." a green, blue, and white starburst apparition descended from just above camera view into view above the head of her grave, camera rolling. The starburst pulsated larger and smaller with my words, in cadence, as if to acknowledge what I was saying. I was so focused on my farewell, I didn't pay attention to the starburst, thinking it was a product of sunlight glare or reflection from the left (East). The Sun had risen to just above the trees; it was 8:31 am.

When I finished my farewell, the starburst transformed into an angelic figure, and moved its wings up and down several times. This can be clearly seen on the video, and in review and analysis rules out any interpretation of the anomaly as lens flare or light reflection. Something told me to turn off the camcorder. In hindsight, I wish I hadn't, because what happened next would have been recorded as a divine epiphany.

Keep in mind that the Sun was to the far left, barely 10 degrees above the horizon. Suddenly a brilliant light the size and color of the Sun's disc shone through the middle of the overhead cloud, and sent a wide "sunbeam" down to Bonnie's grave. My mother, standing behind me and to my right, exclaimed, "Oh, how beautiful. The Sun is sending a sunbeam down to Bonnie's grave." But this "sunbeam" was no ordinary shaft of light. It was an estimated 8-10 feet wide. It stopped about two feet above the ground, and light shafts that looked like sparkling ice cycles fell off the sides of this "sunbeam." When they hit the ground, they broke into pieces, which bounced on the ground until they disappeared. The action was like that of water bouncing on a hot plate.

The blue, green, and white angelic starburst that had been hovering over Bonnie's grave was now inside the "sunbeam." I watched in disbelief as it rose up inside the beam and disappeared into the second sun in the overhead cloud. We could see the green color of Bonnie's spirit ascend in that golden column of light. The "sunbeam" retracted quickly into that second sun before it blinked out.

At the time this happened in January 1991, I was still unfamiliar with paranormal events, divine occurrences, and UAP sightings (the TV series Supernatural began in 2005). My education would not continue until I got back to New York.

My first UAP sighting would not occur until 9 June 1992, and my education into alleged alien abductions not until the miniseries, "Intruders," appeared on television in May 1992. Thus, I had no reason until more than a year later from the events over Bonnie's grave to review the video with a different mindset and purpose.

After reading some of Jacque Vallee's books on UAP sightings (e.g., Passport to Magonia: On UFOs, Folklore, and Parallel Worlds, 1969; Messengers of Deception, UFO Contacts and Cults, 1979; Dimensions: a Casebook of Alien Contact, 1988; Confrontations: A Scientist's Search for Alien Contact, 1990), I became familiar with how UAPs sometimes hide in clouds, and how beams of light had been seen originating from UAP lights. It was not until I recognized that I had been abducted from y wellsite on 1 October 1981 (Appendix I), in front of my night tower (shift) drill crew and mud loggers that I began recovering childhood memories of abductions. Then I became curious about the light over Bonnie's grave, and the light beam that came down from an overhead cloud. I was prepared to review that video with different eyes.

As I reviewed the video, I saw something briefly appear out the left side of that clo0ud as my mother and I drove into the "Garden of Texas Liberty." I would have to wait until I digitized the analog tape on my computer in the late 1990s before I could enhance the images. I recognized the unusual green and purple lighting on the outside of the Temple columns as highly anomalous, especially because the colors alternated on the columns as if the Temple were a merry-go-round turning. Bonnie loved merry-go-rounds, and I made that symbolic connection. The Temple hologram over the corn field (described and illustrated above) was illuminated with the same color green. I also recognized the same kaleidoscope of colors moving around the Temple columns as we drove into the "Garden," which was a different angle or view that again ruled out sunlight reflection from the rising Sun to the East.

When I did get the "Garden of Texas Liberty" digitized and captured frames for enlargement and enhancement, the cloud over the Temple revealed a huge ice-white frosted or colored object briefly visible sticking out of the left side of the cloud. It was clearly visible for less than as a second, and required capturing the right frames to see it. With a little contrast enhancement it resembled a large disc-shaped craft with a dome on top. There was a double rim around the middle, and a nearly flat bottom. Its overall shape resembled that of the craft photographed by Apolinar (Paul) Villa in 1963, but the cloud craft was much larger (see pictures in Good, 1998: photo #6). Based on the size or dimension for the "Garden," the craft is estimated to be at least 200 feet in diameter, and 20 feet high at its center (scales used for comparison: the triangular "Garden" is about 333 feet wide, and the central Temple is about 20 feet high).

As soon as the craft was visible, a cloud-like veil rose out from the cloud to conceal it again. The exposure was apparently no accident, and was intended to be discovered later after I had my many sightings near Pine Bush. It also required computer technology to evolve with adequate software programs available to the public. Once I realized that the light in the cloud was coming from a UAP hiding in that cumulous cloud, I had to deal with the heaven-shaking possibility that the spirit of my wife was somehow connected to the phenomenon. Bonnie talked about UAPs and abductions only once when we were together, and that was indirectly from stories she heard from acquaintances during the late 1970s.

At no time did she indicate that she was an abductee or experiencer, or had she had a UAP sighting in her 40 years of life. And yet she was a hypnotherapist who with the help of her business partner, the late Jim Forberg, had been hypnotized many times and explored her own memories and subconscious mind, including past life memories. She was very gifted psychically, and her gifts became enhanced after each of three near death experiences before her final death. The video of the starburst over her grave certainly implied that it represented her spirit responding to my farewell message. I had answered a request from an unseen entity in 1981 to help a woman in need. I thought it was God asking me for my help,. But we know that the entities behind the UAP phenomenon sometimes claim to be gods or God, and they are masters of illusion (Communion, A True Story, by Whitley Strieber, 1988).

It is now clear that the marble Temple at the center of the "Garden of Texas Liberty" is what was being imaged over the corn field on 17 July 1996 near Red Mills Rd. in Orange County, NY. And if the UAP lights in the background are connected to the corn in the field, a reference to corn ET or Cornet was also being made. I spent seven years to the day with Bonnie (12 January 1984 to 12 January 1991), and I helped save her life and provided for her during that period of time. Was a connection between UAPs, aliens, and Bonnie being made by the hologram of the Temple, or was my unselfish generosity and caring for Bonnie, and eventual marriage to her on 18 June 1986 (on her 36th birthday) being referenced? Or was there something else that will be discussed in the final chapter of this book? Were Bonnie and I brought together (see The Love Bite, Alien Interference in Human Love Relationships, by Eve Lorgen, 2000)? Were our reproductive cells taken years earlier in 1967 to produce offspring, who I have met, who now live with the ETs, and who are responsible for my Pine Bush experiences and sightings?

There are very few researchers or investigators who find themselves subjectively immersed in the UAP phenomenon, sometimes over their heads, such as Eduard Albert "Billy" Meier, born 3 February 1937, who is a 79 year old, one and a half armed, Swiss farmer. He claims to have been in contact with extra-terrestrials named the "Plejaren," formerly as the "Pleiadians." He has taken hundreds of controversial photographs of the Plejaren craft in Switzerland. These encounters with our Visitors beg the questions: Where is this saga going? How will it end?

The Vortex Photo Revisited

On 28 April 1993 the Vortex photograph was taken as a UAP slowly climbed south over the Indian Mound. Directly over the mound a cluster of white, red, blue, and green lights suddenly expanded into a spiraling cone or funnel-shaped pattern. The History Channel thought this time exposure is so spectacular that they did a 30 minute documentary on it and related photographs, which appeared on "UFO Hunters" on 26 March 2008. The show was called "UFO Vortexes." The rushed nature of the shoot for the show prevented a thorough investigation of the area and its magnetic significance. Thus, the importance of the Indian Mound, its magnetic anomaly, and the pyramid hologram photo went undocumented. If further research and exploration of the mound is not done before the completion of this book, it is imperative that GPR be done to see what lies below the surface. As the center of the area UAP hotspot, because of the sheer numbers of UAPs seen emerging from and disappearing around this location, there must be something very important buried there, as there is for a similar magnetic anomaly under the Jewish cemetery on Rte. 52.

The question becomes, is this symbol meant to be interpreted as indicating 1) a portal or Stargate is located in that area, or 2) a parallel dimension or world is thinly veiled at that location, or 3) there is underground activity, i.e. an alien base, located beneath this location, or 4) some combination of these possibilities?

The picture of the Vortex shows the forest below clearly, because the time exposure was 72 seconds long. Within that forest on the left a strange bluish gray building can be seen that appears to be intentionally illuminated. That structure does not exist in that forest. Is the light making it visible through a thinly veiled parallel world, such as is described in Janet Bord's book, Fairies: Real Encounters with Little People (1997)?

I have already described how a Manta Ray spiraled down to the ground and flew directly into that forest without seeming to collide with any of the trees (pp. 55-57). Was that craft landing on an airfield on the other side of that veil? Were we being given information that would indicate our visitors were not coming from some distant planet in our galaxy, but from a parallel world at a slightly different subatomic vibrational level?

Concepts in physics were rapidly evolving to include the possibility of a ten-dimensional universe, and those new concepts of parallel worlds were coming out in recent books designed to inform and educate the public (e.g. Davies and Brown, 1992, Superstrings: A Theory of Everything?; Hawking, 1988: A Brief History of Time; Kaku, 1994: Hyperspace, A Scientific Odyssey Through Parallel Universes, Time Warps, and the 10th Dimension). But it would take another 14 years before the concept of hyperspace, wormhole travel across the universe (Stargates), and parallel worlds accessible through vortexes would leave the

Hollywood screen of fiction and become subjects of History Channel documentaries.

Even Dr. Ellen Crystall states in her book, Silent Invasion (1991), that she did not believe in parallel dimensions at the time she was witnessing the ships suddenly disappear or dive into the ground during the early to mid-1980s. She thought the aliens could open up portals (which she describes seeing) into an underground base, but why would they have an underground excavation if they could simply sidestep into a parallel dimensional space? Are we on the verge of discovering our true place as humans in the universe as the year new millennium unfolds? Were these performances by non-human visitors and their advanced technology designed to wake us up to their reality in a non-threatening manner before we destroy ourselves in global tribal warfare?

Near the beginning of this book I introduced the reader to the evidence for intentionally-created illusions directed towards specific individuals or targets on the ground. Crystall (1991) describes how she would capture images of craft and aliens on the ground with her camera, while her companion, Harry Lebelson (Editor for Omni Magazine), would have blank film despite standing next to Crystall and using an expensive Leica camera. I describe an experiment where two observers with two cameras saw and captured very different images even though they were standing next to one another. It is clear that the alien technology is so sophisticated that it can project different images to each person in a group standing on the ground, leading to confusing and conflicting eye witness accounts. But just as they can mislead and confuse witnesses, so too can they target specific individuals with information they want them to have.

Crystall was allowed to see and photograph ships and aliens on the ground only in 1980. She was given signs and clues to possible underground alien activity in 1985. This information led her to develop the hypothesis of underground alien bases in her book

(1991). But wonder if those clues to underground activity were only intended to get a person like myself, a geologist, involved? Through mapping a larger area with a magnetometer, I discovered at least two areas with such anomalous magnetic activity that implicated underground alien activity. A triangular object was identified as the source of magnetic signals being focused out into space from under the Jewish Beth Hillel Cemetery on Rte. 52. And it would take a geologist familiar with Earth history and the ages of rocks to suspect that an alien probe may be entombed in Ordovician-age black shales, one that had come to Earth about 420 million years ago and landed on an island in a primeval epicontinental sea. Since geologists are scientists, and most scientists follow official government propaganda that ET and UAPs on Earth do not exist, our visitors had very few geologists from which to choose.

On 23 June 1994 the Encounters TV film crew accompanied Crystall and Cornet into the field to shoot them watching the sky at night for UAPs. The Producer of the show wanted to film Cornet using the Proton Precession Magnetometer in the field earlier that day, and had agreed to rent an instrument from E.E.G. in California, and have it delivered to Cornet 12 days earlier.

I took the opportunity to do a survey of the UAP hotspot center at the Indian Mound locality during the afternoon of 11 June 1994. I took measurements every 25 paces (~80 feet) over a course of 74 measurement stations covering about 5,900 feet distance, looping across the fields to the west of the Indian Mound and up onto that mound, which is eight feet high and about 300 feet wide, with a moat or water-filled trough circling the mound, extending its diameter to 365 feet. After I finished at 3:45 pm, I went back to some of the previous stations or places I had taken measurements one hour earlier to confirm previous measurements, but discovered a significant increase in magnetism. The first readings gave measurements between -541 to -547 mg (miligammas). The readings I got later were 200 to 650 miligammas higher and climbing rapidly, indicating that something unnatural was occurring underground. I decided it was unsafe to remain there at the UAP hotspot center, and left the area.

What I discovered was extraordinary, and indicated a highly unstable magnetic field there with changing values by the minute in the same fields and forests I had photographed UAPs, Black Triangles, and the Manta Ray either originating from or disappearing into the ground.

I had done a traverse across the fields on the west side of the Indian Mound, crossed the tree circle located on the west side towards the mound (see Appendix I and pictures on page 203), and reversed my traverse from the mound back the way I had come. To my amazement the first set of measurements with the proton magnetometer were not duplicated on the return traverse. They should have been duplicated if they measured the underlying rock magnetism. However, all values from before became elevated. Not only that, but with each new measurement location I paced off, the magnetic value was higher than the previous location I had just measured, as if something were about to happen below ground. By the time I had reached the farthest farm field from the mound, the overall strength of the local magnetic field had increased by as much as 650 miligammas. This suggested to me that a portal or Stargate was about to open, or was opening, even though I did not see anything unusual happen above ground. In my photographs taken at the same time I did capture something moving very fast high above those fields.

"If some of the chambers do mark the locations of a doorway to a parallel reality, then it is possible, in theory, that when the "door" swings open, a burst of energy might flow into our universe. The same might happen when the door is shut closed. It's like a door that opens from the outside on a cold winter day. When the door is opened, a burst of cold air enters the home; when the door is closed, a slam can be heard. Perhaps when the door in this other dimension is opened or closed, two different types of energy are released into our world. We mentioned earlier that evidence suggest the UAPs might be using certain chambers to travel from their world to ours. [The chamber would represent a type of Stargate for origination or destination universal coordinates.] This would explain why there is an outbreak of paranormal activity before and after a series of UAP sightings. The energy released by the opening and closing the dimensional doorway may be the power source to trigger the paranormal events. In theory, part of the energy should be generated as an electromagnetic pulse (EMP). This pulse should be picked up on a radio receive that has the ability to scan a multitude of frequencies in a short amount of time.

A similar increase in energy (EMP) was recorded by Imbrogno and Horrigan at the entrance to a stone chamber (#2) on Rte. 301 at Kent Cliffs, NY, in September and November of 1997:

Stone chamber, Rte. 20
right smaller one of two
2 August 1992

Above image: Ellen Crystall standing before one entrance to a double stone chamber at Putnam Valley, NY, across from Putnam Valley Elementary School on Route 20. Picture taken during a field trip on 2 August 1992 led by Philip Imbrogno.

"During the fall of 1997, we were able to borrow a receiver that was able to scan frequencies between 16 MHz and 2,000 MHz an oscilloscope and a chart recorder were also used. We picked the oval chamber on Route 301 since it had a history of paranormal events. We decided to monitor the frequencies at night, so we began the experiment at 11 pm. We continued to scan for two hours, then at 1:15 am on September 21 we picked up a pulse that started around 16.50 MHz, and peaked at 1,675 MHz in the microwave region. The pulse continued and started to drop in signal strength after 1,700, and continued to drop until we lost it at 2,000 MHz We could find no explanation for the energy burst. Was it a doorway to another universe opening and closing? Much to our dismay, we received no reports of UAP sightings or any other type of paranormal activity in that area, but that does not mean they did not take place. It would mean that they were just not reported." (Imbrogno and Horrigan, 2000: p. 144).

A second experiment was done on November 1, 1997 (Imbrogno and Horrigan, 2000: p. 145). "At 4 hours and 38 minutes Universal Time (11:38 EST) at the same chamber on route 301, Kent Cliffs, NY, another EMP was detected. The recording indicated a strong electromagnetic pulse at 1,675 MHz Thus twice over a period of 11 days two EMPs were detected between 11 pm and 2 am at the same location." This type of instability and energy activity was also recorded at the Indian Mound on 11 June 1994, where many dozens of UAPs have been sighted emerging and disappearing from the fields and forests around that mound. Paranormal activity has also been observed and recorded nearby on West Searville Rd., just a mile north of the mound by people sky watching near the intersection of Hill Avenue and West Searville Rd. That activity included luminescent orbs doing loops and making erratic movements on film and video, to orbs surrounding a skeleton figure coming out of an adjacent field (captured on film). Another sky watcher there, who snapped flash pictures in directions where he felt something strange was occurring, captured an image of a human skull.

On the night the Encounters TV film crew was there (26 March 1995), 21 lights were tracked and recorded by me between 8:12 pm and 10:15 pm, a period of two hours. Six lights in the night sky appeared anomalous and showed characteristics inconsistent with conventional aircraft, such as flying just above the tree tops, travelling too slow (35 mph) and looking like a triangle or arrowhead as it hovered above us silently, or circled around us silently without banking. This UAP activity started at 8:44 pm and peaked around 10:10 pm, and represented 29% of the total aerial activity that we observed that night.

Encounters: UFO video footage (June, 1995)

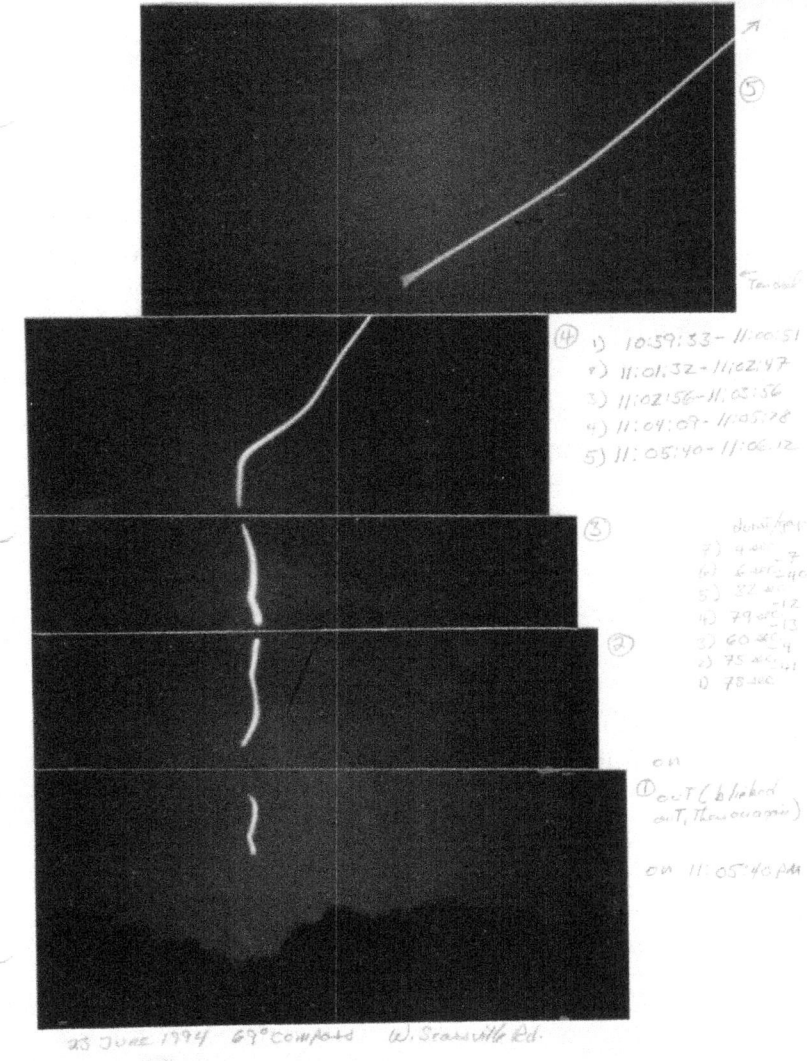

The series of time exposures below were taken on 23 June 1994 just after the above sightings between 10:59 pm and 11:06 pm from West Searsville Rd. A golden light rose up from over the Indian Mound, and after climbing about 150 feet began moving off to the South. Its movement was sinuous and not straight, and the light dimmed and even went out a couple of times, only to come on again slightly higher.

More than a Coincidence

So, by our Visitors involving Crystall, and exposing her to her first UAP-alien encounter in California, she became interested in the Pine Bush phenomenon when she moved back to live with her parents in New Jersey in 1980. And by feeding her specific empirical information in the field that would lead her to suspect an underground alien base, she would write a book, which I would read, and would get me involved. I could have chosen an area for my family to live when we moved from Houston, TX, to New York much closer to the Lamont-Doherty Earth Observatory in Palisades, NY, where I worked, but instead I chose an area in which to live an hour's commute away! That area was Middletown, NY, less than 10 miles from the Pine Bush UAP hotspot! I had no idea at the time (August 1989) that this was the center of paranormal and ET activity, and I had no interest in UAPs. I was too busy working for a post-doc at Columbia University on the geology, paleobotany, and palynology of the Newark Supergroup basins along the East Coast.

When I did become involved in 1992 after reading "Intruders" by Budd Hopkins, and "Silent Invasion" by Ellen Crystall, I put my geologic skills to work. Crystall thought there might be an alien base underground. I could use a proton magnetometer to detect any anomalous electromagnetic activity below ground. From my geologic investigation I developed new hypotheses for why so much alien activity was concentrated in such a small area (less than 20 square miles). Crystall was led to highlight the Jewish cemetery as an area of paranormal activity above ground; I was able to detect anomalous signals coming out of the ground there. I was also attracted to the dogleg bend along West Searsville Rd., which was one of Crystall's favorite observation stations. Over a period of five years I observed many dozens of UAPs landing and taking off from a set of specific farm fields. If it were not for that Vortex picture taken in April 1993, I may not have suspected something unusual existed at the center of the UAP activity until we flew over the area in a helicopter in May 1994. It was then that we discovered the "Indian Mound" and tree circle, implicating the possible special nature of this locality to early Native Americans (e.g. Mohawk and Iroquois).

With each new discovery, we took another step towards understanding the significance of what was happening in this valley and why. Imbrogno and Horrigan's work (2002) on the Stone Chambers east of the Hudson river, and their associated paranormal and UAP activity expands the mystery to include most of the Hudson and Wallkill river Valley.

Chapter 7 The Awakening

In early 1992 I became interested in the UAP phenomenon after seeing the television miniseries, Intruders (based on the work of the late Budd Hopkins and late Dr. John Mack). I thought I had never seen a UAP, thinking that I had missed seeing one at the Bailey No. 1 wellsite in 1981, because my crew had not come to wake me up when a UAP reappeared for the second night over a cut forest next to the wellsite. I was intrigued enough by the subject matter to want to do more research. I went to Dalton's Book Store in Middletown, NY, and found the book, "Intruders" by Hopkins (1987). The bookstore manager saw me looking through that book, and suggested that I also purchase "Silent Invasion" by Ellen Crystall (1991), which had come out recently. Upon reading Silent Invasion I was shocked to discover that Crystall's research focused on UAP activity around where I lived. The subject of UAPs was taboo among scientists, and I was surprised that a show on the subject of alien abductions, no less, had been approved for prime time television. I did not realize it at the time, but I had been abducted by aliens in front of my drill crew at the Bailey No. 1 wellsite on 1 October 1981 near Richmond, VA. It took me another three years to suspect that I had been abducted (based on telltale clues given in "Intruders" and "Missing Time" by Hopkins) and to get up enough courage to go to a hypnotherapist to be regressed back to that abduction. In 1994 my wife and I met Dr. John Mack in Boston, and we became interested in his work on the regression of abductees to find information blocked by hypnotically-induced screen memories created by the aliens. In May 1995 with the help of Fred Max, the hypnotherapist who first regressed Betty Andreasson-Luca (The Watchers and The Watchers II, by Raymond E. Fowler, 1990 and 1996 respectively) I relived that abduction, and discovered that what my crew had told me they saw happen that night was true. That regression became the subject of a web site at

http://www.sunstar-solutions.com/AOP/Abduction/abduct.htm.

Beginning with a scientist's natural curiosity, I was suddenly faced with the reality that I lived in a UAP hotspot area. During the previous four years living in Middletown, I had never seen anything unusual in the night sky. As a geologist I was more accustomed to looking down at the Earth, not up into the sky,. I worked for Columbia University at the Lamont-Doherty Geological Observatory (now called the Lamont-Doherty Earth Observatory), and commuted one hour each way to Palisades, NY. Even the name of that institution has taken on a more space-oriented perspective of our planet.

I was able to get Ellen Crystall's phone number from a friend. I called her and introduced myself. I suggested that we might be able to work together, because she was a musician, and her last name was Crystall; I was a geologist, and my last name was Cornet. She laughed, and told me she had earned a minor degree in geology at Rutgers University. I told her I had played the trumpet in my high school marching band. I asked her if she would be willing to show me more evidence for the claims in her book that an alien base existed below the Wallkill River Valley. She agreed, and said she would pick me up the next evening just before sunset on her way into her fieldwork area. She was taking several people out into the field to sky watch that evening.

As a scientist, I did not know what to expect. Crystall and the people she routinely took with her into the field were not scientists either. None of them had been trained in the scientific method, which as its primary objective attempts to falsify hypotheses and concepts through collecting more data via observation and experimentation. Perhaps out of ignorance or naivety I did not realize how volatile this subject was with my colleagues or peers at Columbia University, because I had never engaged in a conversation about UAPs before with my academic associates. Scientist tend to avoid and even openly reject subjects where the evidence cannot be brought into the laboratory for examination. The most notable exception is astronomy, because the stars do not actively try to avoid being studied and photographed by astronomers using telescopes. Had I known the problems my curiosity would create ("curiosity is what killed the cat") I might have chosen to abandon this interest, and run with all haste in the opposite direction.

Something about the Intruders miniseries bothered me. The aliens depicted in that series were strangely familiar to me. I did not know why, but the show rang true to me. I could not write it off as fiction, and began searching my memory for anything that was familiar about that show. I purchased another book, Missing Time, by Budd Hopkins (1981), and found clues that I might have been abducted. I read my diary or log of drilling activity for the Bailey No. 1 well in September-October 1981, remembering that the crew had reported seeing a UAP hovering over the adjacent field (that had been cleared of trees by lumbering) on two consecutive nights at around 2:00-3:00 am in the morning – the witching time for alien abductions (Hynek et al., 1998: An Analysis of the Situation, Chart II). I then recalled that Rich and Karen, two night tower (shift) mud loggers, told me they saw small figures taking a larger figure in a "beam" of light from the mudlogging trailer to the UAP, a disc-shaped object with as many as 30 multicolored lights blinking on and off. They said they went into the trailer to look for me, because I had specifically asked them to get me if the UAP returned. They found me missing! When I remembered that account, images of alien faces came to mind and I realized I may have been abducted.

From that moment on I realized that my involvement in studying the UAP phenomenon might be much greater than I thought. It also meant that I could never be completely objective in my study. It would not be until 29 May 1995 that I would discover what actually happened to me that night (go to for a transcript of my hypnotic regression).

http://www.sunstar-solutions.com/AOP/Abduction/abduct.htm

As I continued to search my memory with additional clues and associations, I uncovered abductions in 1954 at age nine, indicating a previous abduction in 1953; another in 1962 in my high school band dressing room, right in front of many other band members who became "frozen" in position (30 minutes missing time); and another in March 1967 on a trip to Buffalo, NY, when a sperm sample was taken (one hour of missing time). Details of these abductions and memories will be recounted in a subsequent book involving the meeting of my hybrid human children. Since my active involvement in studying the Pine Bush phenomenon, I have had two additional abductions in November of 1993 and 1994. These two abductions, like the wellsite abduction, occurred in Virginia. The 1993 abduction took place on top of Massanutten Mountain overlooking the ski resort by the same name, and also involved my wife Pat Huff, who was also abducted.

(http://www.sunstar-solutions.com/AOP/Massanutten/Massanutten.htm).

Ironically, Pat's mother owned an A-frame house in Massanutten Village in the valley below, which was located on Hopkins Lane (cr. Budd Hopkins, the abduction researcher). Both abductions in Virginia involved the "wrinkled brow aliens: reported by Poly from Texas (Dr. Karla Turner, Masquerade of Angels, 1994).

With this history of abductions by extraterrestrials in mind, it should not be surprising that I had a deep seeded curiosity of the phenomenon, that just needed the conscious lid to hidden or suppressed memories popped open.

On 9 June 1992 I accompanied Crystall and two other sky watchers to the Jewish cemetery along Rte. 52 east of Pine Bush. We saw two UAPs take off from the field behind the cemetery (Appendix I). Those sightings convinced me that something strange indeed was happening there, and I jumped into an investigation with both feet. I borrowed an expensive geological magnetometer from Lamont-Doherty Earth Observatory, wanting to look for traces of unnatural magnetic activity below ground. That study would ultimately take three years to complete, and involved 1,800 different stations for measurements over a 24 square mile area. The results of that study are published in Chapter Eight (see also Appendix II). But what I discovered in the first 10 days would convince me that the entire valley as under siege, and alien bases underground were the least of my concerns.

Pat Huff and I went out into the field on the first weekend after my night out with Crystall and friends (Sunday, 14 June 1992). I had an EEG Proton Precession Magnetometer that could measure very small magnetic differences in the ground. It was used to map magnetic deposits and formations below the surface. The measurements were recorded in a small computer that was held from the neck. Measurements were recorded in magnetic "gammas" rather than in gauss or milligauss scales. We began a test run for measurements along Rte. 52 from just east of Flury Road, past the Jewish Cemetery, to the hill east of the cemetery. I recorded the measurements at ten pace intervals, and wrote them down in a pocket notebook. The computer also stored the same measurements, and gave each one a unique identifying number. Pat wrote down the measurements as I paced off the traverse along the shoulder of the road.

It soon became apparent that something was wrong. Either the magnetometer was malfunctioning, or there was a major magnetic interference or signal nearby. Instead of getting values for the magnetism of the ground (in the region of 54,000 gammas or 540 milligauss), I was recording spikes of magnetism in the tens of thousands of gammas above normal for the rocks in the valley! Most of the spikes were in the 65,500 gamma region, 11,500 gammas above background.

I stopped and recorded measurements for a while in one place, and saw these spikes recur about every 10 or 11 seconds, with much lower values in between. We drove east to Albany Post Rd., and I measured a traverse along that road to the north of Rte. 52. I was hoping that the interference was local, and I could obtain normal background values in another location. But I continued to record the same types of magnetic spikes 2-3 miles away from the Jewish cemetery. At that point I told Pat we couldn't continue until I got the magnetometer checked out at the lab. Disappointed, we returned home, not realizing that we had recorded an extraordinary event that would not be repeated during the following three years of taking measurements. The entire area was alive with signals coming out of the ground.

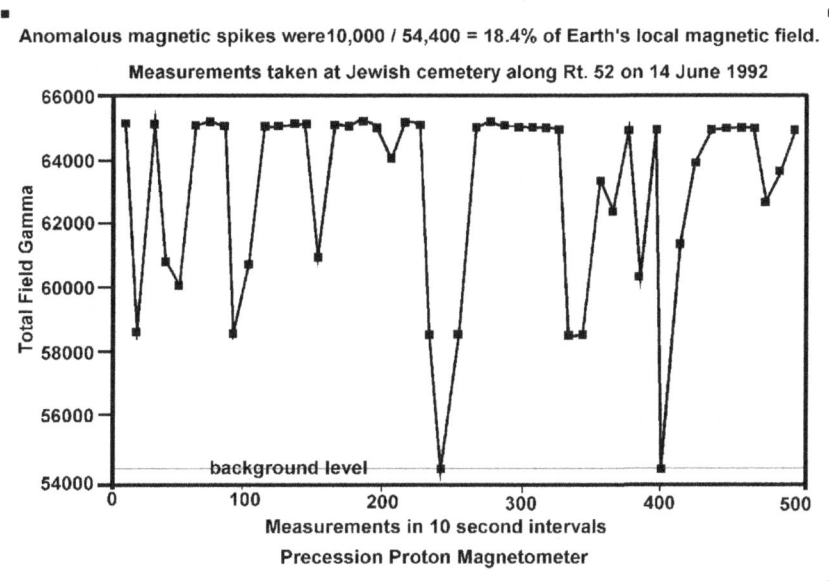

The next day (Monday, 15 June 1992) I had the magnetometer checked out by a technician at the lab. He calibrated it and said there was nothing wrong with it! So I took it back into the field, wondering what those strange spikes represented, and wanting to document them further. Were they the proof that I was looking for? Were they the "smoking gun" evidence of underground alien bases? Had I stumbled upon Earth-shattering data that would rock the scientific community? Not so fast.

That afternoon after work I went back to Rte. 52 near the cemetery and began taking measurements again. To my surprise there were no magnetic spikes. All measurements were normal and stable. I realized that either I had stumbled into the area unexpectedly, and caught something highly anomalous happening, or the anomalous spikes were the bait to interest me and hook me into doing a more intense and ambitious survey than I had originally planned. After all, collecting data for three years whenever I had time after work or on weekends was an enormous undertaking. It would require hiking many miles a week for months on end to collect 1,800 map points for magnetic contours. It would require going back to a predetermined dozen locations each week to confirm that those locations continued to give the same magnetic values. It would require taking measurements during the same daylight hours to minimize the magnetic influence of the Sun on my data. The absence of magnetic spikes meant that I could begin my survey. But I would soon learn that I was being observed and monitored.

On 18 June 1992 I went out after work to collect magnetic data along roads surrounding the UAP hotspot. Later that evening I met Ellen Crystall and Vince Valenti at the dogleg bend on West Searsville Rd., and we saw three UAPs after 10:00 pm. While taking measurements along the shoulder of Albany Post Rd., 0.2 miles south of Bruyn Turnpike, adjacent to a residential house, I was shot by a very unconventional device from across an expansive field. The time was 8:03 pm, and the Sun was low towards the horizon. I had my magnetometer with its computer box hanging from my neck and the magnetic sensor at the end of a long aluminum pole. My truck was parked far enough away that its metal would not interfere with my measurements.

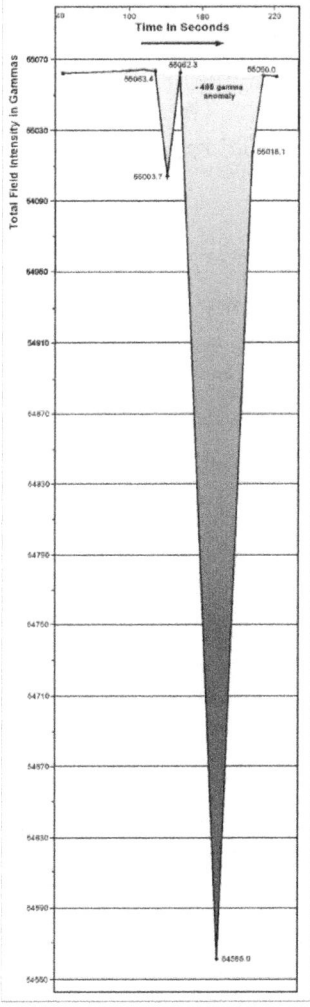

Normally I would take five measurements at each location or station, and use the mean or average measurement for the data point on my map. I had taken four normal measurements. The fifth measurement, however, was 60 gammas below normal. Because it was not normal, I decided to take more than five measurements until I had five close to the same value. Most measurements were within 2 or 3 gammas of the average. I recorded each measurement in a pocket notebook, even though the computer recorded each measurement and gave it a unique reference number, which was also recorded in my notebook.

The sixth measurement was normal. The seventh measurement, however, was a whopping 495 gammas below normal. As I pushed the record button on the rectangular computer control panel, there was a loud report that sounded like a powerful hunting rifle being fired in my direction. Based on my experiences hearing the sounds of hunting rifles fire, it sounded like a thirty-aught-six rifle being fired. I was looking down at the control panel and LED display on the computer when I heard the noise and instantly felt a sharp sting on my lower left chest at the bottom of my rib cage. I saw my shirt move in and out as if hit by a blast of air, but I saw no hole in my shirt or blood. Because I felt no immediate pain or threat of injury, I continued to take measurements, wanting to see what was happening magnetically after that anomalous reading. The thought I had in my mind was that a kid with a BB gun was hiding in the bushes near the house, and he had shot me. But a BB gun does not sound like a thirty-aught-six rifle discharge, and I saw no figure behind the bush. In hind sight I now realize I had been given a subliminal thought to dissuade me from suspecting I had been tagged by aliens for tracking purposes. I had been shot by a magnetic weapon, perhaps analogous to a rail gun that shoots magnetic bullets. The weapon had been sighted, then fired, leaving an unmistakable record of its action on my magnetometer.

The eighth measurement was still negative by 42 gammas, indicating that a residual magnetic field was decaying. The ninth and tenth measurements were normal. The duration for these ten measurements was 90 seconds. There was no movement or activity in the field before me, and no lights were on in the house to my left. I put my equipment back into my Ford pickup truck and drove back home, wondering what had just happened. No pain was evident, so I did not immediately check to see if there was any damage. When I got home I took off my shirt and undershirt.

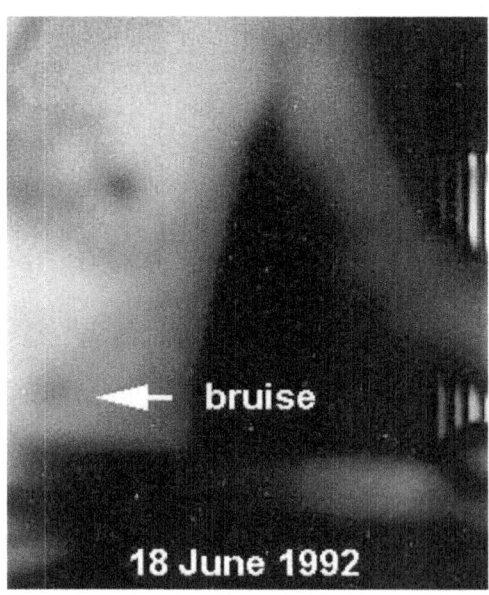

There on my chest where I felt the sting was a large bruise the size of a quarter. The bruised area was concave or indented, and there was a small red dot at the center of the concave indentation, representing broken capillary blood vessels. The bruise healed quickly, but ten days later I began passing blood in my urine, and my left kidney ached. Clearly something physical had penetrated my skin and entered my body.

I called my doctor in Suffern, NY, a 40 minute drive from my condominium. He told me to come to his office immediately. He said he would wait for me to get there. It was after work at Lamont-Doherty Earth Observatory, and I had just arrived home to discover blood in my urine. When I arrived at his office, he took a urine sample and analyzed it. He said I had a very bad kidney infection and gave me samples of a strong antibiotic to take until I could get his prescription filled the next day. I did not tell him what had happened to cause the infection, but it was clear to me that whatever hit me had the force to damage my left kidney and permit an infection to start.

Was I implanted with some type of tracking device or substance? Whatever it was, it was very tiny, perhaps nano-sized. I do not know what it was, and later x-rays showed no anomalous object in my kidney or near it. I do know that when I drove home that night, I was met by a UAP at the Harriman exit #16 off of the New York Thruway, and it paced me all the way home (Appendix I: 18 July 1992). The craft hovered in front of parkway lights that illuminate the toll plaza. I saw one bright plasma light and a small red light, with no visible structure connected them. The lights paced me from the toll plaza all the way home, and the craft kept pace with his truck, but would not allow me to get closer enough to it to see structure. It flew low over trees to right side of highway. As we approached Orange County Airport aircraft crossing highway from left to right, the UAP rose above the air traffic, changed its lighting configuration to resemble that of airplanes below, then dropped back down, returning to a single bright light after it passed over the aircraft. The object continued to pace me home, where it parked in night sky to the northwest. It parked itself in a location where no bright star existed at that time and place on a star map.

I got the distinct impression that whoever shot me was concerned for my health, and wanted to show his concern in a non-threatening manner. The next day (2 July 1992) I would see this same type of UAP again, but up close (less than 30 feet away) with Fred Brock, editor of the Wall Street Journal. We were driving on Rte. 52 west out of Walden, NY. After we had passed over the Wallkill River bridge, the road turns sharply north. A bright golden-white light with small red light behind it soon passed over Brock's car and took up a position directly in front of us over the road. It was no higher than the tops of telephone poles. No visible structure could be seen, only a bright white front light and a small red light in back, less than 15 feet apart. It rapidly became clear that I was being tracked. After the shooting, they even came to Red Bank, NY, after I moved there in late 1992. For a more complete description of this encounter, go to page 303. Simulation given below, showing the toll plaza lights at the top of tall poles:

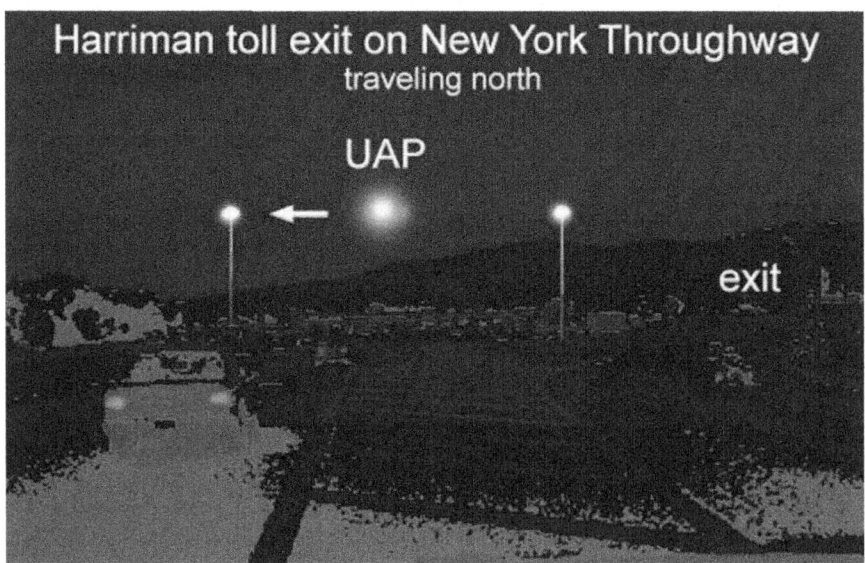

Whatever shot me wasn't a conventional weapon made by humans. No conventional weapon, aside from a possible magnetic railgun, could produce a magnetic anomaly like the one that shot me (cf. Sci Fi railgun in Eraser movie with Arnold Schwarzenegger, 1996). The projectile was microscopic, and did not visibly break the skin surface – just minor tissue damage and broken capillaries along a microscopic projectile path. It is a shame that scientists and doctors were so closed-minded towards alien technology at the time. What an incredible opportunity that encounter would have been for science to examine the anatomy of an alien tracking device and evidence for its high-tech delivery system, especially if the implant was a microscopic nanoprobe transmitter. Is it still in my body? Possibly. But if it lodged in my kidney, it may have been passed during the healing process, if it was nano-sized (one billionth) and small enough.

From March 1992 when my curiosity about UAPs and aliens was stimulated after seeing the miniseries, Intruders, on television, to being shot by an unknown magnetic device on 18 June, to being tracked and followed closely by a largely invisible or transparent UAP with a white and red light on 1 and 2 July, memories of my own abductions surfaced. Those memories stimulated me even more to want to discover why I was involved. Having repeated sightings in June and July of 1992 might be explained if I had somehow been involved with the visitors most of my life. That realization would only increase as contact continued in the months and years ahead. Both Crystall and I had been presented with mind-expanding evidence of ET existence and presence at the beginnings of our awakenings.

Mimicry in Black

Reports are common in UAP folklore of Men in black making their presence known, and appearing to people who had recent UAP sightings, warning them with threats to remain silent and to not share their encounters. Their vehicles would be described as vintage sedans, all black, but polished and looking new. Yet the vehicles were decades older than current cars. It was as if human-like ETs (Scandinavian or Nordic in appearance) had sampled human culture and technology years before, but had not realized just how rapidly clothing and car models changed. The same holds true for their craft, some of which they apparently modified to mimic our aircraft for the time or period of sampling. Thus, in hind sight, it no longer seems strange that they would modify their craft to resemble our aircraft, but for a period in our past. The following descriptions of a Boeing 707-like alien craft (The Boeing 707 first went into service in 1954, 38 years earlier.) highlight the visitor's attempt to blend in with our aircraft in the sky. It must have been a shock for them to discover that their attempt at mimicry or camouflage was already out-of-date in the 1990s.

On 23 June 1992 Crystall and I drove to the dogleg bend on West Searsville Rd., accompanied by her friend, Rich Pascorella. At 11:03 pm we saw our first UAP fly towards the "Indian Mound" from the southwest. It flew low to the trees silently. Only a pair of white lights could be seen. When it got over the fields next to the mound, it dropped down behind the tree row on the other side of the field in front of us, and disappeared. At 11:14 pm another UAP was spotted coming from the west over the low ridge. It too flew low, just above the tree tops, silently, and disappeared into the same field about 3,000 feet away (Appendix I: 23 June 1992). We knew these were not conventional aircraft.

Then at 11:22 pm Crystall spotted a cluster of lights approaching from the northwest on the other side of a low ridge (Thompson's Ridge). These lights were following the flight path for commercial aircraft flying to Stewart International Airport on the opposite side of the valley. Those flights would turn east and the red beacon on the ridge top just south of our location and fly towards the main runway at the airport.

I turned my camera around on the tripod, and opened the shutter for my first time exposure at 11:22 pm. I wanted to record every set of lights in the sky, not assuming the identity of any of them. It had been only two weeks since I began accompanying Crystall into the field at night, and I did not know what to expect. After the set of lights, which did not look like those of a conventional aircraft (no colored wingtip lights), had moved the estimated distance across the film frame for my zoom 300 mm lens, I closed the shutter and rotated the camera counterclockwise for the next time exposure. No sooner than I opened the shutter for the second picture, than something startling happened.

All of a sudden the lights began unfolding and moving apart. Some lights rotated outwards at the tips of wings, which were oriented vertically. Tail stabilizers unfolded at the rear of a cigar-shaped fuselage. Then we heard for the first time a sound, like that of a jet aircraft. This was not a conventional aircraft, because it continued to fly straight with its wings pointed up at the sky and down towards the ground. Had it been a fixed-wing aircraft, it would have fallen out of the sky like a rock.

A) West Searsville Rd. 11:14pm 6-23-92. Travelling W→E; silent, flat; same flight path (Treeline≈230°)

B) 11:19pm East of picture A, craft levels off after descent, dropped below Treeline at 132° (Treeline≈180°)

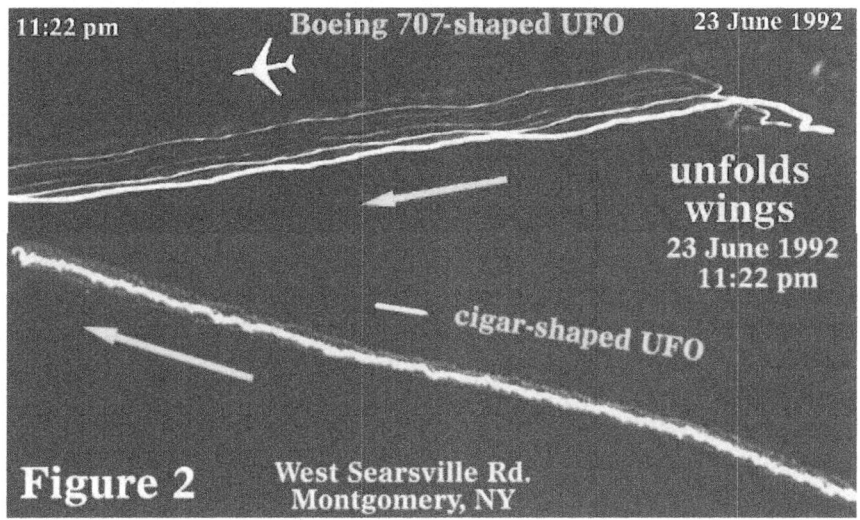

Figure 2 West Searsville Rd. Montgomery, NY

When the UAP-plane got to the red beacon on the ridge, it banked like a conventional aircraft, and turned east towards Stewart International Airport. I closed the shutter, realizing I may have just captures something no one had ever captured on film before – and I had: A UAP transforming into the overall shape and size of a conventional jetliner. Then Rich broke our dumbfounded silence by yelling, "No way!" as he bolted for his car parked on the shoulder of the road. His words and emphasis revealed that he was unconvinced it was a conventional aircraft, and that he was not fooled. Crystall yelled at me, "Come on," as she ran with her camera towards her car.

Please put this into perspective: I was a seasoned scientist who had for the first time in my life accompanied non-scientists into the field at night looking for UAPs. At the very least, I was blown away by what I had seen going into the field with Crystall just three times. I never expected to see what I saw. Planes don't do what I had just seen UAPs do flying in and out of farm fields and forests at night. I was convinced by then of Crystall's observations and many of her interpretations expressed in her book. But now I faced a response by Rich and Crystall I did not expect. I grabbed my camera and tripod and rushed to Crystall's car, jumping into the back seat. I barely got in and secured my camera on the seat when Crystall floored her car to catch up with Rich's car that was speeding down a winding narrow country road at night! He was chasing the UAP in his car! What had I gotten myself into, I thought.

We drove south to South Searsville Rd. and turned hard left. To my amazement Crystall said the UAP was right up ahead of us when I asked if she could even see it. We sped down South Searsville Rd. at over 50 mph, swerving and squealing back and forth as we encountered curves in the road. At that moment I had my first serious doubts about the venture I had gotten myself involved in. I thought we would end up in a ditch or in a car wreck. Because I was not in the front seat, I could not see the UAP lights above and in front of us. Then suddenly Crystall jammed on her brakes as we approached the intersection with Albany Post Rd.

Crystall got out of her car and ran to Rich, who was already out of his car looking up into the sky. I got out, and asked where it was. They both pointed up over the intersection. There at an altitude of about 1,500 feet, below a bank of high clouds backlit by the Moon (producing a silver-screen effect) was what looked like an all-black Boeing 707. The only problem is that it had stopped in mid-air and was hovering silently. We witnessed it rotate "on a dime" 180 degrees and then fly back over us in the direction from which it had come. As it flew directly over us, we could see its outline clearly. Indeed, it looked like a Boeing 707, but there was no indication of engine pods on its wings – or for that matter, on its fuselage. It flew back west silently and disappeared from view.

Imagine yourself as a scientist in my shoes. You had just witnessed something that few humans have ever witnessed, something that had the potential of shattering your concept of reality. Who would believe you? Even with two other witnesses, most people would have trouble believing that planes could fly without wings, or with wings oriented towards the ground and sky. And if you firmly rejected the idea that Earth was being visited by beings with advanced technology, how could you explain what you just saw? Then I got my film developed, and there in Black & White prints was irrefutable evidence of what we had seen. The second time exposure showed the lights moving apart as the craft unfolded its wings and stabilizers! Without that hard evidence, all I had is a story, an anecdote. I now had photographic proof that I was not delusional or deceived and mistaken, whereas most of my scientific colleagues and skeptics were in denial and disbelief. Hynek et al. (1998) in their book expressed the same sentiments towards the myopic vision of a scared scientific community. Hynek was originally a skeptical scientist, until he began reviewing case after case of sightings by others.

The sighting of a UAP that resembles an all-black Boeing 707 without external engines on 23 June 1992 was but a warm up rehearsal of what I would see and record on 24 September, just 86 days later. I would get to see this phantom Boeing 707 up close during daylight, and have it fly over me twice so that I could capture its existence on multiple time exposures. In the meantime, I would have 17 additional UAP sightings, and see two more Black Triangles and the Manta Ray twice (Appendix I). I was not only being given clear unmistakable evidence that our skies were being invaded by extraterrestrial technology, but also that our visitors were capable of **mimicry** and **camouflage** in order to remain in the shadows of human conscious recognition and acceptance. The Mayans prophesied that the gods would return again near the end of their last Long Year near the year 2012, and so they have (2012, The Year of the Mayan Prophecy, by Daniel Pinchbeck, 2008).

It would also be during this period of time that my attempts to educate my scientific colleagues at Lamont-Doherty Geological Observatory (later renamed the Lamont-Doherty Earth Observatory) would encounter a brick wall of skepticism that would ultimately turn to censorship and loss of my Associate Research position in 1993. It then became clear to me how the3 government was controlling the thousands of scientists across the country through intimidation, fear of ridicule, loss of reputation, and withdrawal of NSF funding if they failed to comply with their unofficial ban on UAP research.

Have you ever been followed home by a Boeing 707?

http://www.sunstar-solutions.com/AOP/Boeing/boeing07.htm

On 24 September 1992, late Thursday afternoon, I was driving home from Lamont-Doherty in my two-tone brown, Ford F100 pickup truck. I would take back roads to the New York Thruway, and travel north until I reached the Harriman Exit #16 to State Route 17, which I would take west to Middletown where I lived. We had just completed drilling a series of core holes in New Jersey, and I had been describing cores stored at the Lamont-Doherty core repository on campus at Palisades, NY. It normally took me an hour to commute to home. It was still daylight but near sunset when I was approaching the Harriman exit. There was a lot of traffic on the New York Thruway. The highway sign for the exit was up ahead. Then I saw it. A large dark aircraft was flying towards me on the right side of the highway. It was a couple miles away, and was flying no more than 300 feet above the highway. That was determined by it being below the tops of the mountains (e.g. Bear Mountain) to the east side of the thruway, which rise no more than 300 feet above the valley floor.

About a mile away the aircraft flashed its headlights three times in a row. That caught my attention, making me think the pilot was trying to warn drivers that he was in trouble and was preparing to land on the highway. I assumed the aircraft had just taken off from Stewart International Airport at the next exit (#17) about ten miles north of the Harriman exit. I had to watch for my exit coming up, keeping one eye on this plane as it got closer and closer. Then right in front of me in full daylight what superficially looked like a jetliner banked to its right and crossed the highway only about 90-100 feet above the ground! It crossed just to the south of the overhead highway signs for the exit ahead (the simulation below shows its position over the thruway). I watched in amazement as this large aircraft slowly crossed over the highway, reaching the other side at about the time I passed it (truck position #1 in map below). It was all black, with no markings on it, but even more strangely, it had no visible fuselage windows, even in front for the cockpit. I heard the distinct sound of jet engines, but I could see no engine pods on its wings or fuselage. And yet its shape and size was distinctly that of a Boeing 707.

Equally disconcerting was its apparent speed: It crossed the highway no faster than the vehicular traffic on the highway! I would later during the same trip clock its speed at 50 mph or less, about 100 mph below stall speed for a real Boeing 707. I was relieved that the pilot made a turn without crashing on the highway, and may have been trying to make it back to Stewart International Airport. Then I took my exit.

After paying the toll, I proceeded west on State Route 17 towards Middletown, NY. The Sun was very low in the sky, almost to the horizon, and directly over the highway in the distance. About 1.4

miles from the toll booths I was shocked to see the same black Boeing 707 again. This time it was flying just above the trees and crossed the highway about 900 feet (1.7 miles) ahead of me (truck position #2 in map below) traveling from left to right. Even though it was heading towards Stewart airport, I couldn't imagine it making that distance at such a low altitude with hills rising above it on either side. I fully expected to see a fireball rise off to my right any second, but none occurred.

I drove about a half mile, and then to my surprise I saw it again (truck position #3 in map below), now crossing the highway again about 900 mile ahead of me traveling from right to left! This time the Sun was on the horizon and directly behind the UAP. It took several seconds for it to cross the road, indicating it was still traveling at an impossibly slow speed for a real Boeing 707. I lost track of it behind trees on the left side of the highway as it traveled south, away from Stewart! I assumed the pilot was still looking for a non-populated area to crash land between the low hills and residential areas around Harriman, NY.

I drove about another half mile, and again saw this craft slowly cross the highway just above the trees about 900 feet ahead of me (truck position # 4), but this time from left to right. The Sun was fading below the horizon at 6:49 pm, and soon it was dark. Over a distance of about 3.4 miles and a period of five minutes, I had seen the same craft four times.

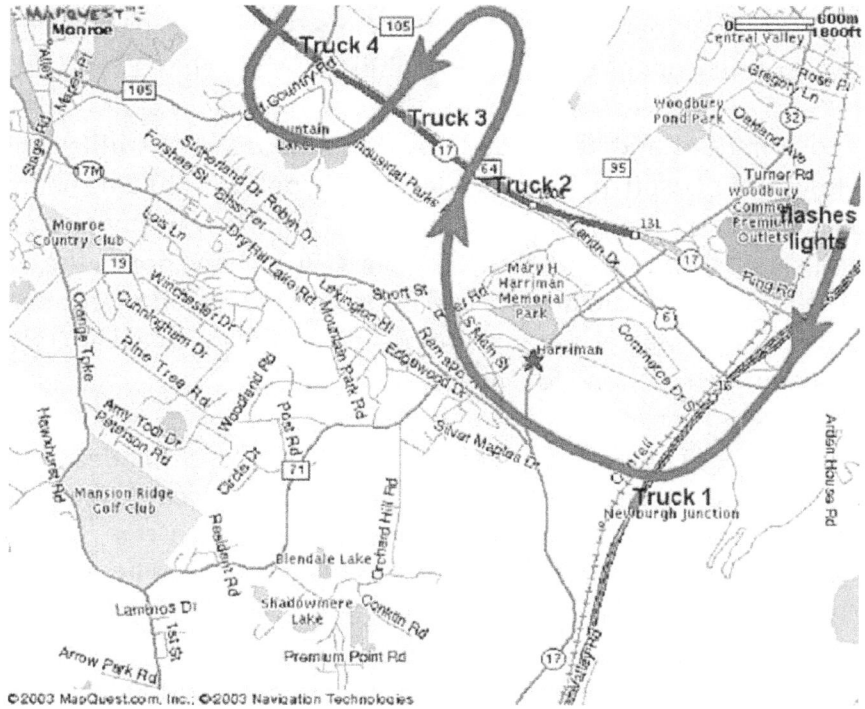

It had taken a sinuous flight path back and forth over the highway, and was clearly targeting me. Only those drivers headed in my direction either directly ahead of me or behind me would have seen the same spectacle. I wondered what other drivers who saw the UAP-plane were thinking. It would take me another 20 minutes until I reached my exit for Route 211 East at Middletown. Nothing happened again until I reached that exit 25 minutes later.

As I approached my exit at around 7:15 pm, I saw an aircraft approaching my location over Middletown on the left side of the highway (truck position #5 on map below). It was dark, and the aircraft had all its navigation lights and headlights on. That was not unusual for this area at night, because Route 211 East was under the flight path to Stewart International Airport 21 miles to the east. Many commercial aircraft fly over this road in this area each day at about 1,500 feet altitude. But as this aircraft approached, I had a very strong gut feeling that it was the same UAP-plane. Part of that feeling came from its unusually low altitude, well below 1,000 feet!

I took my exit and descended down the ramp to the traffic light at Route 211 East (truck position #6 on map below). As I did I heard the sound of jet engines overhead and assumed it was that aircraft turning and lining up with Route 211 East. But when I got to the rid light and stopped, the aircraft had not appeared ahead of me, and its sound remained above me. Impossible if this was a real aircraft. It had stopped with me for the light!

When the light turned green, I turned right and headed east down the road. The black Boeing 707 emerged from above my truck and turned also to the east. As I drove down the road, it slowly moved ahead of me. I accelerated and kept up with it until I was slowed back down by traffic ahead of me on the road. I clocked it at 45 mph! I watched in amazement as the UAP-plane disappeared in the distance, flying no more than 300-400 feet above the ground (truck position #7 on map below).

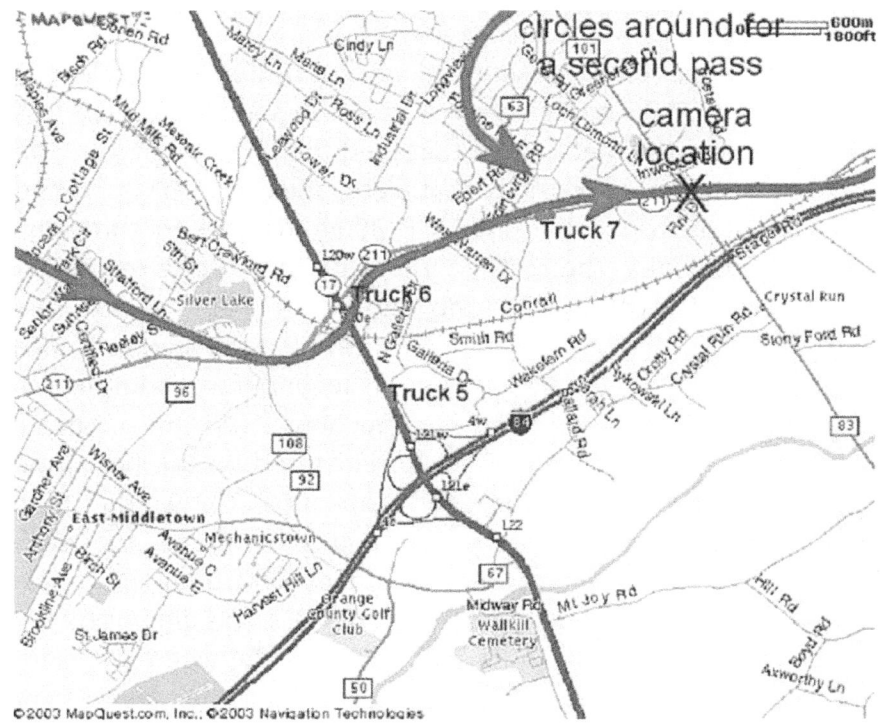

My condominium at Hillside Village was only 1.5 miles from State Route 17. It took me about five minutes to reach the parking lot next to my condominium (X on map below). Sensing that something was up (literally), I ran into my home and retrieved my

camera and tripod set-up. Outside again, I walked to the lawn in front of the condominium complex along State Route 211 E near the intersection with Goshen Turnpike (State Rte. 101). Since this craft followed me home, I thought it might circle around and come back down Rte. 211 E again, but this time towards me. I didn't know why I thought that, but I did. I waited for about fifteen minutes.

I had a chopper device assembled in front of the 300 mm lens of my Minolta XG7 35 mm camera. A pressure-sensitive switch was installed below the metal plate that held the chopper motor. The camera was mounted on top of that plate, and it in turn was bolted to the top of the tripod. In front of the 300 mm lens were three pie-slice-shaped blades. The motor turned the blades one revolution every three seconds. The movement of lights would be  intermittently blocked once a second for about half a second from being registered on the film (time exposure). This would create a dashed trace for each steady light of an aircraft or UAP. If some dimension for distance between two light traces could be measured or estimated, the speed of the object could be calculated. The chopper device (see Cornet using the chopper in image below) was a scientific tool engineered and built by Bob Wisch, a professional and expert in optics who accompanied us into the field numerous times (Appendix I).

Soon my wait paid off. In the distance I saw a set of aircraft lights flying over Rte. 211 E towards me, not far above the street lights. Because of its low altitude, I knew this was the black Boeing 707 returning. I waited until its wings nearly filled my viewfinder and opened the shutter at 7:39 pm. I then swung my camera around and took a picture of it leaving. I took four time exposures as it flew over and past me. I turned on the chopper device for the third and fourth photographs.

I calculated its speed based on distance traveled (406 feet in about 11 seconds) at 25 mph. I used the wingspan of a conventional Boeing 707 (145.6 ft.) for measuring the distance traveled on the time exposure light traces. During the middle of that time exposure I activated the chopper device. The third and fourth pictures of it were taken as it flew past the parking lot and disappeared in the distance as it traveled east. The calculated speed for the second time exposure was again 25 mph, but the speed for the third time exposure was 17.4 mph (425 feet in about 17 seconds). For the calculated speed to have been greater, the wingspan would have to be increased proportionately. Consequently, it can be safely concluded that unless one wants to use a wingspan measurement of 800 feet or more (which is preposterous), this aircraft was probably flying at less than 1/5th the stall speed of a Boeing 707.

After the UAP-plane flew over me and disappeared into the night sky past the tree line, I waited for about 15 minutes, again thinking that the event was not over. Then at 7:55 pm I saw the same craft lights again over Rte. 211 E heading towards me from the west. This time I wanted to capture an image of it just before it reached me without the chopper activated, so I waited to turn on the chopper half way through the time exposure. I waited patiently, focusing my camera as I poised my finger on the sensitive chopper switch. When the craft was in range I opened the shutter.

I braced myself as the aircraft slowly approached as it took the exact same route it had taken before (over Rte. 211E). As its image filled the lens of his camera, I opened the shutter. But to my surprise the aircraft momentarily stopped in midair, backed up, and then proceeded forward in a slight saltatorial motion (jumping or bouncing: Its movement resembled a drunk stumbling along as he walked). I saw this happen, but forced myself to concentrate on getting a quality image. See the labeled image of Figure 4 below. Its movement is not something a conventional aircraft is capable of doing, and I caught it on film. I estimated when the object would be in the middle of the picture frame and gently pushed the button to the chopper device. The latter half of that image became segmented, and from that data I was able to calculate additional velocity information. All totaled, I took three time exposures of the UAP during its second pass over. Note the decrease in apparent speed during each fly-over:

Fly-over	Calculation for first image	Calculation for second image	Percent decrease
7:39 - 7:40 pm	25.0 mph	17.4 mph	30%
7:55 - 7:56 pm	53.0 mph	32.7 mph	38%

Note also that the white lights visible from a frontal view become golden colored when the UAP-plane is viewed from the rear. The light pattern is different from that of conventional jetliners in that the lights that simulated landing lights (wing-mounted lights on either side of the fuselage) become smaller and golden colored. On conventional jetliners the landing lights are shielded so that they cannot be seen from the rear. Also, a small white light was situated on the belly just behind a larger white light, which is redundant. That light has an unusual bluish tint to it. Usually the belly light is red, not white or bluish white. In addition, the forward white belly light in time exposure #5 became several closely-spaced smaller golden-colored lights in time exposure #6. Lights on conventional aircraft do not change like this.

And finally, as the UAP-plane approached me in time exposure #5, the set of lights intended to simulate the red and green navigation lights on the wingtips are not shielded from a set of white lights usually located directly behind them on conventional jetliners. These sets of lights are always shielded on conventional aircraft so that pilots can determine if an aircraft is approaching them or moving away from them. Therefore, both sets of wingtip lights should not be visible in time exposure #5. Both sets can be seen clearly only when a jetliner is directly above the observer. The intensity of each set rapidly changes as the shielding between them unblocks the rear set as the forward set becomes blocked with changing angle of view. Was I being used for quality control in our visitor's attempt to perfect their camouflage and mimicry?

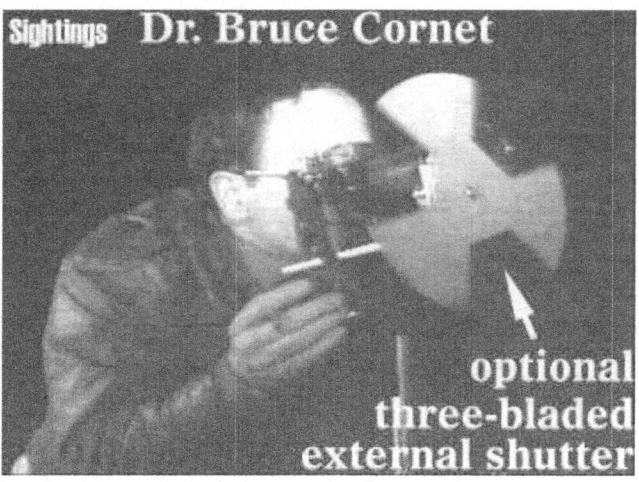

Figure 1, 7:39 pm. Note enlargement showing pulsing energy reflections off fuselage.

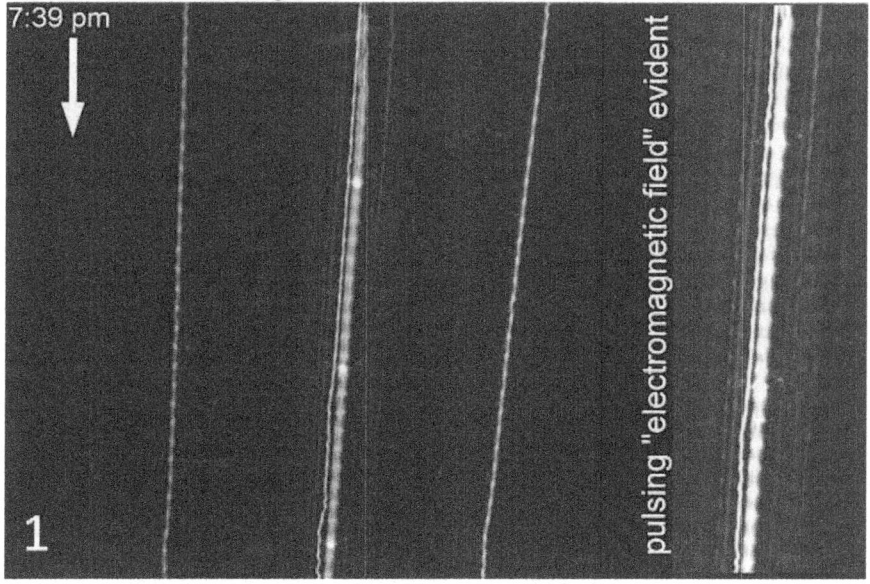

Figure 2. Chopper used for entire time exposure, causing segmentation.

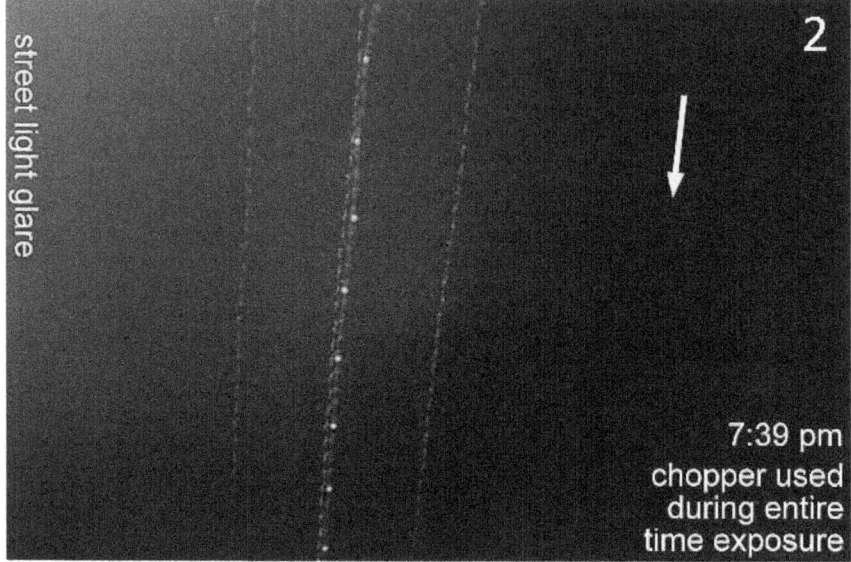

Figures 3 and 4, 7:40 pm. Note how craft swerves to avoid street light.

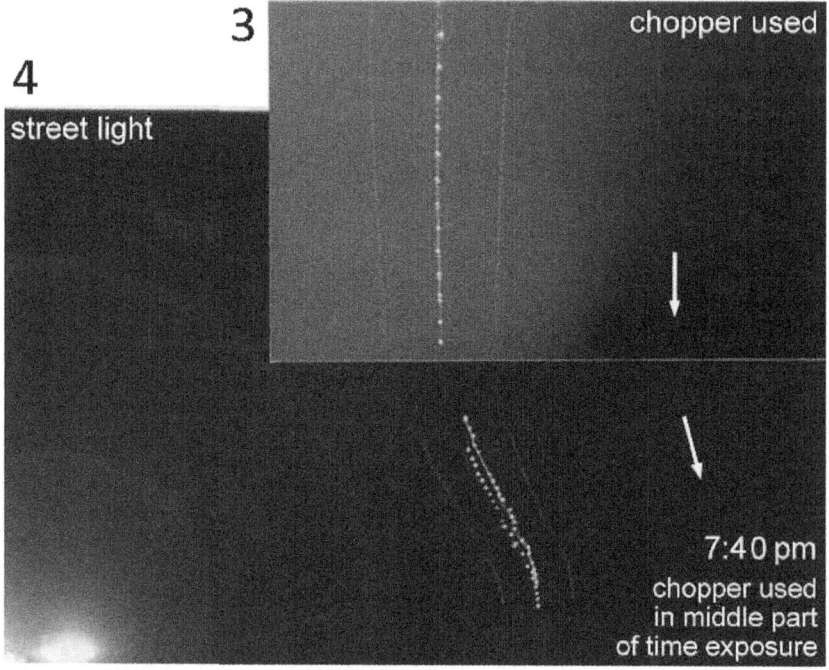

Figure 5, 7:55 pm. This image is explained below. Chopper used at end of time exposure.

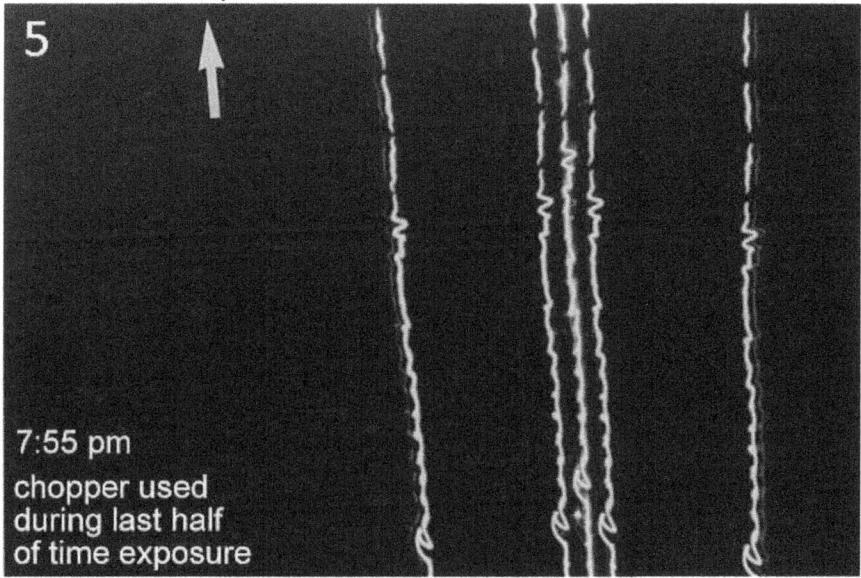

7:55 pm
chopper used during last half of time exposure

Figure 6, 7:55 pm. Note sinusoidal movement of lights and multiple strands of lights.

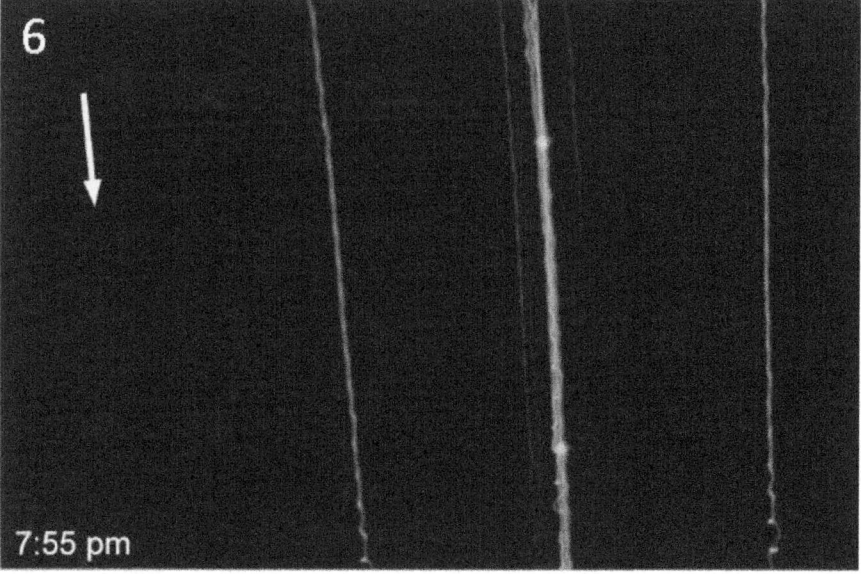

7:55 pm

Figure 7, 7:55 pm. Chopper causes light traces to break up into segments.

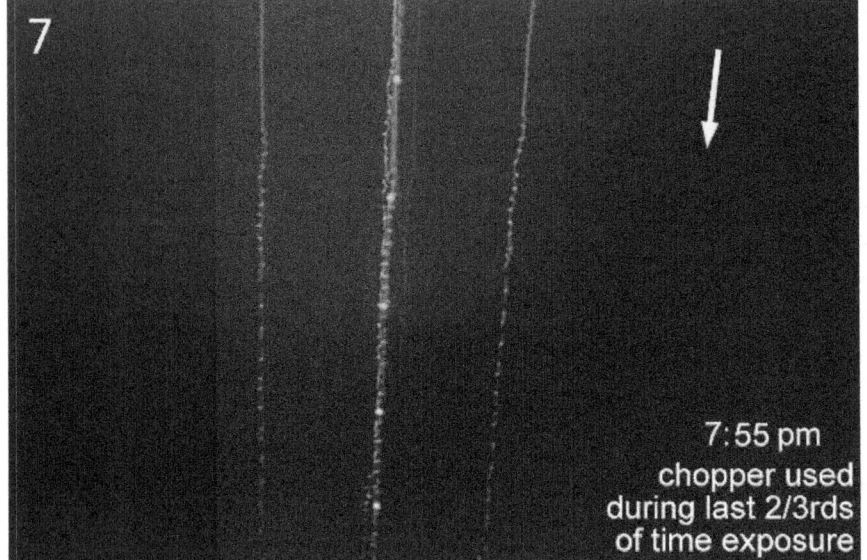

Time exposure #4 with labels to show how chopping the image allowed me to calculate speed, given that width of outer lights was equivalent to the wingspan length of a Boeing 707 or 146 feet. Note how the lights at beginning of time exposure (bottom) show that the craft stopped, backed up, and then moved foreward doing alternating loops through the air, something conventional aircraft cannot do, and also something that manipulating the camera mechanically cannot do.

The first time exposure (#1 below) of it traveling directly over me on the first pass at 7:39 pm also showed something very strange. The entire belly of the cigar-shaped fuselage was illuminated or glowing, as if the lights reflected off of a metallic surface. However, the bars of light on the fuselage indicate a pulsing action as the reflections grew brighter and dimmer, implying some sort of alternating or pulsing power source. The sound coming from the UAP-plane was similar to that of a jet aircraft, but nowhere near as loud as that of a real Boeing 707 flying no more than 300-400 feet overhead (clearly well below F.A.A. minimum altitude of 1,500 feet).

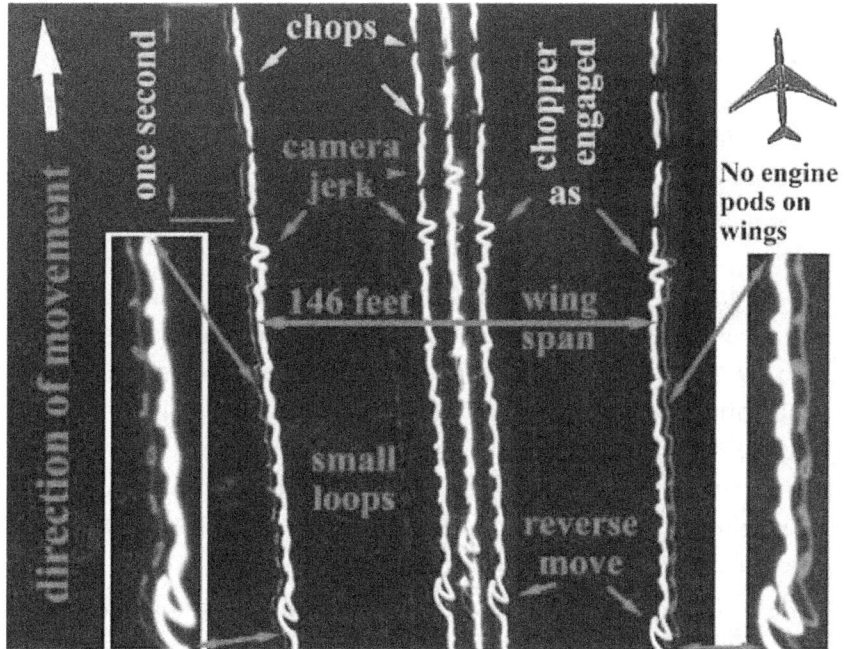

24 September 1992, Middletown, New York.
Time exposure of lights on Boeing-shaped UFO over Rt. 211E travelling north at ~ 300 ft. elevation.
Calculated speed: 53 mph

© B.Cornet 1997

Other UAP-Plane Sightings

This is not the only type of UAP plane I saw near Pine Bush. Crystall (1991) reports that she saw the "Big One" come down for a landing in a field, and she and her friend Cathy witnessed a small winged UAP at close range fly over a house and tree (see page 82). On two separate occasions I saw a T-tail aircraft that resembled a DC-9, but it had no engines near the back of its fuselage or on its wings. The first time I saw this UAP was on 18 July 1992 as I was driving north on Albany Post Rd. I was traveling at about 40 mph and heard this noise of a jet behind me. I looked to my right and saw a large all-white jetliner slowly pass my car about 200 feet above the ground and trees. Its body seemed to glow dimly, which is why I could see it at night. I could see no engines on its wings or fuselage. I did see what looked like dark cockpit windows, but no windows along the side of the fuselage, and there were no identification markings of any kind on it. The distinctive T-tail, like that of a DC-9, was the last thing I saw as it slowly disappeared in the distance, traveling no more than 60 mph.

On 25 January 1997 John Macedo Jr. and I went to South Searsville Rd. just east of the beacon on Thompson's ridge, which is used by commercial aircraft to align with the runway at Stewart International Airport on the other side of the valley. Below I reproduce the account of three sightings from my web page to illustrate the unusual nature of these sightings, and how some of them resemble UAP-planes. In this account I call UAPs: AOP, which stands for Anomalistic Observational Phenomena. After the three craft flew over us, we followed them north and watched them one by one circle and land in an area where Lake Osiris is located just north of Walden, NY.

The Landings

http://www.sunstar-solutions.com/AOP/Landings/landing.htm

Macedo and Cornet initially parked on the shoulder of South Searsville Rd. between two farm fields (location 1X on map below), a location Ellen Crystall had taken Cornet on several occasions (S. Searsville Station). The Moon was full and bright, giving added illumination for photography that evening. Visibility was excellent. The time was about 8:00 pm.

At 8:17 pm the first set of lights (AOP No. 1) was spotted just to the right (north) of the red beacon on top of the ridge, but they were no more than a few hundred feet above the ridge as they crossed it. From that point onwards the lights climbed in altitude as they came directly towards Cornet and Macedo. Cornet captured excellent video of the craft as it passed overhead, and then flew nearly due east until it was almost to the other side of the valley (where Stewart International Airport is located). The only problem they initially had with this object being a commercial jetliner is that it dropped out of sight before getting to Stewart. Most jetliners on approach to Stewart can be seen descending slowly until they get to the airport. AOP No. 1 disappeared well short of and to the north of the runway.

When AOP No. 1 flew over Cornet and Macedo, they could not see a fuselage. All they saw was a deltoid-shape wing with what appeared to be two booms sticking out front, which could be seen because a large pair of lights at the front of the wing illuminated them. Reflection off those booms is what led a number of skeptics to question Cornet's interpretation, but as will be seen and detailed on a web page for this AOP, there is a large amount of contradictory data that cannot be explained by a commercial jetliner identification.

At 8:24 pm another set of lights (AOP No. 2) was spotted approaching their position from behind the ridge to the west. This set of lights crossed the ridge a little further to the south (see map and picture below). As it flew over them, Cornet saw the distinctive shape of a jetliner with a T-tail, similar to that of a Boeing 727 or DC-9. However, no engine pods were seen attached to its fuselage or wings. DC-9's commonly land at Stewart airport, but this jetliner was not flying towards Stewart. It was flying in a northeasterly direction towards Wallkill, NY. There are no airports in that direction. AOP No. 2 also climbed in altitude as it approached their position, which is unusual for commercial aircraft flying in this region and in that direction.

As AOP No. 2 flew slowly away, Cornet captured on video something highly unusual. At 8:25:02 pm the AOP lit up its belly.

That's right, its belly glowed white for two seconds, which can be seen clearly on the video. There was no searchlight beam coming up from the ground, which might be the first suspicion of a skeptic. In addition, the strobes on its wings, belly, and tail fired semi-randomly and out of sync, something that conventional aircraft strobes do not do. More data in support of these statements are given on a web page for this AOP.

Cornet and Macedo decided to drive to West Searsville Rd., but after driving a very short distance (about 30 seconds driving time) west on South Searsville Rd. they spotted a third set of lights approaching them from the west. The lights were at about 500 feet altitude (well below the minimum 1,000 ft. FAA-approved altitude). Macedo quickly pulled to the shoulder of the road and stopped (2X on location map below). They got out rapidly and Cornet began videotaping almost immediately. The time was 8:26 pm.

AOP No. 3 approached over the saddle in the ridge where South Searsville Rd. goes over the ridge (north or to the right of the previous approaches). As it approached, it also climbed in altitude, but more steeply than the previous AOPs. Its lighting pattern was unlike that of a conventional jetliner. A bright light (as bright as a landing light) was positioned at its port (left) side, while no complimentary or symmetrical light was positioned at its starboard (right) side. A small light was perched above the plane of the craft off to one side – not where a fuselage would be. But what was most unusual is that as AOP No. 3 approached, it put on a spectacular colored light show, flashing multicolored strobes all over its belly and leading edge. In other words, it illuminated lights in places where conventional aircraft do not have lights, and it flashed them in a random sequence. A web page has already been devoted to this sighting: The Light Show, which includes a detailed animated gif. More information is given here on a web page for this AOP.

The shape of AOP No. 3 was unlike that of any conventional jetliner. Its shape resembled a kite, which is a shape Cornet has seen and photographed many times in the valley. He calls the AOP with this shape the Manta Ray.

Map below shows initial approaches and directions of travel.

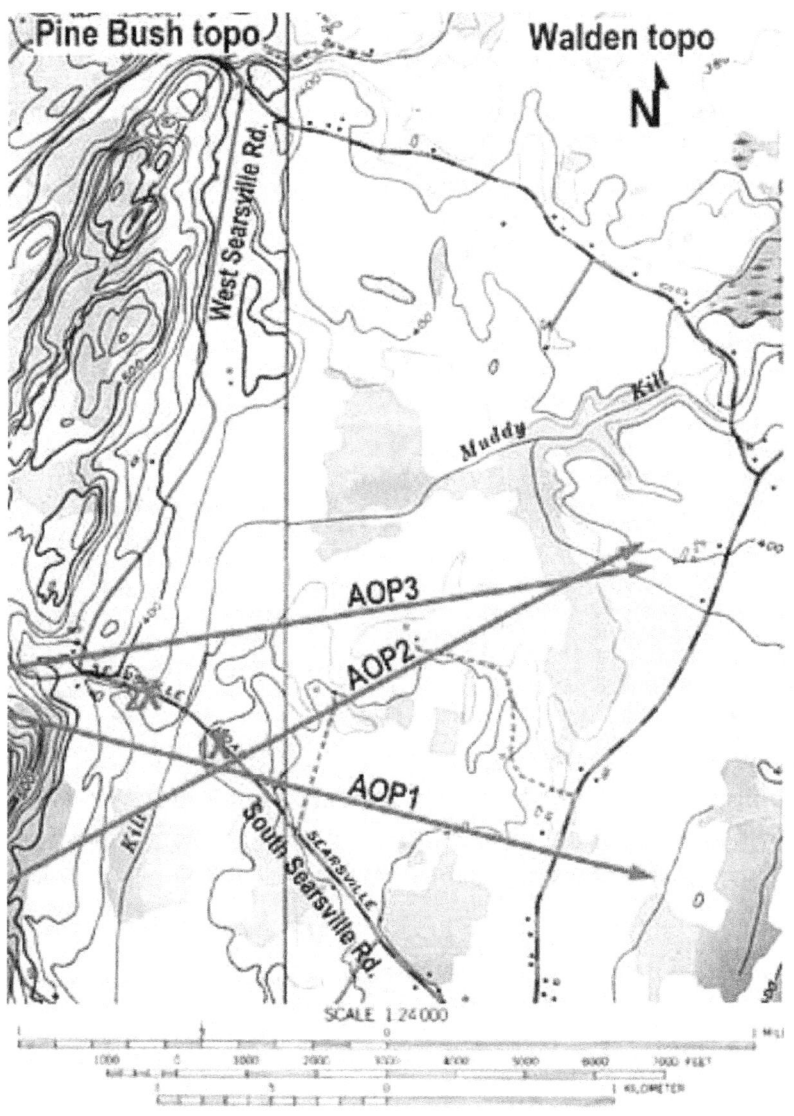

All three AOP produced muffled jet-like sounds as they flew over Cornet and Macedo. Skeptics beware: Sounds can be faked and broadcast. Some of these same AOP or similar ones have been recorded producing various types of sounds, and have been recorded producing no sound at all. See frequency spectrogram for AOP No. 1.

Photograph below showing direction of approaches of the three AOP. Thompson ridge is visible in the distance. There is a red beacon positioned on top of that hill, which directs commercial aircraft to Steward International Airport. Note that the altitude for all three AOP was below 1,000 feet (minimum FAA altitude = 1,500 feet).

Below pattern of strobes on the T-tail UAP-plane shows abnormal strobing of its wingtip and belly lights, out of sequence with one another and unequal in brightness. The timing and duration of the belly glow is also shown, proving this was no FAA-approved aircraft.

Unconventional Aerial Phenomena

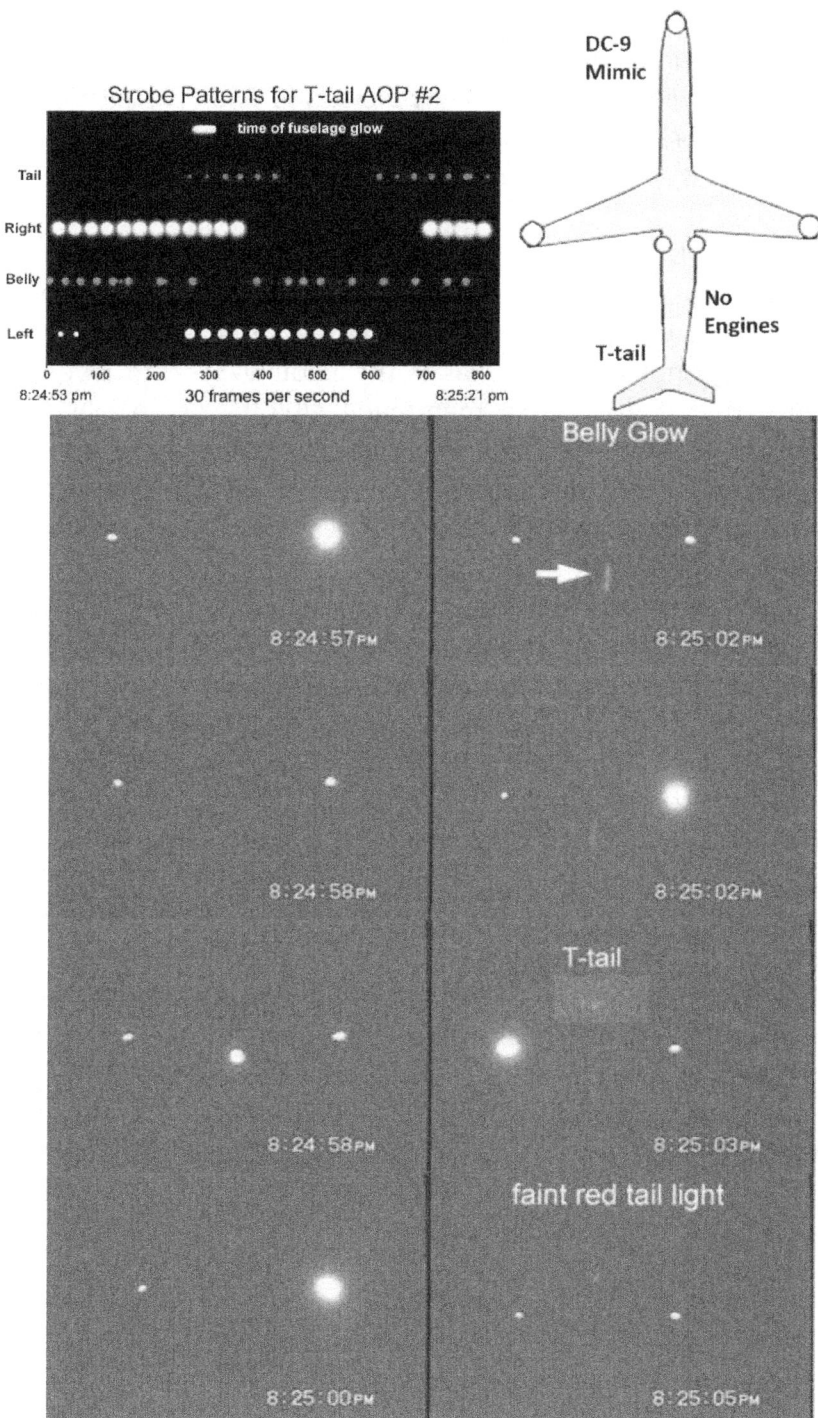

343

Above: Individual frames from video showing uncoordinated brightness of wingtip strobes, belly glow, and T-tail (barely visible with computer contrast enhancement), but with no external engines. This is an excellent example of alien mimicry.

As the T-tail UAP-plane flew away, it illuminated the length of its underbelly – a sight to behold. The Illumination lasted only for a couple of seconds, but was clearly something that a conventional jetliner could not do. It was as if the pilot wanted to give me evidence that it was not a man-made aircraft, or he was just signaling an acknowledgement to us: "Hi," or "Surprise!" The video record clearly shows the outline of the plane-like object, its wings, and its T-tail. Conspicuously absent were any engine pods on its fuselage or wings.

In summary, on four separate occasions UAPs in the shape of man-made aircraft made their appearance between June 1992 and January 1997 (Appendix I). One was shaped after the 1960s version of the Boeing 707, while the other was shaped after the contemporaneous DC-9 used by American airlines out of Stewart International airport in Newburgh, NY. The best images of the overall shape of the DC-9 UAP were captured on video on 25 January 1997 near the end of the second wave of UAP sightings in the region. The time exposures of the Boeing 707 UAP captured in June and September of 1992 provide conclusive evidence of advanced alien technology and attempts at mimicry, and would not have been provided unless the visitors wanted to reveal themselves to a selected audience (cf. leaky embargo hypothesis of Deardorff, 1986).

The Transformation

http://www.sunstar-solutions.com/AOP/SOW/transfor.htm

Events for the night of 24 September 1992 did not end with the Boeing 707 UAP flyovers. Something told me to wait, so I patiently waited by my camera on the lawn at the corner of Goshen Turnpike and Route 211 East for three hours. At 10:57 I spotted a bright golden white light moving from left to right (south to north) very low over the trees and turn sharply towards me when it reached Route 211 East. I immediately focused my camera on this light, which was larger and brighter than any aircraft light I had seen, and opened the shutter. I took a series of seven time exposures as this UAP came towards me, climbing in altitude as it approached, and flew directly over me. When it flew above me it no longer resembled a giant ball of plasma light, but had wings and navigation lights and strobes like that of a small private aircraft. In addition, I heard the distinct but quiet sound of a propeller-driven plane (Appendix I: 24 September 1992).

All the time it approached me it was changing and becoming more and more like a small aircraft, based on its lights. Initially it was just a single bright light that progressively dimmed and grew smaller in size. Then outboard lights became visible (or turned on). They were red on the right and bluish green on the left. The strobe on the left side strobed twice in a row each time it flashed, while the strobe on the right fired only once each time (except once when it strobed twice), and sometimes did not fire in sequence (missed a beat). It was not until I got the pictures developed that I could make out spectacular details of light movement that were not evident while I focused on taking the pictures. Because the UAP climbed in altitude and slowed down as it approached, details of its lights were "stretched" out on my film, details that would not have been recorded had the craft maintained a constant altitude and sped past me.

What became evident when I got the negatives enlarged, and then photographed the negatives under enlargement on a light microscope, was the spiral and looping form of all the lights. The difference in strobe pattern for port and starboard sides (left and right wingtips) was very obvious, indicating that the strobes on the right side were either malfunctioning or that there was a deliberate attempt to falsify a conventional aircraft interpretation by the pilot. The strobe discrepancy by itself is not the "smoking gun" proof for alien technology. It is what all the lights were doing (i.e. tight loops) as the UAP moved towards my camera. They were all looping in unison in an alternating pattern, first clockwise, then counterclockwise (backwards), then clockwise again and again. Humans could rig an aircraft with lights that moved like that, but it would be very expensive and for what purpose? In addition, the rate of looping can be seen on the enlarged negatives to decrease from a very high rate of spiraling when the UAP first turned towards me over the road (Appendix I: color photographs of spiraling lights).

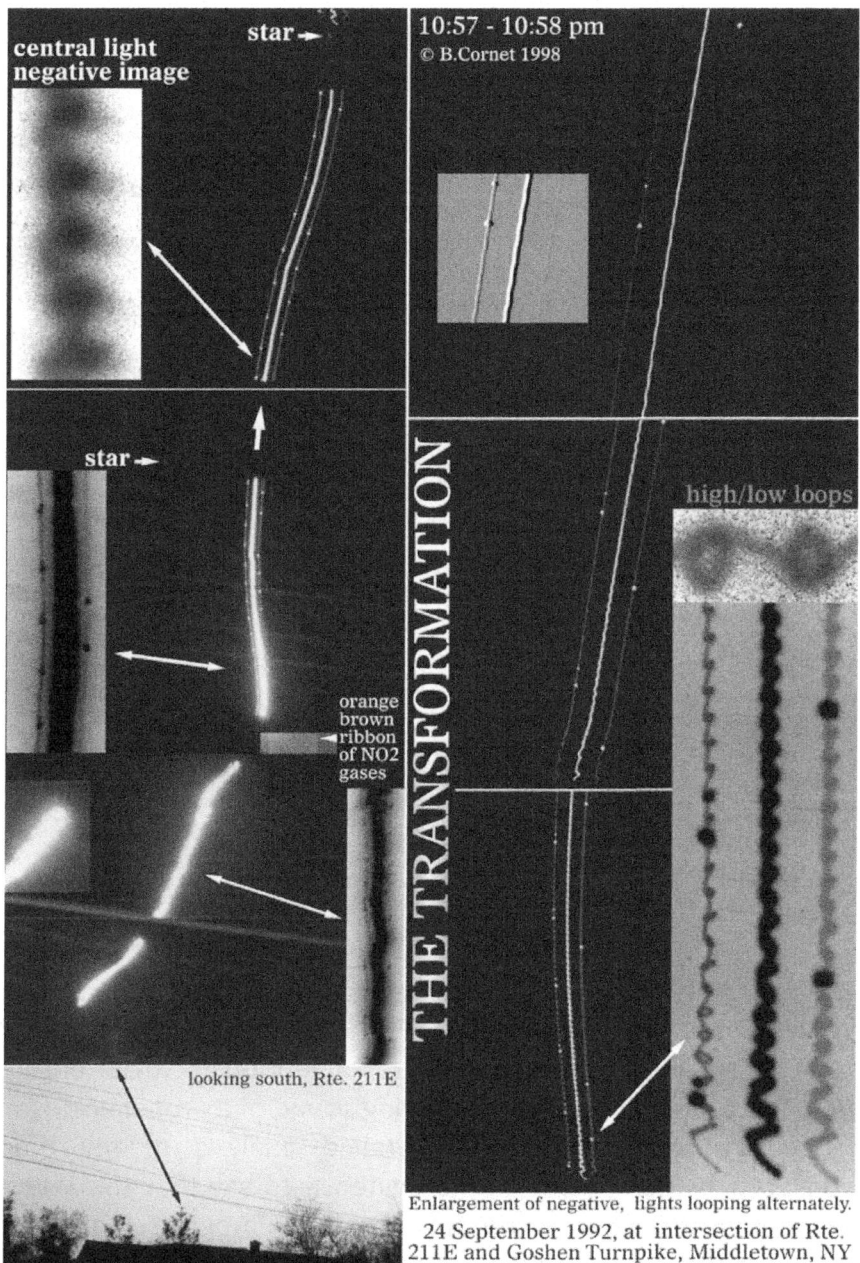

Enlargement of negative, lights looping alternately.
24 September 1992, at intersection of Rte.
211E and Goshen Turnpike, Middletown, NY

24 September 1992 10:57pm
TRANSFORMATION

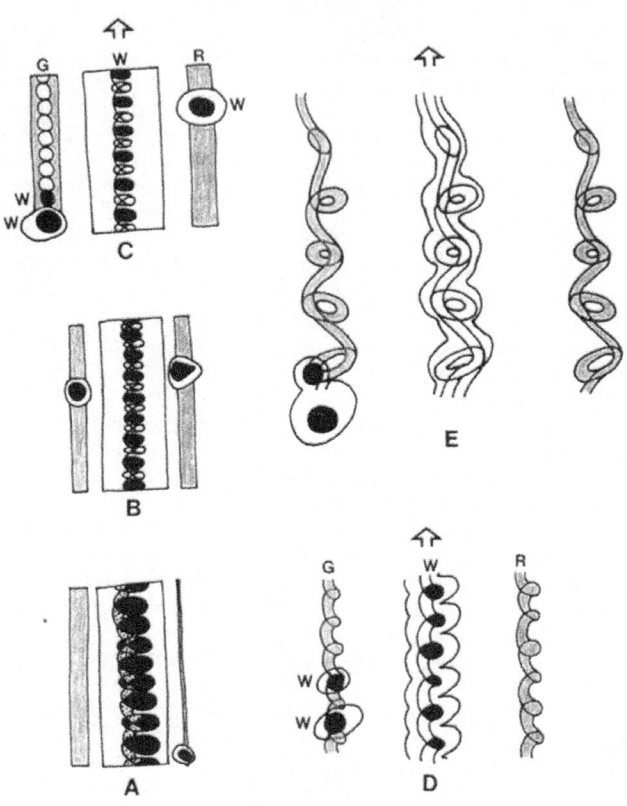

The encounter began with my spotting a brilliant ball of light turning and coming towards me. By the time it got to me, it had transformed into the lights of a conventional single engine airplane, complete with accompanying sound of a propeller-driven engine. Without careful and detailed analysis of my time exposures, I would have had no evidence to say that this was a mistaken identification. As the transformation progressed, the rate of looping and spiraling decreased until the lights did not loop or spiral at all. At that time the craft was almost directly over me (last time exposure), and my camera was pointing almost vertically. It was at that time and that time only that the sound of an engine could be heard. Before that, there was no sound.

Below is a diagram of the distribution and relative spacing of the lights, as well as an estimate of the width of the outer lights. Single engine propeller-driven private and military aircraft (WWII) have wingspans between 32 feet and 55 feet, although the Temco (Globe) Swift 125 has a wingspan of only 29 feet four inches (A Field Guide to Airplanes, Completely Revised and Updated, by Montgomery and Foster, 1992). The estimated width of the outer lights of the UAP is only half those wingspans, if that estimate is in the ballpark.

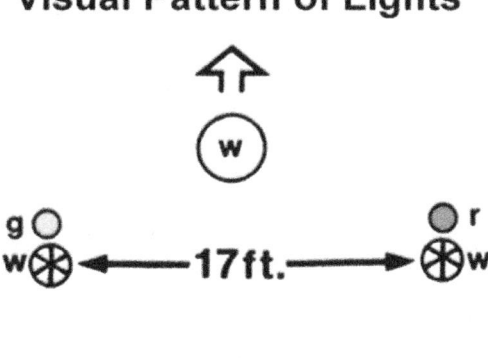

Another important detail evident in the enlarged photos was that before the central light decreased in brightness, almost to the intensity of the outboard lights, a ribbon of brown gas can be seen paralleling the craft just beyond the red port (left) navigation light. This ribbon is wider than the spread of lights on the craft, and resembles nitrous oxide gas in color. It can be seen on the above collage of prints where a brown glow initially follows the brightest central light, and has a sharp outward termination (computer enhanced and labeled in one photograph). My interpretation is that the ribbon of gas represents nitrous oxide generated when the central plasma light was very bright - due to high temperature conversion of nitrogen and oxygen in our atmosphere in a very strong electromagnetic field. This gas was diverted to one side of the craft through a pipe that dispersed it along its length, much like a crop duster airplane dispenses its insecticide out of spray nozzles along tubes sticking out of either side of the aircraft.

Even if someone had gone to enormous effort and expense to rig a crop duster with spiraling lights, it is highly unlikely that plasma lights would have been used, or that the engineers would have wanted to simulate nitrous oxide gas shunting. That would not have been a detail human hoaxers would have thought of, in my opinion. And because the looping of lights intensified just after each time exposure started, how would a human pilot have known the precise timing of my camera shutter: As indicated in Chapter Five (p. 115) on telepathy, I have confirmed that the pilot was reading my mind and thereby knew by my thoughts when I opened my camera shutter.

Anyone seeing a plane-like object at night that looked like a Boeing 707 jetliner probably would have not questioned or even noticed the absence of external engines if they heard a jet-like sound. Most people who saw the silhouette of a small single-engine airplane late at night and heard a sound similar to that made by a propeller-driven aircraft probably would not have paid any further attention to it if it had red and green strobing navigation lights. They might have thought it was very late for such a plane to be flying at night (local small airports typically close at 9:00 pm), but gone about their business.

What was demonstrated to me that September night is a prima facie example of a leak in the embargo with evidence for extraterrestrial visitation. That leak was demonstrated to a targeted individual who had the intelligence, training, and interest in recording, analyzing, and interpreting scientific data. Rather than present these data to any researcher of UAPs, who might take the evidence to a skeptical scientist, the data were presented to a scientists directly, one who was not easily fooled or inexperienced in the phenomenon. "...by not providing sufficient evidence to make their reality totally obvious to [all] scientists and society in general, the ETs are following a strategy or program that avoids inflicting catastrophic shock to society as a whole, which any overt contact could cause, while preparing us for eventual open contact. This could say something about their level of ethics." (Deardorff et al., 2005: p. 48).

By leaking significant evidence to a mainstream scientists (geologist and paleontologist), they were probably testing how our scientists would react (especially for the more open-minded and controversial scientists) to limited and controlled overt exposure.

Chapter 8 Why a Geological Investigation?

Dr. Ellen Crystall had a Ph.D. in music, but she also minored in geology as an undergraduate at Rutgers University in New Jersey. Even though her sightings drew her attention to the skies, other evidence drew her attention to the ground. She recognized patterns of UAP movement, and identified areas where craft frequently originated and disappeared on the ground and in forests. Hynek et al. (1998 also recognized areas on the ground, many containing stone chambers, where craft would first be seen and last be seen. Imbrogno and Horrigan (2000) investigated numerous abandoned iron ore mines in the Mount Carmel area of New York, where UAPs and aliens have been seen; some aliens were reportedly seen by witnesses emerging from old mine entrances or from stone chambers.

Crystall (1991) did not believe in interdimensional portals or Stargates. She did not believe craft disappeared through portals or by passing to invisible parallel dimensions, but either turned on a cloaking or invisibility device, and went through a physical door or opening into an underground base. "For example, one night I accidentally passed a field where drilling was occurring about midnight; no lights were being used, of course. The drilling was so loud that I heard it with my [car] windows rolled up and the air conditioning on." (Crystall, 1991: p. 91). She continued, "I clearly heard something like a pneumatic drill from the dense forest behind a small field. I got out of my car with my flashlight and crossed the bushy border into the field, It was eleven o'clock. I was alone and afraid in the dark, but determined, even as my knees began to shake.

"I kneeled and stupidly shined my flashlight into the woods. The drilling stopped. I held my breath. Nothing happened. For an hour I sat on the ground without moving, but I heard nothing more. The meaning was clear: I was not welcome to observe their operations, let alone their installations.

"...We also believe that aircraft are being housed in these installations. On many occasions we have seen ships come down to the ground level and not rise for the rest of the night. On other occasions, at dusk we have seen ships ascending from the ground in the areas we suspect have entrances (although we recognize they would have landed invisibly).

"I'm not alone in my conclusions that there are underground alien bases (see Timothy Good, 1998: Alien Bases). I have received reports from around the country about underground drilling in the woods during the night and electrical generators operating in areas where no one is doing construction." (Crystall, 1991: p. 91-92).

Crystall told me that by the mid-1980s she stopped hearing underground noises, and by the late 1980s she was writing her book, "Silent Invasion," and the chapter on "Aliens Underground." I too have seen and photographed ships or craft disappearing into fields and forests, or originating from them. Most of these occurrences were in the fields and forests around the "Indian Mound" or center of the UAP hotspot. Even though her hypotheses for underground bases led me to become involved and do a 24 square mile magnetic survey, which took me three years to complete, other evidence now suggests that these locations contain portals or Stargates into other dimensions (see chapter Six on The Vortex Revised). Both of these concepts (i.e. underground base and portals or Stargates) could be true, and they are not mutually exclusive. There is even a third possibility of ancient robotic probes being entombed in solid rock representing ocean muds that covered them and entombed them hundreds of millions of years ago. Chapter Eight, Evidence for an Underground Magnetic Focusing System, and Appendix II deal with the evidence for this hypothesis.

Thus the noise and lights indicating drilling activity at night in fields and forests may not represent mining or underground base excavation, but could be either a ploy to convince Crystall of underground activity so that her research would attract a geologist such as myself to investigate, or it could be an attempt to access ancient probes embedded in solid rock, specifically for information they may contain or to reactivate them so they could be sued as beacons for communication and/or coordinates for Stargate operation. Whatever the reason for the noises, lights, and air-blasts Crystall felt coming out of the ground very near the Jewish cemetery, they stopped by the time I became involved in June 1992. Was that because the purpose for the underground activity had been achieved, or was it because it was just a ruse to get a geologist involved? If it were the latter, why? What would I discover or uncover that required such an earthly investigation?

On 2 August 1992 Phil Imbrogno took a group of people to the stone chambers in Dutchess, Putnam, and Westchester counties, NY. I attended that field trip with Ellen Crystall, and took my borrowed Proton Magnetometer with me. I assisted Imbrogno by doing a detailed magnetic study around three stone chambers in southeastern New York. Those results are published online at

http://www.sunstar-solutions.com/AOP/chambers/stone_chambers.htm,

and presented in their appendix in his book Horrigan, Celtic Mysteries (2000).

Imbrogno and Horrigan (2000) describe the visions of a psychic named Loretta Chaney, who lives in Connecticut. Her visions may bear on what is going on underground and in the areas where ships appear and disappear. Imbrogno and Horrigan write: "Before Loretta was taken to Ninham Mountain, we took her to a number of stone chambers, and the Balanced Rock in North Salem, New York. Months before, with the assistance of Dr. Bruce Cornet, we were able to do extensive magnetic anomaly studies at several chamber locations. These studies along with graphs are presented in Appendix 2, page 135. Loretta was able to find the center of the magnetic anomalies at each of these locations. She was so precise that at first it was very hard to believe – but then again, birds and certain animals are sensitive to changes in magnetic fields, so then why can't certain people? Scientifically it is possible, but it is a theory that has never been proven. This was enough proof to us that Loretta was indeed able to sense various types of energy fields. In the summer of 1997, we decided to take Loretta and her husband Scott to Ninham to see what she could tell us about this sacred place of the Wappinger tribe." (Imbrogno and Horrigan, 2000: p. 84-85).

Loretta's first impression was of the ice age 18,000 years ago and thick ice sheets moving down from Canada. "Loretta's next impression was that she was being told that this was a location 'where the earth breathes.' She said that extraterrestrials are also attracted to this place for that reason. All of a sudden a rush of images and information came to Loretta, a rush so great she felt dizzy and had to sit down. She then saw Atlantians working side by side with the Indians a very long time ago. She told us that the Atlantians were a hybrid race, part human and part extraterrestrial, and they had a more advanced civilization than the rest of the world. [The extraterrestrial Ahlahn told the White Sands Proving Ground engineer, Daniel Fry, a very similar history in 1954: Good, 1998, p. 69-70.] She saw that the spots that now have [stone] chambers on them once had small pyramids. As Loretta moved forward in time she then saw ancient Phoenicians in the area and she claims that they were the ones who built the first chambers. After the pyramids weathered away the Indians marked the sacred ground with standing stones. The Phoenician traders and explorers built stone temples over these high energy areas and the later chambers were built by the Celts with strong Druid influence." (Imbrogno and Horrigan, 2000: p. 86).

Here we have independent information that may help explain why a pyramid image appeared on top of the "Indian Mound" at the UAP hotspot, and why that area may be a location for portals to another world or universe (cf. the Vortex image discussed on pages 176, 192, and 212). Loretta said that an energy pulse of the right frequency will open the portal, and Imbrogno and Horrigan (2000) documented EMPs occurring at one test chamber on Route 301 in Kent Cliffs, New York. They also relate that the Druids believed there were certain locations that are "Holy ground," which are windows or doorways that open into an alternate universe or dimension. The images of a building structure showing up in the forest next to the "Indian Mound," a structure that does not exist in that forest, during UAP performances on 28 April 1993 (Appendix I), can be explained if the veil or barrier between dimensions is thin there. Perhaps a certain frequency of light generated in that alternate dimension can pass into our dimension, making visible to us whatever object or structures are illuminated on the other side. A thin veil would also explain how ships and craft such as the Manta Ray could transition from one dimension to the other by changing their atomic or subatomic (i.e. "string") frequencies. The time exposure of the Manta Ray passing through the forest and into the ground (see page 67) no longer seems impossible, and explains why ET wanted a geologist, especially one with access to a sensitive magnetometer, to do a detailed study of the valley. ET was giving cl to humans where to look to find answers for why so much UAP activity exists in the Hudson and Wallkill river valleys. The embargo was leaking. Now let's take a look at in more detail what my 24 square mile magnetic survey uncovered. Apocalypse is a Greek word meaning 'to uncover.'

A Magnetic Focusing System from an Underground Robotic Probe

http://sunstar-solutions.com/AOP/Appendix_II/APPENDIX_II.htm

Although some of the information given below has been presented earlier, I am now going to give a much more detailed account of the Triangular Alien Probe and what its significance may be for the UAP Corridor in New York.

Dr. Ellen Crystall claims in her 1991 book, Silent Invasion, that she detected evidence for an underground alien base near Pine Bush, NY.

In 1992, I began a three year, 24 square mile, magnetic mapping project in the Wallkill River Valley of New York in order to test her hypothesis.

Magnetic measurements were made at 1,800 locations or stations within the mapping area.

Several maps were generated which show underground geologic features, and various types of magnetic anomalies.

Some of these anomalies appear to be unnatural and technologically based.

Due to erosion, some of the ancient islands have been exhumed or exposed in the southern part of the Wallkill River Valley, NY. They stick up in odd shapes reminiscent of pyramids, and have been given symbolic names by the early European settlers in the valley.

Dr. Bruce Cornet www.OrionWorks.com/bcornet/

The magnetic map that I generated for Orange and Ulster counties is displayed below. It shows the distribution of over 15 magnetic anomalies distributed along the axis of the valley. These anomalies are buried and most are not visible at the surface. They represent iron-rich granite and metavolcanic rocks that were part of an ancient island system or archipelago. Mt. Adam and Mt. Eve are a major exception to the south between Pine Island and Warwick, NY (see image above), because they have been nearly completely exposed due to erosion of surrounding softer shales. One island in the mapped area is represented by a magnetic anomaly, and is not completely buried. It is located at the northern end of Red Top road, due west of the City of Wallkill. Brown, weathered metavolcanic rock pokes up through a sea of surrounding black Ordovician shale (Austin Glenn Formation). Prospecting pits in that rock are still visible, and were areas where iron ore was probably being sought.

It became clear during my initial geologic investigation that these islands were being identified on the magnetic map below as magnetic anomalies. Most of these anomalies are stationary, meaning their magnetic values do not change.

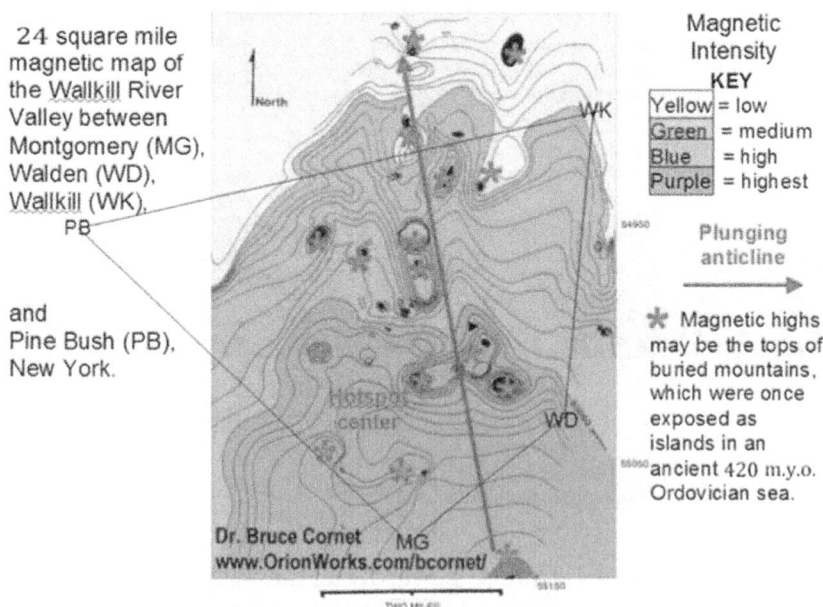

24 square mile magnetic map of the Wallkill River Valley between Montgomery (MG), Walden (WD), Wallkill (WK), PB

and Pine Bush (PB), New York.

Magnetic Intensity
KEY
Yellow = low
Green = medium
Blue = high
Purple = highest

Plunging anticline
⟶

✷ Magnetic highs may be the tops of buried mountains, which were once exposed as islands in an ancient 420 m.y.o. Ordovician sea.

Dr. Bruce Cornet
www.OrionWorks.com/bcornet/

TWO MILES

As my research on UAPs in the valley continued, and after I had made my first map of magnetic anomalies in that 24 square mile area, I realized that most if not all sightings of craft at night were occurring directly over or near those magnetic anomalies.

There is a strong correlation between AOP activity (green circles) in the valley and magnetic anomalies (red and clear ellipses), indicating some type of relationship.

Question: Is this relationship due to a preference of AOP for iron-bearing igneous formations underground, or is it because the natural magnetic anomalies provide a measure of camouflage against easy detection of artificial magnetic activity underground?

The thought that the Visitors were using natural underground magnetic anomalies to disguise their technology underground was my first suspicion. That explanation would seem to confirm Ellen Crystall's hypothesis of an underground base. As a geologist, I knew that the rocks causing the magnetic anomalies were of igneous origin, and as such would be much more competent and stronger structurally than the surrounding ancient marine shales, made up of claystone and siltstone. The ancient islands, if like the rocks exposed in Mt. Adam and Mt. Eve further south, were made up of granites and metamorphosed volcanic rocks, where the mineral crystals would be interlocked and much harder than clay minerals. Those islands would make better areas for underground bases, which could be carved out and excavated, forming underground caves. Crystall provides evidence for digging activity in the woods, which supports her idea of underground tunnels and caverns.

It was not until I discovered the underground alien probe beneath the Beth Hillel Jewish Cemetery that an alternative explanation came to light. Because that probe was sending three magnetic beam signals out into space, I was able to locate with my magnetometer exactly where those beams were located when they came out of the ground. A triangular array of three similar beams defined the triangular shape and size of that buried object. Because that probe was resting on top of a larger and more extensive magnetic anomaly, the possibility of the signals coming from an ancient alien probe that had landed on an island when this area was an ocean seemed more reasonable than underground activity coming from tunnels and caverns. And the question of why the signals could only be detected when the Constellation Booties was overhead had to be explained.

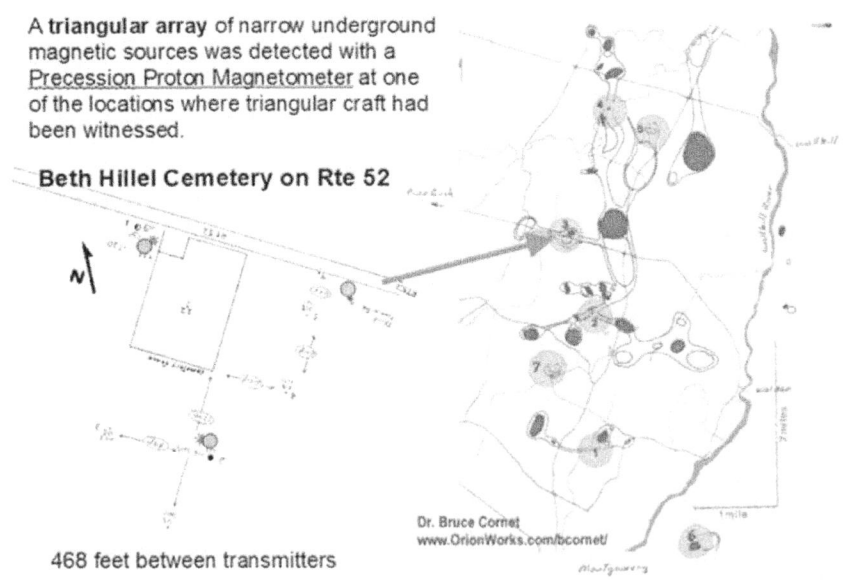

A **triangular array** of narrow underground magnetic sources was detected with a Precession Proton Magnetometer at one of the locations where triangular craft had been witnessed.

Beth Hillel Cemetery on Rte 52

468 feet between transmitters

Dr. Bruce Cornet
www.OrionWorks.com/bcornet/

As initial or early questions were answered, more questions arose as a consequence of this discovery:

- Why would the Visitors be sending signals out into space from underground when they could exit their underground base and transmit through our atmosphere?

- How could magnetic beams pass through solid rock if the probe was entombed in black marine shale?

- Were there other energy beams coming out of the ground that were not magnetic, but which other people had observed when they generated light?

- Were other areas where I detected similar signals coming out of the ground nearby also coming from alien probes that had landed on other islands of this ancient archipelago?

- What was the alien probe trying to signal or communicate with out in space?

- Could these signals be intercepted by other alien probes out in space, and that is why the area has attracted so many visitors coming to Earth to investigate the source of that signal? Was ET trying to call home?

- How could these probes still be active and functioning after 420 million years after being entombed and even buried deeply at one time in the Earth's crust?

On three separate occasions anomalous magnetic spikes were recorded at the source locations. The spikes occurred every 5-10 seconds, and were typically 3,000 to as high as 11,000 gammas (~56 to 205 mG) above or below background levels.

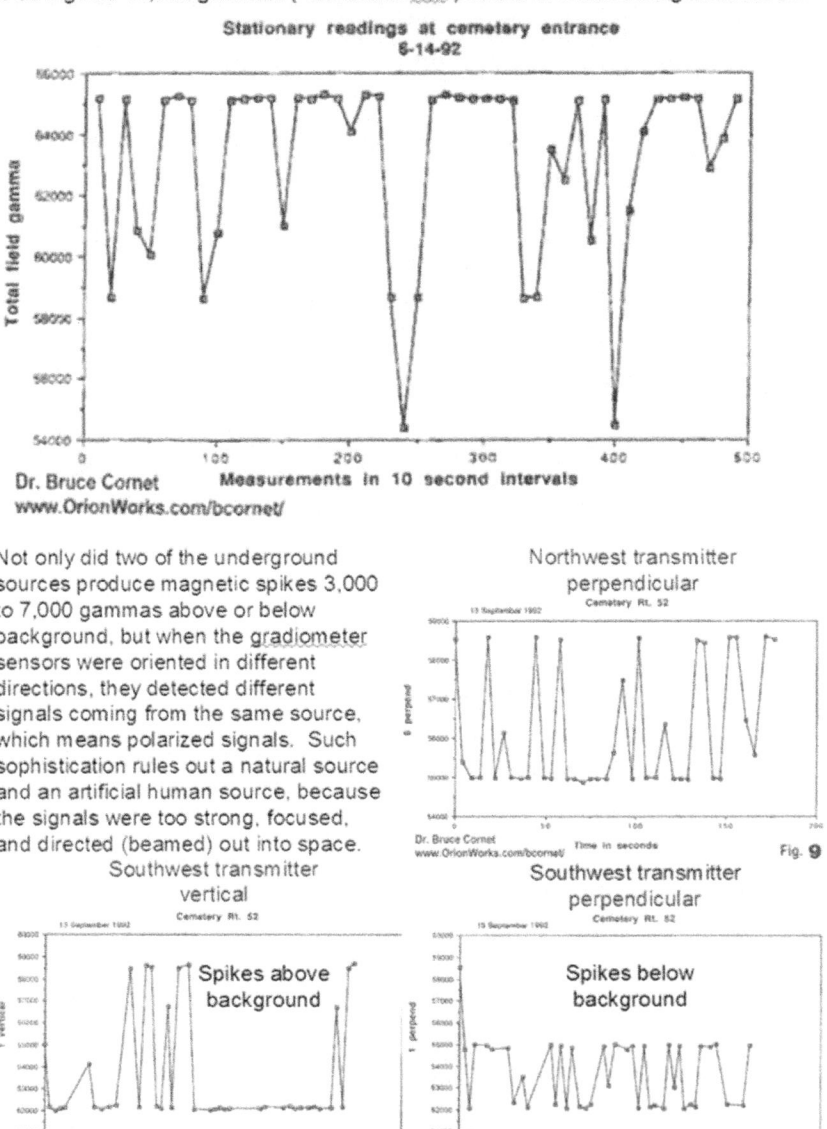

Not only did two of the underground sources produce magnetic spikes 3,000 to 7,000 gammas above or below background, but when the gradiometer sensors were oriented in different directions, they detected different signals coming from the same source, which means polarized signals. Such sophistication rules out a natural source and an artificial human source, because the signals were too strong, focused, and directed (beamed) out into space.

Seven notable anomalies have been observed over and around the triangular magnetic array:

- A very large triangular craft was observed hovering directly over the array (and cemetery) one evening. Image below from Silent Invasion by Ellen Crystall (1991).

- Strange sounds and rapid drops in air temperature have been reported at the cemetery and near the transmitter array.

- The author has experienced such drops in temperature and popping sound (cf. vacuum seal being broken) coming from the magnetic transmitter located at the cemetery entrance.

- Sue Mann reported feeling vibrations and a distinct humming coming out of the ground within the cemetery one night.

- Plants and animals have grown or behaved strangely in and around the cemetery and magnetic sources – including plants that grew downwards and towards one source, and worms/insects that allegedly glowed at night.
- The ground around the magnetic source array is sometimes unusually warm – enough so that it rapidly melted snow banks only along the road in front of the cemetery. Pictures below taken by Phil Martin in March 1994.

- Witnesses (Phil and Lynn Martin) observed a tracking beam of light arising from between the three magnetic sources, which tracked a low flying commercial airliner as it passed over the cemetery, and then disappeared as quickly as it appeared.

Location of magnetic source (transmitter) next to Beth Hillel Cemetery fence.

Wild Onions growing back towards the ground.

Location of magnetic anomaly

Scot Stride, engineer for JPL in Pasadena, CA, raised some important points when I shared my discoveries with him. Scot helped design the UHF radio hardware for the Sojourner Mars Rover. He has flight experience with EM signals.

1. As far as being a signal, for it to transmit into space it needs to be electromagnetic. The SETI definition of signal is a very narrow-band (< 10 Hz) CW signal detected at any given microwave frequency.

2. Pulsed magnetic fields can be used to focus particle beams and electromagnetic fields; that explanation is more plausible than a magnetic transmitter.

3. The strength of the detected signals implies artificiality, more than regular pulses or geographic spacing.

4. The three equilaterally spaced magnetic sources could be part of a beam focusing system. With three magnets, a beam of charged particles can be steered and pointed into space. Four or more magnets could be used, but controlling them gets complicated.

5. Therefore, the three magnetic anomalies detected are probably not actual signal beams. Rather, they are indirect evidence that an unknown centrally-located signal (between the three magnetic sources) was being generated when the magnetic spikes were detected. In support of that interpretation, **the Martins saw a beam of light originating from the cemetery near that central location**.

6. As for the beam going through solid rock, the particles would need to be very tiny and energetic, like neutrinos or weakly interacting dark matter.

7. If an active alien transmitter exists underground, and sends signals out into space from time to time, it represents a unique opportunity for monitoring and study.

8. All three magnetic transmitters detected with a <u>Precession Proton Magnetometer</u> are located on private property just outside a local cemetery.

9. A seismic survey might be able to detect the depth of the devices producing the magnetic signals, as well as their sizes and shapes.

10. Monitoring equipment could be positioned at each magnetic source location so that a continuous record of activity can be made and studied.

11. Determining the characteristics of the magnetic fields focusing the signal, and the type of carrier beam (e.g. charged particles, neutrinos, weakly interacting dark matter, etc.) could help us understand why SETI has been unsuccessful in detecting ETI communications.

12. If this is alien technology, it means ET is already here and transmitting data to a distant location in space (shades of the movie, *The Arrival*).

13. Why is this technology located underground?
    - Camouflage and inaccessibility, or
    - An ET probe arrived here long ago and landed on an island in a shallow Ordovician sea. It was gradually buried in marine mud as sea level rose. With time and continued burial the enclosing muds were turned into rock. Erosion and glaciation have nearly exposed the probe, and it is trying to signal "home".

14. If this is a buried probe, why is it transmitting today?
    - It may contain a long-lived power source, have been engineered and designed to seek out planets with

primitive life, and transmit signals when it detected the evolution of intelligent life.
- Triangular craft are commonly spotted flying over the area, and some stop and hover directly over the buried probe as if drawn to its beacon.
-

A Triangular Probe descends over an Island Archipelago and lands on one of the islands.

420 mya

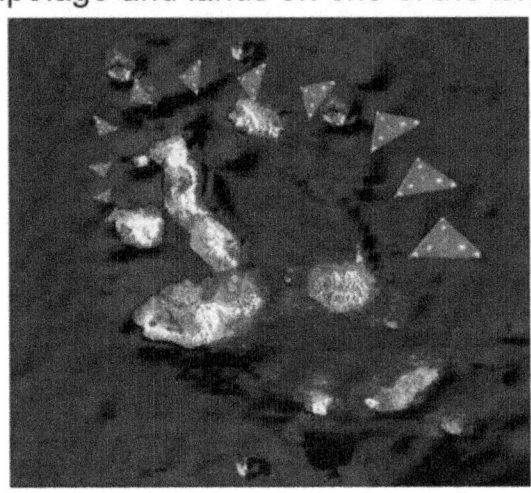

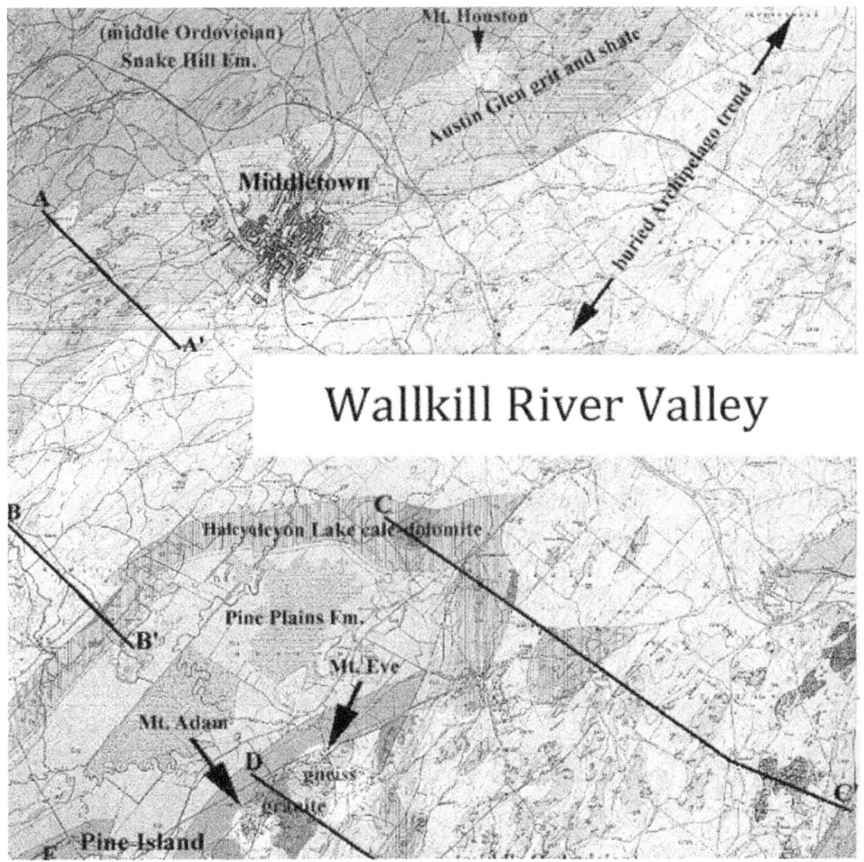

HISTORY CHANNEL TESTS HYPOTHESIS

On 8-9 November 2007 the History Channel brought magnetometers and a ground penetrating radar system to the location of the anomalous magnetic signals.

The fence around the Beth Hillel cemetery had been extended west and south. It now encloses the tree and signal source that had been outside the cemetery fence before 2005.

They discovered that the large tree next to the signal source was magnetized: 1700 milligauss, compared to 55 milligauss background level.

Magnetic readings next to the cemetery were fluctuating in three different directions simultaneously.

Ground Penetrating Radar (GPR) along the south side of the cemetery encountered a buried object seven feet below the surface, surrounded by Ordovician shales of the Austin Glen Formation – exactly where Cornet's hypothesis predicted it would be found.

The object is more than 17 feet thick.

GPR data extends only down to 27 feet, and does not show the bottom of the object, nor what it is resting on.

History Channel Film Shoot

Bill Birnes (UFO Magazine) with History Channel film crew.

Autumn Humphreys, Producer

Below the History Channel film crew is pictured with Dr. Bruce Cornet (on right), across from the Beth Hillel Cemetery along Rte. 52 between Pine Bush and Walden, NY.

Beth Hillel Cemetery, Rt. 52

signal location   magnetic tree   fence extended in 2005

Ground Penetrating Radar Device

Above picture was taken behind the Beth Hillel Jewish Cemetery in November 2007, showing the GPR device used to image the ground down to 27 feet with the UAP Hunters for the History Channel. The images below show the initial results of that survey, along the grass skirt behind the cemetery fence. It shows the corner of a solid structure at the east end of the fence that is no more than nine feet below the surface.

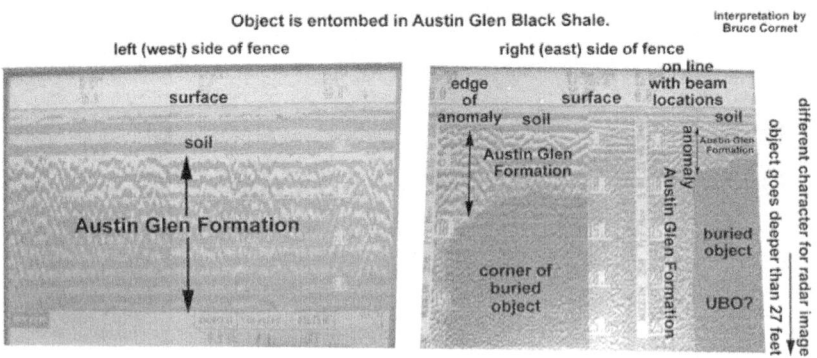

Above: Raw Data for Ground Penetrating Radar Images of UBO

Below: interpreted GPR data.

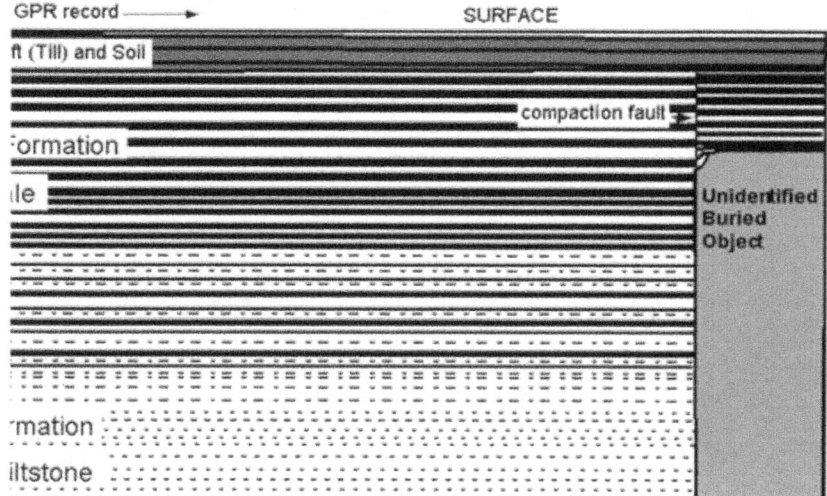

The satellite image below of the cemetery adjacent to Rte. 52 shows the positions of magnetic beam generators at the corners of a triangular object and a central signal generator, which has been witnessed projecting a beam of light into the atmosphere. Significant is the size of this object resting on top of a granitic island, and the fact that it has magnetized the tree in the upper left corner of the picture. That magnetism was measured by the UFO Hunters team at 1.7 gauss, which is greater than the total magnetic field of the Earth. A Triangular craft was seen hovering over the cemetery by Dr. Ellen Crystall (1991: drawing), while other witnesses have heard a humm coming up out of the ground, or felt the temperature change over the cemetery, but not elsewhere. Phil and Lynn Martin even witnessed the snow on top of the cemetery melting and running into the road on a cold winter day, indicating that the ground over the Triangle had been heated enough to cause that melting. I have witnessed and photographed plants growing back down towards one of the magnetic transmitters next to the magnetic tree, indicating that these wild onions "thought" a source of energy (i.e. sunlight) was comng up from below them. Strange popping sounds have also been heard coming from the corner transmitters. In 1992-1993 I monitored those magnetic beams using a Proton Procession Magnetometer. Those results (presented in a talk at the Society for Scientific Exploration in Las Vegas on 5 May 2004) show that this buried alien probe sends powerful magnetic beams out into space every time the Constellation

Booties is overhead. This discovery was reported in the November issue of UFO Magazine. This probe landed on Earth 420 million years ago in the Late Ordovician.

Vol.22 No.11
Issue #140
Nov. 2007

Anomaly at Pine Bush

by Bruce Cornet

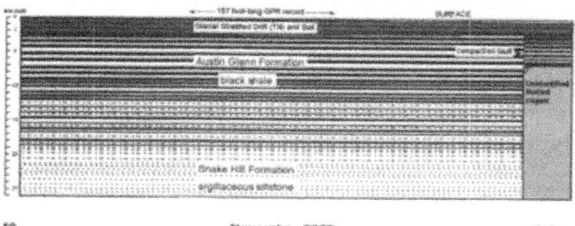

Thanks to
Bill and Nancy
Birnes, prod. & ed.

UFO Magazone, Who is winning the war for our souls?
Vol. 22, No. 11, Issue #140, November 2007.

Chapter 9 Summary and Conclusions

The three UAP crashes or landings in Roswell, New Mexico, in 1947 (LZ1, LZ2, and LZ3) and the Kenneth Arnold sighting are considered the beginning of the UAP phenomenon in the United States, but a crash retrieval occurred earlier in 1941 at Cape Giradeau, Missouri, where Dr. Harley Rutledge taught physics at the SE Missouri State University (Rutledge, 1981)(L.M. Howe, 2017). The Night Siege (Hynek and Imbrogno, 1986) about the Hudson Valley flap that occurred in New York between 1981 and 1986 of huge UAP seen by thousands of people reached public awareness after Dr. Ellen Crystall had started her investigation of the Pine Bush phenomenon in 1980 (Crystall, 1991). This investigation of that phenomenon began in June 1992 after Cornet read Crystall's book.

It has been generally believed by UFOlogists that the phenomenon involves structured craft piloted by biological entities from other worlds in our galaxy and universe. That is why UFO has become synonymous with ExtraTerrestrial Vehicles, or ETVs, even though UFO means Unidentified Flying Object. During his 11 year field investigation into the Pine Bush phenomenon, Cornet began to accumulate evidence for Crystall's theory that the phenomenon hid in underground bases in the area was confirmed by a 24 square mile magnetic map. He also found that the center of the UAP Hotspot was located over a previously unknown archaeological site representing the outline and mound (Unchi) for a Native American Sweat Lodge that predated the invasion of the valley by Europeans in the 17th century. In other words, this phenomenon may have not only been observed by those natives, but they may have considered the night lights to be sacred Earth spirits, with whom they tried to communicate in their sweat lodge. That would mean the UAP phenomenon predated Roswell and Cape Giradeau by more than 300 years.

Cornet also determined that the structured vehicles that he saw and photographed were trying to mimic the wing shapes of conventional and military aircraft that traversed the valley daily as they flew in and out of the Stewart Army & Air Force base, and International Airport near Newburgh, NY, on the opposite side of the valley. He frequently recorded light patterns and sounds coming from these craft that resembled the navigation lights at night and sounds of turbofan jet engines and propeler engines, indicating a level of camouflage that was rarely reported by other investigators, who saw mostly silent discs, spheres, and cigar-shaped craft.

It became obvious to him that in order for these Earth entities to exist and operate in the Wallkill River Valley, they had to blend in and be mistaken for conventional aircraft. During 1992 and 1993 he was treated to dozens of performances by the pilots or intelligence operating the craft, with the clear intent of testing his ability to distinguish their craft from manmade vehicles. He became convinced that he was being used as a type of quality control for their camouflage and deception. He was able to document telepathic communication with the pilots and conducted experiments which proved the craft operated like projection platforms, which could send separate beams of information to human witnesses on the ground standing only two feet apart.

It became clear to Cornet that he was dealing with a highly sophisticated technology that had been on Earth possibly longer than humans on Earth. It was during his three year, 24 square mile magnetic survey that he discovered an entombed alien probe that had landed on Earth over 400 million years ago. It was encased in Ordovicial black shale that buried a granitic island in an epicontinental ocean. A Ground Penetrating Radar survey paid for by the History Channel revealed that this huge triangular probe, 465 feet on a side, was only about nine feet below the surface, and continued to send information signals out into space whenever the constellatioSn Bootes was overhead. He determined that it was nuclear powered due to higher-than-background gamma radiation coming from the areas where magnetic beams shot out into space.

This discovery put a whole new spin on the question of whether humans are alone in the Universe. It answers the question of whether ET has visited the Earth, and when ET got to Earth. It rendered moot the astronomer's skepticism of ET making the light-years jouney to Earth by indicating that ET (or its AI robotic probes) has been on Earth for almost as long as complex multicellular life has been preserved as fossils on Earth.

Recent disclosures by the Pentagon and To The Stars Academy in November and December 2017 of Navy pilots giving chase to luminous ETVs over the ocean, nicknamed Tic Tacs, began a series of acknowledgements that UAPs not only exist but are flying in our atmosphere and oceans. Dr. Bruce Maccabee put this and subsequent disclosures in 2019, by the Navy into perspective.

On May 29, 2019, in a Facebook post, Dr. Bruce Maccabee wrote:

DISCLOSURE CREEP - you heard it here first. the major media are freaking out as more and more Navy personnel come "creeping" forward with unbelievable UFO/UAP stories that cannot be easily rejected by the "tradition" that all sightings are either failures by the witnesses to correctly identify phenomena, hoaxes or delusions. (See The Legacy of 1952, the Year of the UFO for information about the "UFO tradition.") Over the last two years Navy UAP data releases have crept along at a more or less steady pace and the media have to treat this seriously - no more "woo woo music - because the Navy is clearly serious about investigating sightings and informing the public. The "self-cover up" is being broken down and it appears that what I called the "Emperor has no clothes" event may soon come about: someone (Navy?) in the "crowd" will stand up and yell "THEY'RE REAL" at which time everyone else will say "Yes, Real .....we knew it all along!"

On June 5, 2019 in a Facebook post, Dr. Bruce Maccabee wrote:

We are told of the newly discovered amazing characteristics of the UFOs..But these are new only to those who are "UFO newbies" who don't know UFO history. Consider the following: ESSENTIAL ELEMENTS OF INFORMATION (EEI)

Essential elements of information are lists of particular characteristics of some technological device(s) about which an intelligence agency wants information. The EEI are generated at the headquarters of the agency and are then sent to the agents to tell them what to look for as they are collecting intelligence information (spying! Gulp!). One particular Intelligence Collection Memorandum lists some EEI that apply to gathering information about flying objects that have the following characteristics. It reads in part:

- Extreme maneuverability
- Ability to hover
- Absence of exhaust trail
- Ability to appear suddenly without warning as if from an *extremely high altitude* (my emphasis)
- The ability to quickly disappear *by high speed* or by complete disintegration
- Exhibits evasive action

Comparing the information in these two documents we see they are quite similar in their lists of characteristics, particularly with regard to speed and maneuverability. In the list of EEI is the suggestion that they can appear suddenly by dropping at high speed from a high altitude. What is the document which lists these EEI? It's an Air Force intelligence document generated under the authority of General George Schulgen of the Intelligence Requirements Division. This EEI document was created in the fall of 1947! (e.g. , see The FBI CIA UFO Connection. Pg 31)

Makes one wonder, have we learned anything in the last 72 years? Bruce Maccabee: In conclusion, ET does not have to be ExtraTerrestrial if it has been living underground and/or on Earth's surface longer than humans have existed. If we did not evolve from primates on Earth, or if we are hybrids created from the combining of primate and non-human genes, then WE may be the true ExtraTerrestials, and many of those we are calling ETs may be the original natives or inhabitants of Earth. If THEY created us in their image and likeness, then perhaps THEY are already walking among us and guiding our civilizationsl

Bob Tarantino, an electrical engineer by education and trade, in a post dated June 18, 2019, summarized the recent news (since November 2017) and commentary regarding those luminous ETVs chased by Navy fighter pilots over the Pacific Ocean:

Concerning the Tic Tac, it is said that five observables have been made:

1) Anti-gravity lift.

2) Sudden and instantaneous acceleration.

3) Hypersonic velocities without signatures.

4) Low observability, or cloaking.

5) Trans-medium travel.

We know that they are found throughout ufology, and its just the beginning. There are many more traits that are associated with UFOs. And, of course there are many shapes.

UFO Shapes

http://blog.presentandcorrect.com/ufo

Even the term "flying saucer" referred originally to the motion of a UFO, not its shape. Kenneth Arnold claimed he saw nine objects that were skipping like a "saucer on water." So, even from the early days, we can see that these objects fly erratically. The objects he saw were more crescent shaped.

Concerning metamaterial, I do not understand how this would work, but I assume that the thought is that it would coat the exterior of the object. Then, a thin boundary layer around the object would cause destiffening of space time, and cause the object to "fall" in the direction of curved space time. By falling, there is acceleration, but no force.

Some photographs seem to show distortion of the UFO as if light is bending around the object. There are reports of a force field around the object, where tree limbs were seen to move out of the way. Paul Hill cites a case where a rock was thrown at the object, but it bounced off an invisible field surrounding the object.

There are multiple cases where a car was levitated beneath a UFO. And, Travis Walton was knocked down so hard by the tilting of a UFO that it may have killed him, only to be revived by aliens on the ship. Clearly, the evidence suggests a much larger boundary region, and a downward propulsive force. Further evidence of this is water spouts that develop when a UFO is over water. Also, water seems to move out of the way, before a UFO enters.

Given the EM nature of these objects, that are known to shut off car engines, it also appears to me that these objects use resonant frequencies, perhaps in the amplification of gravity waves to cancel the effects of gravity. The side effects may be microwaves and other types of EM radiation that are known to cause burns on people near by. Multiple amplifiers may be necessary to create a resonant effect. Small EM effects may be drastically amplified in these cases.

It is clear to me that advanced materials are used on cloaking these objects. The reports are too numerous to mention, as these objects have been seen at close range and are described as having a "glass like " appearance.

So, it appears to me that Hill was right in that instead of a fall in the direction of a curved space time, it is rather the case that the effects of earth's gravity are canceled and even reversed. Now, the method by which this is done is unknown, and indeed perhaps metamaterials could be used. Perhaps gravity waves would need to be generated and amplified.

There are many types of UFOs and some may use very different methods of propulsion. But the smallest UFOs may use a very different approach to propulsion.

But I'm NO expert. I'm just a saucer head. That's my two cents.

-Bob

THAT'S ALL FOLKS

References

Allen, V., 2015: UFO's and the Paranormal Are Normal, Balboa Press.

Berliner, D. and Friedman, S.T., 1992, Crash at Corona: The U.S. Military Retrieval and Cover-up of a UFO, Da Capo Press.

Bord, J., 1997, Fairies: Real Encounters With Little People, Michael O'Mara Books.

Braden, G., Russell, P., and Pinchbeck, D., 2007, The Mystery of 2012, 2009, Sounds True, Incorporated.

Braden, G., Russell, P., Pinchbeck, D., Macey, J.R., Jenkins, J.M., et al., 2009, The Mystery of 2012: Predictions, Prophecies, and Possibilities, AbeBooks.

Brookings Report, 1960, p. 182-184 (commissioned by NASA and created by the Brookings Institution in collaboration with NASA's Committee on Long-Range Studies).

Canada, S., 2006: Bible encoded Crop Circle Gods, authorHouse.

Carter, M.J.S., M.Div., 2013: Alien Scriptures: Extraterrestrials in the Holy Bible, Grave Distractions Publications.

Davies, P.C.W. and Brown, J., 1992, Superstrings: A Theory of Everything?, Cambridge University Press.

Cooke, P., 2005, The Greatest Deception, The Bible UFO Connection, Oracle Research Publishing.

Cornet, B., 2007, Anomaly at Pine Bush, UFO Journal, vol. 22, no. 11, issue #140.

Corso, Col. P. J. (Ret.), 1999, with William J. Birnes, The Day After Roswell, Atria Books.

Costa, C., and Costa, L.M., 2017. UFO Sightings Desk Reference, United States of America 2001-2015, Unidentified Flying Objects, Frequency – Distribution – Shapes, CreateSpace, An Amazon.com Company.

Costa, C., writer for the Syracuse Newtimes.

Crystall, E., 1991, Silent Invasion, The Shocking Discoveries of a UFO Researcher, Paragon House, New York.

Daniel, Alan Dale, 2013: Tracking Ancient Legends, How the biblical Flood, Sky Gods, and UFOs Fit Into Prehistory.

Däniken, E. von, 1968, Gods From Outer Space, Return to the Stars or Evidence for the Impossible, G.P. Putnam's Sons, New York.

Däniken, E. von, 2002, The Gods Were Astronauts, Element Books, New York.

Däniken, E. von, 2009, History is Wrong, The Career Press, Inc.

Däniken, E. von, 2010, Twilight of the Gods, The Mayan Calendar and the Return of the Extraterrestrials, New Page Books.

David, F. R., Ph.D., 2010, The UFO-Christianity Connection: Fact or Fiction, iUniverse.

Deardorff et al., 2000: Abstract.

Deardorff, J.W., 1986, 27, 94-101, A Possible Extraterrestrial Strategy for Earth, Department of Atmospheric Sciences, Oregon State University, Corvallis.

Deardorff, J.W., Haisch, B., *Maccabee, B.* and *Puthoff,* H.E., 2005, Inflation-Theory Implications for Extraterrestrial Visitation.

Dennett, P., 2008, UFOs Over New York, A true History of Extraterrestrial Encounters in the Empire State, Schiffer Publishing Ltd.

Dickerson, K., Live Science, Scientific American, July 9, 2014.

Downing, Rev. Dr. B.H., 1968, 1989, The Bible and Flying Saucers, New York: Avon and Berkeley CA.

Downing, Rev. Dr. B.H., 2015, UFO Revelation.

Fowler, R. E., 1991, The Watchers: The Secret Design Behind UFO Abduction, Bantam Books.

Fowler, R. E., 1996, The Watchers II: Exploring UFOs and the Near-Death Experience, Wild Flower Press.

Good, T., 1998, Alien Bases, Earth's Encounters with Extraterrestrials, Century.

Greco, S. and Gordon, S., 1992, Williamsport Wave, MUFON UFO Journal, No. 290, June 1992: pp. 3-10.

Greer, S.M. M.D., 1999: Extraterrestrial Contact, The Evidence and Implications, Crossing Point, Inc. Publications.

Guerra, J.L., 2018, Strange Craft, The True Story of an Air Force Intelligence Officer's (Major George Filer) Life with UFOs, Bayshore Publishing Co.

Hawking, S., 1988, A Brief History of Time, Bantam Books.

Hoagland R.C. and Bara, M., 2007, Dark Mission, The Secret History of NASA, Feral House.

Hoagland R.C. and Bara, M., 2009, Dark Mission, The Secret History of NASA, Enlarged and Revised Edition, Feral House.

Hoagland, R.C., 1987, The Monuments of Mars: A City on the Edge of Forever.

Holy Bible: Ezekiel describes this object or wheels again in Chapter 10:1-22.

Howe, L.M., 2017, The Unheard Truth Behind Roswell's UFOs| FREE Episode of ...- YouTube.

Hynek, Dr. J. A., Imbrogno, P.J., and Pratt, B., 1998: Night Siege, Llewellyn Pubs.

IHS Jane's Defense Weekly, 17 June 2015.

Imbrogno, P.J. and Horrigan, M., 1997, Contact of the 5th Kind, The Silent Invasion Has Begun, Llewellyn Pubs.

Imbrogno, P.J. and Horrigan, M., 2000, Celtic Mysteries in New England, Llewellyn Pubs.

Imbrogno, P.J., 2008, Interdimensional Universe.

John E. Chitty, 2002: The Broken Bible, Picking Up the Extraterrestrial Pieces.

Kaku, M., 1995, Hyperspace, A Scientific Odyssey Through Parallel Universes, Time Warps, and the 10th Dimension.

Kent and Olsen, 1999. Kent, D.V. and Olsen, P.E., 1999, Astronomically tuned geomagnetic polarity time scale for the Late Triassic, Journal of Geophysical Research, v. 104, p. 12,831-12,841.

Kent, D. V., Olsen, P. E., and Witte, W. K., 1995, Late Triassic-Earliest Jurassic geomagnetic polarity reference sequence from cyclic continental sediments of the Newark rift basin (eastern North America). Albertiana v. 16, p. 17-26.

Korff, K.K., 2000, The Roswell UFO Crash, What They Don't Want You to Know, Dell.

Lorgen, E., 2000, The Love Bite, Alien Interference in Human Love Relationships, ELogos & HHC Press.

Marcel, J. Jr., M.D., and Marcel, L., 2008, The Roswell legacy, The Untold Story of the First Military Officer at the 1947 Crash Site, Weiser.

Martin, R.L. and Wann, O-Q.T., 1995, The Coming of Tan, Historicity Productions.

Martinez, S.B., 2013, The Lost History of the Little People, Their Spiritually Advanced Civilizations around the World, Bear & Co.

Njemanze, P.C. MD, 2015: Igbo Mediators of Yahweh Culture of Life, Xlibris.

Polise, V., 2005, The Pine Bush Phenomenon, Trafford Publishing.

Randle, K.D., 2002, Operation Roswell.

Richard M. Dolan and Bryce Zabel, 2010: A.D. After Disclosure.

Roberts, A.R., 2012, From Adam to Omega, An Anatomy of UFO Phenomena, iUniverse, Inc.

Rutledge, H., 1981, Project Identification: the first scientific field study of UFO phenomenon, Prentice Hall.

Santos, G., 2012: UFO: Angels and the Mayan Calendar, Trafford Publishing.

Sarfatti, Dr. J., Pers. Comm. 2003.

Shriner, S., 2005: Bible Codes Revealed – Analyzing last-days Bible prophecy and current events as revealed in the Bible Codes, iUniverse, Inc.

Strieber, W., 1988, Communion, A True Story.

Sutherly, C., 2001, UFO Mysteries: A Reporter Seeks the Truth. Llewellyn Publications.

Turner, Karla, Ph.D., 1994, Taken: Inside the Alien-Human Abduction Agenda, Kelt Works.

Vallee, J., 1991, Messengers of Deception, UFO Contacts and Cults, Daily Grail Publishing.

Vallee, J., 1991, Revelations: Alien Contact and Human Deception, Ballantine books.

Vallee, J., 2008 (1979), Messengers of Deception, UFO Contacts and Cults, Daily Grail Publishing.

Webster, 1961.

Zimmermann, L., 2013, In The Night Sky, Hudson Valley UFO Sightings From The 1930s To The Present, Eagle Press, New York.

Index

'S'-shaped, 80, 82, 83
Abductee, 186
aerial, 1, 80, 98, 108, 237, 282, 302
Ahlahn, 106, 141, 166, 246, 252, 355
Albany Post Rd., 112, 168, 169, 311, 312, 322, 337
Alien Bases, 42, 105, 138, 139, 166, 276, 353, 385
Atlantis, 141, 142
Barbara Hartwell, ix, 55, 227, 283
Basking Ridge, 149
Bernardsville, 149
Beth Hillel Jewish Cemetery, 60, 67, 72, 100, 113, 120, 135, 160, 264, 360, 372
Biaveh, 144
Biaviian, 143
Billy McNamara, ix, 60, 63, 124, 160
Binary Solar System, 267
black shales, 8, 9, 135, 298
Black Triangle, 19, 28, 29, 124, 125, 126, 133, 135, 143, 149, 151, 157, 165, 174, 183, 188, 190, 197, 198, 199, 200, 204, 207, 209, 211, 243, 248, 264, 273, 279, 282
Bob Pratt, 1, 107
Bob Wisch, ix, 72, 153, 330
Boeing 737, 192
Boeing 747, 50, 57, 60, 116, 195
Boeing 767, 66
Boing 707, 66
Boomerang, 4, 46, 49, 50, 116, 118
Booties, 360, 374
Bruyn Turnpike, 312
Budd Hopkins, ix, 4, 91, 138, 304, 306, 308, 309
C-5 Galaxy, 57, 181, 226
camouflage, 30, 115, 148, 204, 208, 227, 230, 231, 318, 323, 333, 377
Canada, 139, 239, 355, 383
Cape Giradeau, 376
Carmel, 4, 352

Celtic Mysteries, 7, 114, 240, 270, 354, 386
Central America, 5
Cheryl Costa, 10, 120, 122, 124
chopper device, 330, 331, 332
Chris Burns, ix, 100
Close Encounters of the Third Kind, 180
Cogan Station, 4
Colm Kelleher, ix, 128, 131, 133, 143, 266, 269
Constellation, 81, 83, 360, 374
Crampton, 47
craton, 8
Crystall, Ellen, ix, 2, 4, 5, 9, 10, 11, 13, 14, 15, 16, 21, 23, 24, 27, 28, 30, 40, 41, 42, 43, 46, 48, 54, 55, 57, 58, 59, 60, 66, 68, 72, 73, 89, 90, 91, 99, 100, 101, 103, 105, 107, 108, 112, 114, 115, 116, 117, 118, 119, 120, 121, 122, 124, 135, 138, 152, 158, 160, 163, 167, 168, 169, 172, 175, 179, 180, 181, 183, 184, 186, 187, 188, 190, 198, 200, 204, 207, 209, 222, 226, 244, 245, 246, 247, 251, 252, 254, 256, 263, 270, 297, 298, 301, 304, 306, 307, 309, 310, 312, 318, 319, 321, 322, 337, 338, 352, 353, 354, 357, 359, 360, 363, 373, 376, 384
CSETI, 5, 6
Daniel Fry, 105, 106, 141, 166, 246, 252, 355
DC-9 jetliner, 59
diamond-shaped, 50, 58, 60, 119, 138, 153, 160
disc-shaped, 14, 138, 288, 293, 308
ditto, 80, 82, 83
Doppler, 30, 58, 89, 191, 192, 193, 194, 195, 196, 197, 198, 205, 206, 207, 210, 211, 214, 222, 224, 226, 227, 232, 233, 234, 235, 257
Dr. Bruce Maccabee,

269, 378, 379
Dr. Harley Rutledge, 376
Dr. Jack Sarfatti, 191
Dr. Jacques Vallee, 3
Dr. John Mack, ix, 91, 306
Dr. Joseph Burkes, 5
Dr. Samuel Greco, 4
Dr. Steven Greer, 4, 5
Dr. Susan B. Martinez, 18, 141
Dutchess, 2, 13, 49, 240, 354
Edenville, 8
Egyptian, 268
electromagnetic, 40, 72, 156, 170, 175, 184, 185, 209, 263, 300, 301, 304, 350, 366
Empire State, 2, 270, 385
Energy Emission, 13
Eve Lorgen, 295
Ezekiel, 103, 104, 386
FAA, 28, 75, 95, 148, 149, 153, 211, 228, 277, 340, 342
Fairfield, 2, 7, 49
ferris wheel, 102
Fifth Kind, 6
Flury Rd., 114
Franklin Mineral Museum, 9
Fred Brock, ix, 72, 74, 153, 167, 246, 316
Garden of Texas Liberty, 285, 286, 287, 289, 292, 293, 294
Glaciation, 142
granite, 358
Harriman, 41, 44, 168, 246, 315, 324, 326
Harry Lebelson, 13, 112, 167, 297
headlights, 28, 33, 57, 61, 62, 68, 77, 78, 79, 80, 84, 111, 115, 119, 124, 150, 163, 165, 174, 183, 207, 212, 226, 228, 229, 243, 271, 272, 324, 327
Hudson Valley, 1, 2, 4, 7, 10, 13, 41, 42, 49, 55, 107, 108, 119, 126, 133, 135, 147, 151, 152, 159, 185, 227, 228, 263, 267, 269, 376, 388
Hynek, J. Allen, 1, 2, 7, 11, 13, 43, 49, 54, 55, 89, 92, 105, 107, 108, 118, 119, 120,

133, 135, 145, 147, 149, 150, 152, 156, 159, 187, 188, 244, 252, 269, 270, 308, 323, 352, 376, 386
implant, 317
In The Night Sky, 4, 13, 42, 68, 108, 270, 388
Indian Mound, 16, 18, 21, 77, 97, 113, 124, 154, 226, 248, 254, 264, 267, 268, 271, 277, 278, 279, 295, 298, 299, 301, 303, 305, 318, 353, 355
island, 298, 358, 360, 367, 373, 377
jet engine, 54, 89, 188, 191, 192, 200, 206
Jim Forberg, 294
John Macedo Jr., ix, 338
Jumbo or Big One, 116
jumpjet, 72, 194, 244
kaleidoscope, 292
Kent, 4, 82, 300, 301, 356, 387
Kingston, 4, 7, 8, 13, 68, 112
Kodak 400 ASA, 71
L.M. Howe, 376
Lake Carmel, 4

Lamont-Doherty Earth Observatory, 67, 253, 304, 307, 310, 315, 323, 399
lamps, 68, 72, 152, 156, 158, 159, 170, 185, 235
Lemuria, 141, 142
Lewisburg, 4
Linda Zimmermann, 4, 68, 108
Linden, 4
Long Island, 4, 10, 11, 13, 108, 121, 151, 189, 232
looping, 89, 175, 178, 298, 346, 348, 350
loops, 78, 175, 179, 186, 237, 302, 336, 346
magnetic, 2, 7, 8, 9, 21, 38, 40, 66, 67, 77, 80, 82, 83, 84, 108, 112, 113, 154, 236, 240, 248, 253, 278, 281, 295, 298, 299, 310, 311, 312, 314, 317, 353, 354, 356, 357, 358, 359, 360, 361, 362, 363, 364, 366, 367, 369, 373, 376, 377
magnetic survey, 66,

108, 353, 356, 377
Manta Ray, 45, 50, 58, 59, 60, 61, 62, 63, 65, 66, 68, 71, 72, 73, 75, 78, 80, 84, 85, 88, 89, 90, 92, 96, 97, 100, 110, 111, 116, 119, 122, 124, 156, 157, 160, 163, 165, 181, 182, 189, 190, 197, 204, 205, 206, 211, 214, 215, 219, 220, 222, 223, 224, 225, 226, 229, 230, 231, 232, 235, 247, 252, 253, 263, 264, 269, 283, 284, 296, 299, 323, 341, 356
Marc Whitford, ix, 124
Marianna Horrigan, 107
Maryland, 110
Massanutten Mountain, 145, 309
Mayan Calendar, 239, 384, 388
Maybrook, 44, 45, 47
MD-11, 132
Memphis, 13
merry-go-round, 289, 292
metaterrestrial, 245

metavolcanic, 358
Mexico, 5, 6, 7, 106, 166, 246, 287, 376, 399
Mexico City, 6, 7, 245
Middletown, 2, 44, 45, 46, 47, 67, 112, 121, 126, 128, 167, 304, 306, 307, 324, 325, 327
Mike Bara, 260
miligammas, 298, 299
Mimicry, 59, 318
Mini-Flap, 44
Minolta XG7, 26, 68, 198, 241, 330
MisterX, 139
Mohonk, 16
Montgomery, 8, 44, 45, 60, 68, 107, 112, 154, 233, 256, 267, 349
Monticello, 43, 44, 118
Mount Adam, 8
mudstones, 9
Mystery of 2012, 82, 90, 143, 383
National Institute for Discovery Science, 128
New Haven, 2, 49
New Jersey, 4, 8, 67, 82, 92, 96, 117, 124,

125, 195, 304, 324, 352, 399
New York, i, 1, 2, 4, 8, 9, 10, 13, 41, 44, 46, 63, 92, 100, 107, 108, 112, 120, 126, 160, 167, 168, 240, 246, 248, 267, 270, 287, 288, 289, 291, 304, 315, 324, 352, 354, 356, 357, 376, 384, 385, 388
Newburgh, 7, 44, 45, 47, 50, 110, 180, 226, 236, 246, 283, 344, 377
NIDS, 128, 130, 131, 132, 133, 134, 135, 266
Night Siege, 1, 42, 54, 89, 105, 118, 159, 270, 376, 386
NORCANS, 139
North America, 5, 240, 387
NUFORC Database, 133
Object, 376
Olsen, 82, 387
O-Qua Tangin Wann, 143
Orange County, 2, 11, 42, 60, 68, 75, 90, 95, 112, 120, 122, 126, 160, 288, 294, 315
Ordovician, 8, 9, 108, 135, 253, 298, 358, 367, 370, 374
Ottowa, 139
Palisades, 304, 307, 324
paranormal activity, 2, 7, 91, 240, 248, 300, 301, 304
Passport to Magonia, 292
Pat Huff-Cornet, 110, 124
Phanerozoic, 8
phenomena, 1, 2, 3, 4, 5, 14, 15, 28, 34, 42, 91, 99, 105, 130, 227, 248, 378
Phil and Lynn Martin, ix, 364, 373
Philip Imbrogno, 1, 107, 301
Pine Bush, 2, 3, 4, 5, 8, 11, 13, 14, 26, 41, 43, 45, 50, 54, 55, 60, 61, 68, 90, 91, 99, 107, 108, 110, 112, 118, 120, 121, 122, 124, 126, 135, 136, 151, 160, 163,

168, 185, 186, 189, 225, 226, 227, 236, 237, 240, 256, 263, 267, 269, 270, 283, 294, 295, 304, 309, 337, 357, 370, 376, 383, 387
Pine Island, 8, 10, 13, 68, 358
Plasma Orb, 35
Pleiadians, 295
Popocatepetl, 5, 7
Poughkeepsie, 60, 232
Pozzuoli, 46
Preston Dennett, 2, 63, 270
propellors, 117
Proton Precession Magnetometer, 21, 67, 298, 310
Putnam, 2, 4, 7, 13, 49, 240, 301, 354, 384
Pyramid, 268, 277
Ramapo, 16, 153, 284
Randy Morrison, 105
Ray Sandford, 105
Rectangle, 116
Red Bank, 125, 128, 157, 160, 222, 316
Rich Pascorella, ix, 318
Richard C. Hoagland, ix
Richard Dreyfuss, 111
Riley Martin, 143
River Chebar, 103
Robert Bigelow, ix, 132
Rockland, 2, 49
Rte. 17, 41, 43, 46, 121, 126, 168
Rte. 211 East, 47, 48
Rte. 52, 60, 67, 110, 112, 114, 135, 136, 160, 168, 296, 298, 309, 310, 311, 312, 316, 370, 373
Russel Pardi, 43
Rutgers University, 14, 307, 352
Salt Point, 60, 63, 122, 124, 160, 165
Scottstown, 47
screen memories, 306
Sean Donovan, 105
Searsville Rd., 16, 17, 29, 55, 67, 68, 75, 84, 90, 92, 95, 96, 97, 99, 100, 101, 107, 110, 153, 154, 168, 169, 179, 183, 188, 191, 198, 207, 222, 226, 227, 237, 241, 242, 248, 256, 267, 270, 272, 277, 303, 304, 312, 318,

322, 338, 340
Selinsgrove, 4
Sharon Cunningham, ix, 110
shot, 169, 190, 212, 243, 312, 314, 316, 317, 378
Silent Invasion, 2, 5, 66, 100, 101, 105, 114, 116, 270, 297, 304, 306, 353, 357, 363, 384, 386
skyglyph, 77, 82, 83
skywatching, 68, 75
Small airplane, 116
South Searsville Rd., 68, 92, 322, 340
Stan Gordon, 4
Stargate, 248, 296, 299, 300, 354
Stewart Airport, 57, 58, 133
Stone Chambers, 108, 114, 305
Swamp Gas, 92
sweat lodge, 18, 19, 279, 282, 376
Syracuse Newtimes, 10, 120, 384
Telepathy, 166, 174
Temco (Globe) Swift 125, 349
The Coming of Tan, 143, 387
The Experiment, 16, 21, 32, 198
The Larry Hatch Triangle Database, 133
The MUFON Triangle Database, 133
The teacher, 140
Thompson's Ridge, 73, 88, 92, 153, 219, 319
Timothy Good, 42, 105, 138, 139, 166, 353
transversable wormholes, 265, 266
treble clef, 257
Triangle, 7, 29, 49, 50, 116, 120, 124, 125, 126, 127, 128, 132, 133, 138, 157, 174, 200, 204, 209, 210, 273, 276, 287, 289, 373
Trickster, 110
UAP, ii, 1, 10, 11, 13, 21, 23, 43, 44, 45, 46, 47, 48, 53, 75, 81, 92, 96, 100, 107, 111, 120, 149, 150, 151, 154, 168, 169, 173, 189, 198, 227, 231, 236, 237, 256,

261, 264, 265, 266, 292, 294, 295, 299, 304, 306, 308, 315, 318, 319, 321, 322, 323, 337, 344, 345, 346, 356, 376, 378, 399

UAP Corridor, 10, 11, 13, 108, 120, 151, 232, 263, 357

UBO, 372

UFO, 1, 2, 3, 4, 5, 6, 7, 9, 10, 11, 13, 14, 40, 41, 44, 45, 46, 47, 48, 50, 53, 54, 57, 68, 69, 70, 75, 81, 82, 83, 84, 89, 91, 96, 100, 106, 108, 111, 112, 113, 120, 124, 130, 132, 135, 145, 147, 148, 150, 151, 154, 156, 166, 168, 169, 172, 173, 174, 178, 180, 186, 187, 189, 190, 191, 197, 199, 203, 215, 222, 227, 232, 236, 237, 239, 240, 244, 245, 246, 249, 254, 261, 263, 264, 265, 266, 267, 277, 284, 286, 288, 291, 292, 294, 295, 298, 300, 301, 302, 304, 305, 306, 307, 308, 312, 315, 316, 317, 318, 321, 322, 323, 324, 326, 327, 328, 330, 331, 332, 336, 337, 338, 342, 344, 345, 346, 349, 352, 353, 356, 357, 372, 373, 374, 375, 376, 378, 379, 380, 381, 383, 384, 385, 387, 388

UFO Vortexes, 113, 249, 264, 267, 295

Ulster, ii, 2, 5, 8, 44, 100, 121, 122, 357

ultraphysical, 245

Unchi, 17, 18, 282, 376

unconventional, 156, 197, 204, 231, 312

underground, 10, 61, 66, 108, 162, 226, 296, 297, 299, 304, 310, 311, 352, 353, 354, 357, 359, 360, 361, 366, 367, 376, 380

underground base, 61, 162, 297, 352, 353, 359, 361

Vincent Polise, 11, 14, 91, 99, 108, 270

volcano, 5, 6, 7
V-shape, 49
Walden, 8, 44, 60, 67, 68, 76, 90, 107, 110, 112, 135, 153, 160, 168, 189, 240, 254, 267, 316, 338, 370
Wallkill, i, ii, 2, 8, 10, 13, 26, 44, 45, 60, 68, 107, 112, 153, 154, 168, 232, 236, 240, 305, 307, 316, 339, 356, 357, 358, 377
Wallkill River, i, ii, 2, 8, 10, 13, 26, 60, 68, 112, 153, 154, 168, 232, 236, 240, 307, 316, 357, 377
Wallkill River Valley, i, ii, 2, 8, 10, 13, 60, 68, 112, 153, 232, 240, 307, 357, 377
Walnut or turtle shape, 116
Wann, 143, 145, 387
Warwick, 8, 358
Westchester Boomerang, 46, 116, 118
Westchester County, 49
Whitley Strieber, 294
Williamsport, 4, 50, 54, 95, 148, 149, 385
Wisconsinian, 142
Xretsim, 139
Young's Experiment, 39
Zenite 35 mm SLR, 14
Zret, 139

About the Author

Bruce Cornet received a B.A. degree (1970) in biology from the University of Connecticut, a Masters degree (1972) in paleobotany from that same university, and graduated from Penn State in 1977 with a Ph.D. in geology and palynology (the study of fossil spores and pollen, used to age date rocks). He spent 11 years in the oil industry, working for Gulf Research & Development, Exxon USA, Mobil Oil Corporation, and Superior Oil Company, all in Houston, TX. Between 1981 and 1982 he ran his own independent exploration company (Geminoil, Inc.), which drilled for and found oil in eastern Virginia. Between 1988 and 1993 he held a research position at Lamont-Doherty Earth Observatory (part of Columbia University). He is profiled on ResearchGate, and has published many scientific Articles (28), Books (1), Chapters (3), Conference Papers (2), Thesis (1), Technical Reports (4), Research (2), Experiment Findings (1), and Presentations (6). All total (48). He taught classroom geology and botany, and an online geology course for the Raritan Valley Community College in New Jersey for seven years (2002-2008). He also taught physical and historical geology for the El Paso Community College in Texas, and for the Dona Ana Community College, a branch of New Mexico State University. Now retired, he has been writing books and continuing his research into UAP and Alien Abductions.

Made in the USA
Monee, IL
20 February 2021